T0239544

TUNNELS AND UNDERGROUND CITIES: ENGINEERING AND
INNOVATION MEET ARCHAEOLOGY, ARCHITECTURE AND ART

PROCEEDINGS OF THE WTC2019 ITA-AITES WORLD TUNNEL CONGRESS, NAPLES, ITALY, 3-9 MAY, 2019

Tunnels and Underground Cities: Engineering and Innovation meet Archaeology, Architecture and Art

Volume 9: *Safety in underground construction*

Editors

Daniele Peila
Politecnico di Torino, Italy

Giulia Viggiani
University of Cambridge, UK
Università di Roma "Tor Vergata", Italy

Tarcisio Celestino
University of Sao Paulo, Brasil

CRC Press
Taylor & Francis Group
Boca Raton London New York

CRC Press is an imprint of the
Taylor & Francis Group, an **informa** business

A BALKEMA BOOK

Cover illustration:

View of Naples gulf

CRC Press/Balkema is an imprint of the Taylor & Francis Group, an informa business

© 2020 Taylor & Francis Group, London, UK

Typeset by Integra Software Services Pvt. Ltd., Pondicherry, India

Published by: CRC Press/Balkema
  Schipholweg 107C, 2316XC Leiden, The Netherlands
  e-mail: Pub.NL@taylorandfrancis.com
  www.crcpress.com – www.taylorandfrancis.com

ISBN: 978-0-367-46874-3 (Hbk)
ISBN: 978-1-003-03166-6 (eBook)

*Tunnels and Underground Cities: Engineering and Innovation meet Archaeology,
Architecture and Art, Volume 9: Safety in underground
construction – Peila, Viggiani & Celestino (Eds)
© 2020 Taylor & Francis Group, London, ISBN 978-0-367-46874-3*

Table of contents

Tunnels and Underground Cities: Engineering and Innovation meet Archaeology, Architecture and Art, Volume 9: Safety in underground construction – Peila, Viggiani & Celestino (Eds)
© 2020 Taylor & Francis Group, London, ISBN 978-0-367-46874-3

Preface

The World Tunnel Congress 2019 and the 45th General Assembly of the International Tunnelling and Underground Space Association (ITA), will be held in Naples, Italy next May.

The Italian Tunnelling Society is honored and proud to host this outstanding event of the international tunnelling community.

Hopefully hundreds of experts, engineers, architects, geologists, consultants, contractors, designers, clients, suppliers, manufacturers will come and meet together in Naples to share knowledge, experience and business, enjoying the atmosphere of culture, technology and good living of this historic city, full of marvelous natural, artistic and historical treasures together with new innovative and high standard underground infrastructures.

The city of Naples was the inspirational venue of this conference, starting from the title Tunnels and Underground cities: engineering and innovation meet Archaeology, Architecture and Art.

Naples is a cradle of underground works with an extended network of Greek and Roman tunnels and underground cavities dated to the fourth century BC, but also a vibrant and innovative city boasting a modern and efficient underground transit system, whose stations represent one of the most interesting Italian experiments on the permanent insertion of contemporary artwork in the urban context.

All this has inspired and deeply enriched the scientific contributions received from authors coming from over 50 different countries.

We have entrusted the WTC2019 proceedings to an editorial board of 3 professors skilled in the field of tunneling, engineering, geotechnics and geomechanics of soil and rocks, well known at international level. They have relied on a Scientific Committee made up of 11 Topic Coordinators and more than 100 national and international experts: they have reviewed more than 1.000 abstracts and 750 papers, to end up with the publication of about 670 papers, inserted in this WTC2019 proceedings.

According to the Scientific Board statement we believe these proceedings can be a valuable text in the development of the art and science of engineering and construction of underground works even with reference to the subject matters "Archaeology, Architecture and Art" proposed by the innovative title of the congress, which have "contaminated" and enriched many proceedings' papers.

Andrea Pigorini   Renato Casale
SIG President   *Chairman of the Organizing Committee WTC2019*

Acknowledgements

REVIEWERS

The Editors wish to express their gratitude to the eleven Topic Coordinators: Lorenzo Brino, Giovanna Cassani, Alessandra De Cesaris, Pietro Jarre, Donato Ludovici, Vittorio Manassero, Matthias Neuenschwander, Moreno Pescara, Enrico Maria Pizzarotti, Tatiana Rotonda, Alessandra Sciotti and all the Scientific Committee members for their effort and valuable time.

SPONSORS

The WTC2019 Organizing Committee and the Editors wish to express their gratitude to the congress sponsors for their help and support.

Tunnels and Underground Cities: Engineering and Innovation meet Archaeology,
Architecture and Art, Volume 9: Safety in underground
construction – Peila, Viggiani & Celestino (Eds)
© 2020 Taylor & Francis Group, London, ISBN 978-0-367-46874-3

WTC 2019 Congress Organization

HONORARY ADVISORY PANEL

Pietro Lunardi, President WTC2001 Milan
Sebastiano Pelizza, ITA Past President 1996-1998
Bruno Pigorini, President WTC1986 Florence

INTERNATIONAL STEERING COMMITTEE

Giuseppe Lunardi, Italy (Coordinator)
Tarcisio Celestino, Brazil (ITA President)
Soren Eskesen, Denmark (ITA Past President)
Alexandre Gomes, Chile (ITA Vice President)
Ruth Haug, Norway (ITA Vice President)
Eric Leca, France (ITA Vice President)
Jenny Yan, China (ITA Vice President)
Felix Amberg, Switzerland
Lars Barbendererder, Germany
Arnold Dix, Australia
Randall Essex, USA
Pekka Nieminen, Finland
Dr Ooi Teik Aun, Malaysia
Chung-Sik Yoo, Korea
Davorin Kolic, Croatia
Olivier Vion, France
Miguel Fernandez-Bollo, Spain (AETOS)
Yann Leblais, France (AFTES)
Johan Mignon, Belgium (ABTUS)
Xavier Roulet, Switzerland (STS)
Joao Bilé Serra, Portugal (CPT)
Martin Bosshard, Switzerland
Luzi R. Gruber, Switzerland

EXECUTIVE COMMITTEE

Renato Casale (Organizing Committee President)
Andrea Pigorini, (SIG President)
Olivier Vion (ITA Executive Director)
Francesco Bellone
Anna Bortolussi
Massimiliano Bringiotti
Ignazio Carbone
Antonello De Risi
Anna Forciniti
Giuseppe M. Gaspari

Giuseppe Lunardi
Daniele Martinelli
Giuseppe Molisso
Daniele Peila
Enrico Maria Pizzarotti
Marco Ranieri

ORGANIZING COMMITTEE

Enrico Luigi Arini
Joseph Attias
Margherita Bellone
Claude Berenguier
Filippo Bonasso
Massimo Concilia
Matteo d'Aloja
Enrico Dal Negro
Gianluca Dati
Giovanni Giacomin
Aniello A. Giamundo
Mario Giovanni Lampiano
Pompeo Levanto
Mario Lodigiani
Maurizio Marchionni
Davide Mardegan
Paolo Mazzalai
Gian Luca Menchini
Alessandro Micheli
Cesare Salvadori
Stelvio Santarelli
Andrea Sciotti
Alberto Selleri
Patrizio Torta
Daniele Vanni

SCIENTIFIC COMMITTEE

Daniele Peila, Italy (Chair)
Giulia Viggiani, Italy (Chair)
Tarcisio Celestino, Brazil (Chair)
Lorenzo Brino, Italy
Giovanna Cassani, Italy
Alessandra De Cesaris, Italy
Pietro Jarre, Italy
Donato Ludovici, Italy
Vittorio Manassero, Italy
Matthias Neuenschwander, Switzerland
Moreno Pescara, Italy
Enrico Maria Pizzarotti, Italy
Tatiana Rotonda, Italy
Alessandra Sciotti, Italy
Han Admiraal, The Netherlands
Luisa Alfieri, Italy
Georgios Anagnostou, Switzerland

Andre Assis, Brazil
Stefano Aversa, Italy
Jonathan Baber, USA
Monica Barbero, Italy
Carlo Bardani, Italy
Mikhail Belenkiy, Russia
Paolo Berry, Italy
Adam Bezuijen, Belgium
Nhu Bilgin, Turkey
Emilio Bilotta, Italy
Nikolai Bobylev, United Kingdom
Romano Borchiellini, Italy
Martin Bosshard, Switzerland
Francesca Bozzano, Italy
Wout Broere, The Netherlands
Domenico Calcaterra, Italy
Carlo Callari, Italy

Luigi Callisto, Italy
Elena Chiriotti, France
Massimo Coli, Italy
Franco Cucchi, Italy
Paolo Cucino, Italy
Stefano De Caro, Italy
Bart De Pauw, Belgium
Michel Deffayet, France
Nicola Della Valle, Spain
Riccardo Dell'Osso, Italy
Claudio Di Prisco, Italy
Arnold Dix, Australia
Amanda Elioff, USA
Carolina Ercolani, Italy
Adriano Fava, Italy
Sebastiano Foti, Italy
Piergiuseppe Froldi, Italy
Brian Fulcher, USA
Stefano Fuoco, Italy
Robert Galler, Austria
Piergiorgio Grasso, Italy
Alessandro Graziani, Italy
Lamberto Griffini, Italy
Eivind Grov, Norway
Zhu Hehua, China
Georgios Kalamaras, Italy
Jurij Karlovsek, Australia
Donald Lamont, United Kingdom
Albino Lembo Fazio, Italy
Roland Leucker, Germany
Stefano Lo Russo, Italy
Sindre Log, USA
Robert Mair, United Kingdom
Alessandro Mandolini, Italy
Francesco Marchese, Italy
Paul Marinos, Greece
Daniele Martinelli, Italy
Antonello Martino, Italy
Alberto Meda, Italy

Davide Merlini, Switzerland
Alessandro Micheli, Italy
Salvatore Miliziano, Italy
Mike Mooney, USA
Alberto Morino, Italy
Martin Muncke, Austria
Nasri Munfah, USA
Bjørn Nilsen, Norway
Fabio Oliva, Italy
Anna Osello, Italy
Alessandro Pagliaroli, Italy
Mario Patrucco, Italy
Francesco Peduto, Italy
Giorgio Piaggio, Chile
Giovanni Plizzari, Italy
Sebastiano Rampello, Italy
Jan Rohed, Norway
Jamal Rostami, USA
Henry Russell, USA
Giampiero Russo, Italy
Gabriele Scarascia Mugnozza, Italy
Claudio Scavia, Italy
Ken Schotte, Belgium
Gerard Seingre, Switzerland
Alberto Selleri, Italy
Anna Siemińska Lewandowska, Poland
Achille Sorlini, Italy
Ray Sterling, USA
Markus Thewes, Germany
Jean-François Thimus, Belgium
Paolo Tommasi, Italy
Daniele Vanni, Italy
Francesco Venza, Italy
Luca Verrucci, Italy
Mario Virano, Italy
Harald Wagner, Thailand
Bai Yun, China
Jian Zhao, Australia
Raffaele Zurlo, Italy

Safety in underground construction

*Tunnels and Underground Cities: Engineering and Innovation meet Archaeology,
Architecture and Art, Volume 9: Safety in underground
construction – Peila, Viggiani & Celestino (Eds)*
© 2020 Taylor & Francis Group, London, ISBN 978-0-367-46874-3

Tunnel inspection analysis based on the crack progression of the change by Tunnel-lining Crack Index (TCI) for road tunnel

H. Aio, H. Hayashi & M. Shinji
Yamaguchi University, Ube, Japan

T. Nakamura & S. Yamada
Yamaguchi Prefecture, Yamaguchi, Japan

S. Morimoto
Dobocreate Corporation, Ube, Japan

ABSTRACT: Civil engineering infrastructures in Japan were intensively developed from the 1960s to the 1970s, thus there is a concern that rapid growth of aging will progress in the future. Therefore, it is necessary to evaluate the health status of each civil engineering infrastructure and to develop the method for evaluating the order of importance for maintenance based on the result of investigation. Various methods of health evaluation have been proposed on the lining of road tunnels. One of them is the Tunnel-lining Crack Index (TCI) proposed by one of the authors, which makes it possible to quantitatively evaluate the status of crack distribution on a tunnel lining. In this paper, TCI was applied to multiple road tunnels managed by Yamaguchi local prefecture. From the time-history of TCI, the crack progression and the priority of tunnel inspection were analyzed to improve the efficiency of maintenance in the future.

1 INTRODUCTION

About 10% of road tunnels managed by Yamaguchi prefecture in Japan has been 50 years since the start of service, and this proportion will increase rapidly in the future. Along with the aging problems of such tunnels, the conversion from "post-maintenance type" to "preventive-maintenance type" has been reviewed on the maintenance and management aspect. In addition, periodic inspections of road tunnels were regulated at once every five years in 2014. Therefore, in the current Yamaguchi prefecture, the first periodic inspection was completed in all 129 tunnels, and the second one will be carried out subsequently. As tunnels with a service life exceeding 50 years increase in the future, efficiency and economy in maintenance management are important. However, the conventional health evaluation method for tunnel lining is unsatisfactory in which qualitative indicators based on the experience of the inspectors are included. The practice of comparing inspection results in the same tunnel, being a standard protocol in the past, becomes a shortcoming for the present situation, since data from different tunnels do not share the same frame of reference. Continuation of current practice will lead to a severe lack of quantitative data that makes comparisons between tunnels difficult so the prioritizing of tunnels for inspections and repairs remains a challenge.

In this paper, TCI proposed by one of the authors was applied to 10 tunnels at Yamaguchi local prefecture to quantitatively evaluate the wellness of tunnel lining and we analyzed the progression of cracks and crack occurrence factor by changes in TCI.

2 INTRODUCTION OF TCI

The density, direction and width of cracks in rock in rock mechanics field influence the physical properties (strength, deformation modulus, coefficient of permeability). In this study, TCI which is based on crack tensor was used as an indicator that can comprehensively quantify these effects. Figure 1 shows the diagram to calculate the TCI and the Equation (1) is a calculation expression for TCI adopted in this study. This has a feature that makes it possible to quantitatively evaluate tunnel lining health from the width, length and direction of cracks, without depending on the subjectivity of the inspection engineer.

$$F_{ij} = \frac{1}{A} \sum_{k=1}^{n} \left(t^{(k)} \right)^{\alpha} \left(l^{(k)} \right)^{\beta} \cos \theta_i^{(k)} \cos \theta_j^{(k)} \tag{1}$$

A : Area of tunnel lining
n : Number of cracks
$l^{(k)}$: Length of crack
$t^{(k)}$: Width of crack
$\theta_i^{(k)}$: Angle between normal vector to crack and x-axis
$\theta_j^{(k)}$: Angle between normal vector to crack and y-axis
α : Coefficient on weighting of crack width
β : Coefficient on weighting of crack length

It is suggested that the value of TCI (denoted by F0) is obtained as sum of its vertical (F11) and crossing (F22) components. A larger value denotes more cracks, which means that the health evaluation of tunnel lining is relatively low. Here, α and β in expression (1) are assumed as 1.0 based on a previous study.

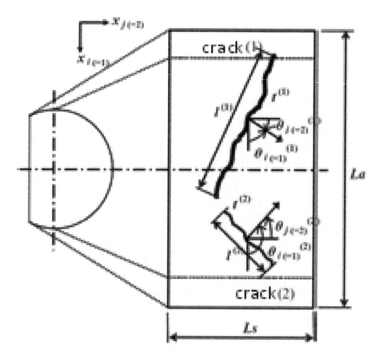

Figure 1.   Diagram to calculate the TCI.

3 ESTIMATION OF TCI ABOUT TUNNELS

Table 1 shows the outline of researched tunnels in this study. They were inspected for 2 or 3 times and were managed by Yamaguchi prefecture. TCI was applied to these tunnels. In this section, TCI calculation result of tunnel B, C and F showing characteristic aging change were charted by the year of inspection and analyzed for the occurrence factors of local cracks.

3.1 *Tunnel B*

Figure 2 shows TCI of each span in Tunnel B. The increase in TCI at Span No.17 is remarkable in Tunnel B. Figure 3 shows the change of crack map on Span No.17 recorded at the time of the last 3 periodical inspections. Although cracks have not been reported in the initial inspection in 1997, crack with a maximum width of 7mm was reported in the second in 2008. Moreover, crack with a width of 8mm was newly reported in the third inspection in 2016. It means that the width of previous cracks has increased, which is considered to have led to a rapid increase in TCI at Span No.17. The geology of this area is pelitic or siliceous hornfels which is hard and dense, thus poor ground strength is not considered to be a problem, however, Span No.17 has the largest overburden of about 55m, and ground investigations report the presence of active faults. For these reasons, it is considered that the overburden or active faults affects the rapid increase in TCI. In addition, TCI increased significantly for each span that is marked by a dashed red circle in Figure 2. Figure 4 shows the relative change in amount of TCI at these spans. It shows the ratio (%) of the change in amount of the TCI to the total change in amount from the second to the third inspection (2008~2016), and the larger the value means that the

Table 1. Outline of tunnels.

	Length m	Span	Construction year	Method	Crack map
A	358	40	1971		1999, 2008, 2016
B	273	29	1971		1997, 2008, 2016
C	318	58	1958	Sheet	2011, 2017
D	178	44	1958	pile	2010, 2017
E	45	6	1958	method	2011, 2016
F	166	17	1978		2012, 2016
G	600	67	1973		2011, 2016
H	334	33	2001		2011, 2016
I	708	68	2008	NATM	2011, 2016
J	587	58	1991		2012, 2016

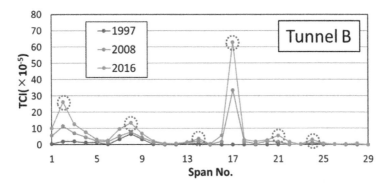

Figure 2. Change in TCI of Tunnel B.

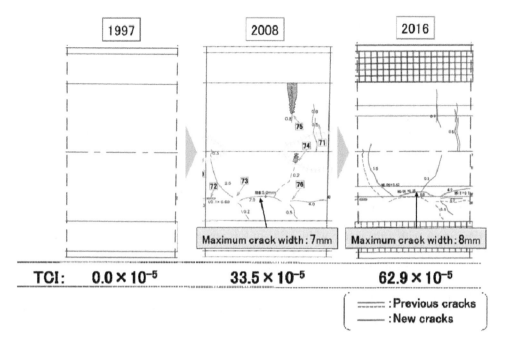

Figure 3.　Crack map and change of TCI at Span No.17 in Tunnel B.

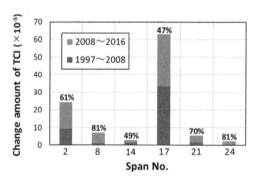

Figure 4.　Change amount of TCI at selected spans in Tunnel B.

progression of cracks is accelerating. From this figure, not only Span No.17 but also other extracted spans can be judged that cracks were progressing rapidly, therefore, it is considered that urgent maintenance is necessary such as in Span No.17.

3.2　Tunnel C

Figure 5 shows TCI of each span in Tunnel C. The cracks are hardly reported in the central part of Tunnel C, however at the time of inspection in 2011, TCI increased sharply from Span No.45 and attained the highest value of 79.1×10^{-5} in Span No.55. Figure 6 shows the crack map of spans No.46 to 57 with large TCI at the time of inspection in 2011 and 2016. From this figure,
It can be confirmed that cracks in the longitudinal direction occurred in the tunnel arch part. After that, flacking prevention work was carried out from span No.46 to 57 in 2015, and it was considered that cracks have disappeared based on the Maintenance and inspection procedure, thus the TCI decreased at the second inspection in 2017. Thereafter particular attention will be focused on locations where flacking prevention was not applied. Appearance of new cracks or progression of existing cracks will signify a necessity to review the repair method. The geology changes from pelitic schist to granite bordering at Span No.45, according to core boring results at the tunnel main body when repair design was conducted in 2012. This parallels the distribution of TCI which suggests a strong geological influence.

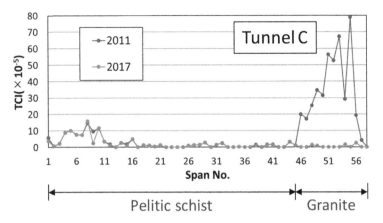

Figure 5.   Change in TCI of Tunnel C.

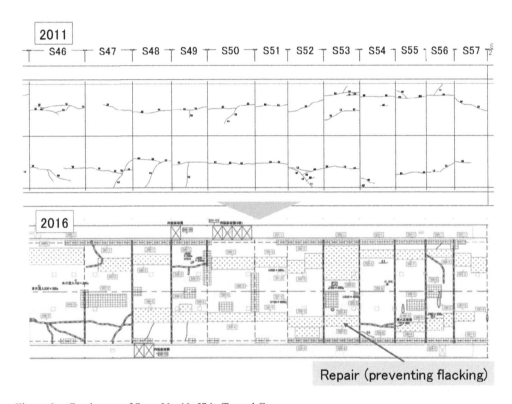

Repair (preventing flacking)

Figure 6.   Crack map of Span No.46~57 in Tunnel C.

3.3   *Tunnel F*

Figure 7 shows TCI of each span in Tunnel F. In addition, Figure 8 shows the crack map of Span No.1 to 5 at the time of inspection in 2017. This tunnel had no significant increase TCI from initial inspection to second one at most spans. For this reason, crack progression is considered to be relatively low. However, from Figure 8, cracks occurred throughout the lining, and the average value of TCI was 2.76×10^{-5} at the time of the most recent inspection for the other 9 tunnels, on the other hand Tunnel F had a large value of 19.75×10^{-5}. In addition to the cracks concentrating at the tunnel mouth, there are also spans where repairs have already been carried out, such as at Span No.5, it is possible to grasp that cracks occurred frequently in the overall tunnel. From the above, Tunnel F is an old tunnel with a service period in excess of 30 years and it does not show the occurrence or progress of local cracks like in

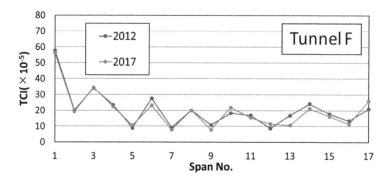

Figure 7. Change in TCI of Tunnel F.

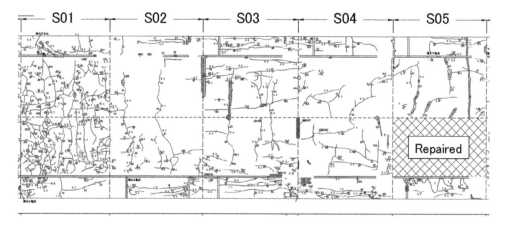

Figure 8. Crack map of Span No.1~5 in Tunnel F.

Tunnel B and C, therefore it can be inferred that the main cause of cracks is not the action of any external force, but due to material degradation of the lining.

Likewise, in the other 7 tunnels, analysis was carried out based on the change in TCI with age. By following the change in TCI with age according to inspection records of multiple years such as in Tunnel B, C and F mentioned in the example, it is possible to estimate the progression of cracks and factor of occurrence which cannot be realized by relying on a single inspection.

4 ANALYSIS OF CRACK PROGRESSION BY TCI

4.1 *Investigation by aging change of average TCI*

Figure 9 shows the change of the average TCI per span in all 10 tunnels. From Figure 9, TCI of Tunnel A, B, E and G increases, indicating that the health evaluation of tunnel lining tends to decline. The TCI of Tunnel H, I and J constructed by NATM remained almost unchanged, and at this stage there is no deterioration in the health of tunnel lining. However, due to a different role played by sheet pile method in lining, it is necessary to monitor how the TCI will change with more years of use from future inspections. In addition, out of 3 tunnels with years of usage exceeding 50 years, Tunnel C and D have been repaired before at some parts, in contrast, Tunnel E not having done so and its TCI tends to increase. Therefore, close monitoring on Tunnel E is recommended. The rapid increase in TCI at Tunnel B Span No. 17 as shown in Figure 3 is a concern for further evaluation, however it was not apparent for Tunnel F as shown in Figure 8. For this reason, we studied further by changing the focus point in the next section.

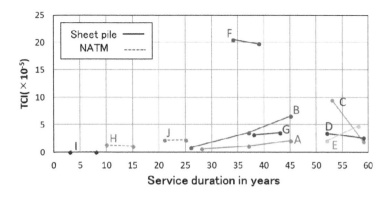

Figure 9.　Average change of TCI of each tunnel.

4.2　*Inspection priority evaluation focus on the crack progression part and crack amount*

Analyzing each of the 10 tunnels, there are cases in which cracks progresses all over or locally. In this paper, all over means the tunnel as a whole, and locally means a case where cracks progress remarkably in a specific span. It is difficult to quantify the progression of such detailed cracks only by the average change of TCI shown in Figure 9. Moreover, we would like to evaluate tunnels with both features such as Tunnel F which has low progression and large amounts of cracks. This can be realized by plotting the local crack progress against overall crack progress, as shown in Figure 10. Local crack progress is evaluated from the slope of TCI at the span with maximum amount of change. Overall crack progress is evaluated from average annual rate of increase. In this case, it was impossible to calculate the increase rate because zero may be included in the TCI of first inspection in the maximum change span, hence TCI slope was adopted as another measure for indicating progression. The average annual increase rate was calculated by dividing the TCI increase rate from the first inspection to the second or third of each tunnel in Figure 9 by the inspection interval years. The reason for using the increase rate is to focus on the change amount of TCI. For example, the slopes of Tunnel B and E in Figure 9 are almost the same as 0.379 and 0.375 respectively, the change amount is larger in the B, and the annual increase rate is greatly different as 43.2% and 13.1% respectively. Furthermore, the span average TCI at the last inspection showing the crack amount is represented by the size of bubbles plotted in Figure 10, and the larger bubble means that TCI is larger, that is, the crack amount is greater.

In the Figure 10, it is considered that the tunnels indicated on the diagonally upper right have higher inspection priority because the crack progresses both all over and locally. Therefore, it can be judged that inspection of Tunnel B should be given the highest priority in all 10 tunnels. Besides, although Tunnel A and G are almost the same in the inspection priority from the viewpoint described above, it is possible to grasp the difference of characteristics such as A is all over and G is locally in terms of crack progressions. Here, even if it is local, if there is a remarkable change, it is considered that it is necessary to undertake urgent maintenance given the importance of safety on traffic vehicles and pedestrians. In addition, since the crack amount is more in Tunnel G, it is considered that Tunnel G should have higher priority than Tunnel A.

Furthermore, it is clear that although the crack amount of Tunnel F is larger than others, it is less progressive and chronic. On the other hand, the crack amount of Tunnel B is less than F, but since the progression is high, it is judged to be acute and thus inspection is given priority. In this way, it is not always the case that the priority of inspection will be higher because the amounts of cracks is large. However, if the cracks concentrate a lot locally, the risk of flaking increases, thus it is considered that the tunnel be closely monitored for inspection. On the other hand, since Tunnel C, D, H, I, J have low crack progressions and relatively few cracks, it can be determined that urgent inspection is unnecessary.

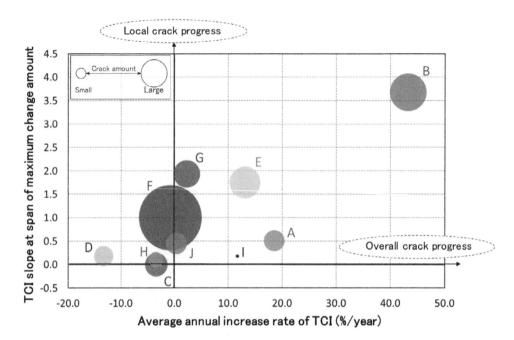

Figure 10. Inspection priority evaluation.

5 CONCLUSIONS

In this paper, TCI is applied to the crack maps which is the inspection result of 10 road tunnels managed by Yamaguchi prefecture for the purpose of improving efficiency of maintenance and quantitative health evaluation of tunnel lining. From the results, crack occurrence factors were estimated and crack progression was analyzed. In addition, the inspection priority determination was applied. The finding obtained by this are as follows.

1) By following the aging change of TCI, it is found that the crack occurrence situation and progression of each tunnel are quantified and various tendencies are shown for each span even in the same tunnel.

2) In tunnel B and C, we focused on the span in which TCI increases rapidly and found ground faults, overburden thickness and geological transition as possibilities of crack occurrence. In addition, in tunnel F, since cracks occurred throughout the tunnel, it is inferred that the main cause of cracks is not due to external loads, but due to material degradation of tunnel lining.

3) By paying attention to whether the cracks progress all over the tunnel or locally, it was possible to ascertain the characteristics of crack progression of each tunnel more. Besides, by including crack amount as well, it was also possible to judge whether the cracks in the tunnel were acute or chronic. The evaluation of inspection priority from these data will support efficiencies in maintenance and management in the future.

REFERENCES

Civil engineering department road maintenance section of Yamaguchi Prefecture. 2016, *Long life and repair plan for tunnels managed by Yamaguchi Prefecture*. Japan: pp.1–15 (In Japanese)

Ministry of Land, Infrastructure, Transport and Tourism JAPAN. 2014. *Regular inspection manuals for road tunnel*. Japan: pp.1–2 (In Japanese)

East Central West Nippon expressway company limited. 2015, *Maintenance and inspection procedure (Structure version)*. Japan: p.145 (In Japanese)

Shigeta, Y., Tobita, T., Kamemura, K., Shinji, M., Yoshitake,I., Nakagawa, K. 2006. *Propose of tunnel crack index (TCI) as an evaluation method for lining concrete*, Journal of geotechnical engineering (F), JSCE. Japan: vol.62, No.4, pp.628–632 (In Japanese).

Tunnels and Underground Cities: Engineering and Innovation meet Archaeology, Architecture and Art, Volume 9: Safety in underground construction – Peila, Viggiani & Celestino (Eds)
© 2020 Taylor & Francis Group, London, ISBN 978-0-367-46874-3

The stresses and safety of the final lining structures in bored tunnels in case of fire

T. Alberini
Italferr S.p.A. Italian State Railways Group, Rome, Italy

E. Cartapati
Department of structural and geotechnical engineering, "Sapienza" University of Rome, Rome, Italy

ABSTRACT: The final lining structures of recently and nearly recently built bored tunnels are essentially constituted by a ring of reinforced (or not reinforced) concrete, conventional excavation or mechanized excavation, bound at its outer contour by the mass of rock/soil crossed by the tunnel. This study wants to focus attention on elements that are important in order to find the stresses, and therefore in order to check the structural safety, of the final lining structures of bored tunnels, in case of fire.

1 INTRODUCTION

In recent years the Italian and European regulatory framework in terms of safety has imposed, including the various aspects and with different purposes, to check the behavior at *high temperature* of the structural elements/final lining of cut and cover tunnels/bored tunnels.

High temperatures mean those that can develop in conditions of generalized fire and then in after flash-over phase. At this stage *fire actions*, represented by the *nominal fire Temperature-time curves* reported in *Figure 1* (RWS-curve, Eureka-curve, Hydrocarbon-curve), are those.

With reference to these *curves* usually the behavior at *high temperatures* of the structural elements/final lining of cut and cover tunnels/bored tunnels, is analytically checked, in terms of *fire*

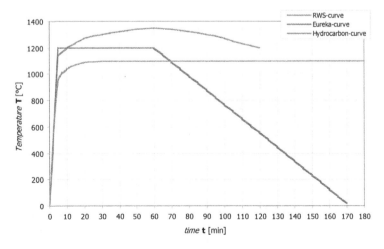

Figure 1. Nominal fire Temperature-time curves usually adopted to check the behavior at high temperature of the structural elements/final lining of tunnels.

resistance R; this analyzing, for the whole required duration, case by case, the mechanical behavior starting from the analysis of the evolution of the temperature inside of the structural sections. This effort is pursued on the basis of national and international regulatory, on results based on scientific research and on calculation methods based on the rules of *Structural Engineering*.

This study, also on basis of experiences, developed by writers in the last 20 years (in the last 20 years in particular in railway) wants to focus attention on elements that are important in order to find the stresses, and therefore in order to check the structural safety, of the structural elements/final lining of cut and cover tunnels/bored tunnels, in case of fire.

2 DESCRIPTION

The final lining structures of recently and nearly recently built bored tunnels are essentially constituted by a ring of reinforced (or not reinforced) concrete, conventional excavation or mechanized excavation (Figure 2).

The behavior at *high temperatures* of the final lining structures is studied, for checking *fire resistance R*, according to the following analytical procedure summarized below:

a) *identification of the action fire*
   depending on regulatory framework and/or performance to be given to the structures (see e.g. the *nominal fire Temperature-time curve* above mentioned).
b) *analysis of the evolution of the temperature in the structural sections*
   are carried out with the determination of the thermal field in the structural sections by the resolution of the problem of the heat propagation ruled by Fourier equation.
c) *analysis of the mechanical behavior of the structures*
   the mechanical behavior of the structures is analyzed, for the combination of the fire load, for all the duration of required *fire resistance R*, taking into account:
   ✓ the reduction of the materials mechanical strength due to the increase of temperature (change of stress-strain relationships of materials with temperature);
   ✓ the effects on stress due to non-allowed thermal expansion, generally, not negligible for the final lining structures (that are *hyperstatic structures*). This also is in accordance with *Section 4 Mechanical actions for structural analysis* point *4.1 General* of Eurocode 1: *Actions on structures - Part 1–2: General actions - Actions on structures exposed to fire* in which (1 (P)) is required that «...*Imposed and constrained expansions and deformations caused by temperature changes due to fire exposure result in effects of actions, e.g. forces and moments, which shall be considered with the exception of those cases where they: - may be recognized a priori to be either negligible or favourable; - are accounted for by conservatively chosen support models and boundary conditions, and/or implicitly considered by conservatively specified fire safety requirements...*».

Figure 2.   Examples of structural types used for *bored tunnels*.

d) *structural U.L.S. (Ultimate Limit State) checking*
   the checks of the fire resistance R are performed by checking that the mechanical strength is maintained for a time *t* corresponding to the duration of required *fire resistance R* with reference to the *nominal fire Temperature-time curve* adopted (point a).

Essentially, the *fire action* alters the governing inequality that expresses the structural verification in specific manner with respect to other actions

$$R \geq S \rightarrow R(t) \geq S(t)$$

where R=fire resistance and S=*Si*+*Se* with *Si*= stresses due to internal stresses; *Se*= stresses due to external loads; *t*=time
   being the phenomenon governed by the *change* (increase) *of temperature T* over *time t* and from point to point (both in the section and in the structure):

✓ the *increase of the temperature T* in the materials causes a decay of their mechanical characteristics → *reduction* of R
✓ in *hyperstatic structures* the *change of temperature T* produces expansions which, being constrained, cause the growth of internal stresses (*Si*) that combine with stresses due to external loads (*Se*) → S
✓ in *hyperstatic structures* the *change of temperature T* also produces a reduction of structural elements stiffness leading to a redistribution of stresses both due to internal stresses (*Si*) and to external loads (*Se*) → influence on *Si* and on *Se*

Therefore, for *isostatic structures* the non-satisfaction of the inequality occurs due to exclusive degradation of materials, as the first plastic hinge arises. For *hyperstatic structures*, like final lining structures in object, the non-satisfaction of the inequality, due to the superabundance of constraints, does not take place as the first plastic hinge arises, but when a number of hinges are formed causing structural lability.
   In the light of the above, in this study particular attention to previous points c) and d) will be paid.

2.1 *Analysis of the mechanical behavior of the structures*

In fire conditions, for a reliable analysis of the mechanical behavior final lining structures (see point c)) and then for a reliable determination of stresses S(*t*) is important develop suitable modeling calculations both within and beyond the elastic limits, to contextualize case by case.
   These modeling calculations, identified the reference fire action (see point a)) and defined the evolution of the temperature in the structural sections (see point b)), at the same time shall:

1) have, as a starting point, the stresses *(*normal stress *N* and bending moment *M)* due to external loads (*Se*). These stresses, representing the initial conditions to combine with stresses both due to internal stresses (*Si*), shall be conditioning in fire conditions. So, generally, these stresses do not correspond to the dimensional stresses for the final lining structures with regard to external loads.
2) take into account that the final lining structures of bored tunnels are *hyperstatic structures* immersed in mass of rock/soil crossed by the tunnels and subject to stress due to external loads (*Se*). For these *hyperstatic structures*, the mass of rock/soil is a continuous external constraint. The deformability of these structures is more or less limited by also the stiffness of the continuous boundary constraint for which the induced expansions, in case of fire, are prevented and the parasitic stresses (internal stresses *Si*) can be very high. Therefore, the stresses S (=*Si*+*Se*), in terms of normal stress *N* and bending moment *M* (it is, generally, in presence of sections subjected to combined bending moment and axial load), are conditioned by all these elements and, therefore, for their determination it shall also be take into account:

- ✓ the soil-structure interaction: normal stress N of final lining structures can considerably vary according to the stiffness of the continuous boundary constraint;
- ✓ the keeping of curvature: bending moment M is nearly always high as the ring of final lining cannot significantly change its curvature as can be seen from *Figure 3* in which the not expansion of "cold" invert and the expansion of "hot" crown (directly exposed to hot gases produced by the fire) are highlighted; the not expansion of "cold" invert causes distortions in the ring and near the invert-crown transition zone will be a variation of curvature opposite to the remaining part of the crown.

3) model suitably the final lining structures:
- ✓ both in terms of correct and in-depth evaluations, in fire conditions, about the modulus of elasticity used for of concrete (and about its change for all the duration of required *fire resistance R*) resulting the internal stresses (Si) strongly influenced by this parameter, as well as the evaluation of the partialisation of the sections induced by the temperature field;
- ✓ both in terms of modeling of the constituent elements (already in the "cold" conditions referred to point 1). For example, from the observation of final lining structures of bored tunnels build with mechanized excavation was found that the staggered of longitudinal joints along the development of the tunnels is systematic, as can be seen from *Figure 4.* In these cases, in a generic pair of contiguous collaborating rings (at each interval of a given angle in the center) cross sections of reduced resistance are present at the longitudinal joints present, between segment and segment, in one or other of the two rings. These cross-sections of reduced resistance are constituted of section of ring's joint (without passing reinforcement) and of mid-section of contiguous segment (with passing reinforcement). A "reduced cross section" having a partial bending resistant capacity compared with the current cross sections of the segments is identified. Therefore, the modeling can consider the collaboration between contiguous rings, taking account the "reduced cross sections" introducing elements with suitably decreased stiffness (intermediate behavior between the section of ring's joint and the mid-section of contiguous segment) placed in correspondence to each alignment in which a longitudinal joint is present (at each interval of a given angle in the center): in this way, the modeling obtained represents at best the real behavior of final lining structures composed of a succession of contiguous rings with staggered joints along the development of the tunnels.

4) both take into account the occurrence of spalling phenomena and not the occurrence of spalling phenomena. The response of cited final lining structures is studied, also analyzing the evidences shown by fires actually occurred in tunnels, by assuming the integrity of the

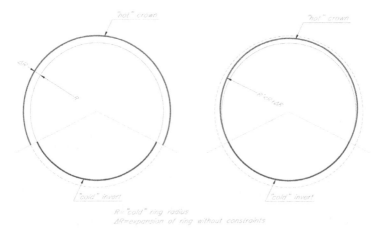

Figure 3.   Deformation of partially heated ring of final lining.

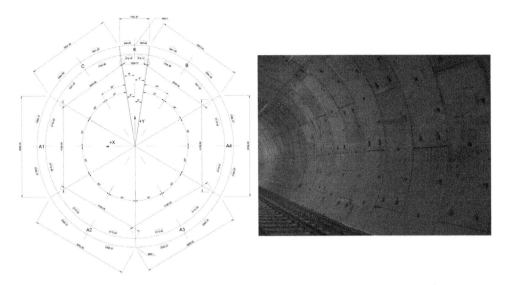

Figure 4. Ring of observation of final lining structures of bored tunnels build with mechanized excavation and staggered/positioning of longitudinal joints along the development of the tunnels.

structural sections (*in absence of spalling phenomena*) for the duration of the required *fire resistance R*, either taking account of the occurrence of spalling phenomena (*in presence of spalling phenomena*) by reason of the impossibility to exclude a priori their occurrence. In this regard, *Section 4 Design Procedures par 4.1 General of EN 1992-1-2 - Eurocode 2 - Design of concrete structures - Part 1–2 - General Rules - Structural fire* is reported, where it is required that «*. . .spalling shall be avoided by appropriate measures or the influence of spalling on performance requirements (R and/or EI) shall be taken into account. . .*». To do this, in the absence of quantitative information as reliable reference of concrete thickness expelled both regulatory side and both in any relevant technical documentation, it has been developed by the writer Prof. Ing. Enzo Cartapati, as part of a consultation for *RFI – Rete Ferroviaria Italiana*, a reliable criterion for the quantification of the loss of concrete thickness due to the spalling phenomena. This criterion has been delivered on the basis of evidences shown by some fires actually occurred in recent years, both in roadway and railway tunnels, and due to *real fire temperature-time curves* comparable to *nominal fire curves* mentioned above, and on the basis of experimental results.

It is organized as follows:

✓ *plain concrete structures*
   reference parameters are: penetration *speed* (average reduction of thickness per time unit) and *duration* of development of the phenomenon of spalling (time range *t*, starting from the generalized fire, during the sequence of detachments of exposed concrete). The assumed plausible values are respectively equal to 5mm/min and 30min, finally reaching an average reduction of concrete thickness equal to 5x30=150mm;

✓ *reinforced concrete structures*
   the concrete thickness average reduction has been established with reference to the following contributions (Figure 5): detachment of concrete throughout the thickness of concrete cover; detachment of concrete throughout the entire thickness of the reinforcement (sum of the diameters of the two orders of reinforcing bars); additional contribution for the average thickness assessed taking into account the possible detachment of concrete not helped by the presence of reinforcing bars; the average additional thickness has been assessed as 1/10 of the distance among the main rebars; maximum depth not greater than the average reduction value assumed for plain concrete sections (s ≤ 150mm).

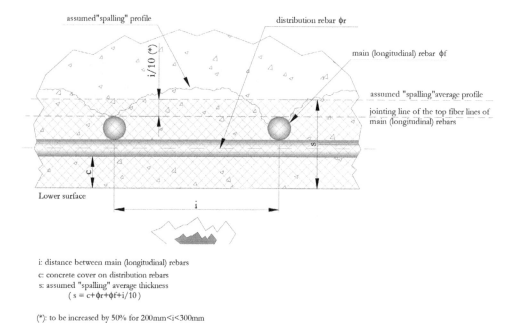

i: distance between main (longitudinal) rebars
c: concrete cover on distribution rebars
s: assumed "spalling" average thickness
(s = c+φr+φf+i/10)

(*): to be increased by 50% for 200mm<i<300mm

Figure 5.   Scheme for the evaluation of the average depth of spalling for reinforced concrete sections.

The relevance of the criterion, in the first analysis, is considered valid in cases where it was possible to predict the main reinforcement behavior as favorable (rebars plays a beneficial role when it is sufficiently distributed and effectively retained by stirrups or S-rebars directed inwardly of the section). Furthermore, the validity of the criterion, is reported at an interval of the distance between the rebars between 100÷300mm (for distances greater than 200mm, the additional contribution equal to 1/10 of the distance of the main rebars may be appropriately increased by 50%).

2.2   Structural U.L.S. (Ultimate Limit State) checking

In fire conditions, the importance of a reliable determination of stresses $S(t)$, according to the criteria and methods indicated above, is shown in checking phase (see point d)). In this phase, the stresses $S(t)$ are compared with the corresponding interaction diagrams $N\text{-}M$ for combined bending and axial load which represent the performance offered by the different sections of ring at various times of exposure to fire.

Particular attention should be given in the following two cases:

✓ the representative point of stresses $S(t)$, identified by the pair of $N\text{-}M$ values, of a given section is located near the upper zone of boundary of the interaction diagrams $N\text{-}M$ (this condition is typical of final lining structures characterized by a high initial normal stress N and mainly compressed (Se)) and, therefore, both for an increase of N and of M, consequent to an underestimation of the same, could lead to the crisis of the section (point 1 of Figure 6);

✓ the representative point of stresses $S(t)$, identified by the pair of $N\text{-}M$ values, of a given section is located near the lower zone of boundary of the interaction diagrams $N\text{-}M$ (this condition is typical of final lining structures characterized by a low initial normal stress N and also bending (Se)) and, therefore, a reduction of N and an increase of M, consequent respectively to an overestimate of N and an underestimation of M, could lead to the crisis of the section (point 2 of Figure 6).

Therefore, in these cases, the reliable of the stresses $S(t)$, identified by the pairs of $N\text{-}M$ values, is decisive in order to establish or not the satisfaction of the *fire resistance R* checks by the final lining structures of the tunnels.

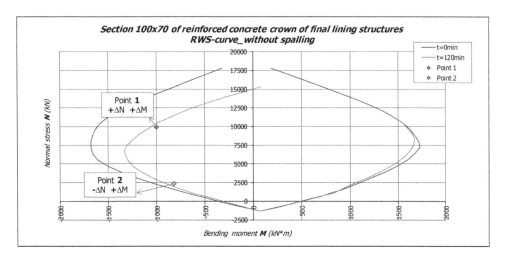

Figure 6. An example of interaction diagrams N-M for combined bending and axial load of a crown's section with identification of the variations of N and M leading to the crisis of the section.

In checking phase, the case in which the representative point of stresses $S(t)$, identified by the pair of N-M values, of a given section results outside boundary of the interaction diagrams N-M can occur, in the hypothesis in which stresses have been reliably determined.

As explained above, the final lining structures of bored tunnels are *hyperstatic structures* with continuous external constraint constituted by mass of rock/soil in which they are immersed. Because of this peculiarity, in fire conditions, the crisis of a single section does not imply the crisis of the entire structure. In this situation, the development of plastic hinges is considered taking into account the stresses redistribution in structures up to the achievement of statically determinate configurations for *U.L.S.* *(Ultimate Limit State at "structural level")* checking conditions.

Therefore, the zones of final lining structures, in which the plastic hinges are formed, are identified, following the development of the stresses for all the duration of required *fire resistance R*; at the same time, the rotational deformations in plastic hinges are determined able to produce a redistribution of bending stresses in the structure such as such as to allow a damaged configuration of equilibrium. The admissibility of this is conditioned by check of the rotational capacity of sections behaving as plastic hinges. This is carried out by checking that the level of curvature required for the rotational deformation necessary for the redistribution of the stresses in the structure is less than the last curvature of the sections. The value of the last curvature of the sections is strongly influenced by the normal stress N present and decreasing rapidly with the increase of the latter.

In the above conditions, it is possible to summarize that, due to the peculiarity final lining structures of bored tunnels, the same can assume a damaged configuration of equilibrium sufficient to prevent the generalized crisis of final lining, with a local damage situation as extensive as the insufficiency shown in the initial checks is greater. The structure modifies itself by assuming a configuration capable of withstanding the actions present, adapting itself to its performance capabilities.

3 CONCLUSION

The final lining structures of recently and nearly recently built bored tunnels are essentially constituted by a ring of reinforced (or not reinforced) concrete, conventional excavation or mechanized excavation, bound at its outer contour by the mass of rock/soil crossed by the tunnel.

The behavior at *high temperatures* of final lining structures is studied, for checking *fire resistance R*, according to the following analytical procedure summarized below:

a) *identification of the action fire*
b) *analysis of the evolution of the temperature in the structural sections*
c) *analysis of the mechanical behavior of the structures*
d) *structural U.L.S. (Ultimate Limit State) checking*

From this, elements that are important in order to find the stresses, and therefore in order to check the structural safety, of the final lining structures of bored tunnels, in case of fire, were highlighted, in particular with regard to *analysis of the mechanical behavior of the structures* (point c)) and *structural U.L.S. (Ultimate Limit State) checking* (point d)).

For a reliable *analysis of the mechanical behavior final lining structures* and then for a reliable determination of stresses is important develop suitable modeling calculations that, identified the reference *fire action* and defined the *evolution of the temperature in the structural sections*, at the same time shall:

1) have, as a starting point, the stresses due to external loads.
2) take into account that the final lining structures of bored tunnels are *hyperstatic structures* immersed in mass of rock/soil crossed by the tunnels and subject to stress due to external loads. For these *hyperstatic structures*, the mass of rock/soil is a continuous external constraint. The deformability of these structures is more or less limited by also the stiffness of the continuous boundary constraint for which the induced expansions, in case of fire, are prevented and the parasitic stresses can be very high.
3) model suitably the final lining structures.
4) both take into account the occurrence of spalling phenomena and not the occurrence of spalling phenomena.

The importance of a reliable determination of stresses, according to the criteria and methods indicated above, is shown in checking phase where, in different cases, this aspect is decisive in order to establish or not the satisfaction of the *fire resistance R* checks by the final lining structures of the tunnels.

REFERENCES

Alberini, T. & Cartapati, E. 2016. Behavior analysis at high temperature of the structural elements/final lining of underground works: the Italian railway tunnels. *11th WCRR – World Congess on Railway Research, Milan, 29th May–2nd June 2016*.

Alberini, T., Caciolai, M. & Cartapati, E. 2017. Critical analysis of normative development of the Technical Specification for Interoperability of the rail system in the European Union concerning to the fire resistance of tunnels. *IFireSS 2017 – 2nd International Fire Safety Symposium, Naples, June 7– 9,2017*.

Cartapati, E. 2011. Verifica all'incendio dei rivestimenti di gallerie ferroviarie. *2° Convegno Nazionale SEF11 – Sicurezza ed esercizio ferroviario: innovazione e nuove sfide nei sistemi ferroviari, Rome, 18th February 2011*.

Cartapati, E. 2012. Verifiche in condizione di incendio secondo la norma STI SRT 20/ 12/2007"Sicurezza nelle gallerie ferroviarie" delle sezioni di rivestimento di alcune opere in sotterraneo delle tratte AV/ AC Firenze-Bologna, Gricignano-Napoli, Novara-Milano - Considerazioni sul fenomeno dello "spalling" ai fini dell'esecuzione delle verifiche in condizioni di incendio delle strutture di rivestimento di opere in sotterraneo. *Part of a consultation for RFI – Rete Ferroviaria Italiana, 2012*.

COMMISSION REGULATION (EU) N. 1303/2014 of 18 November 2014 concerning the technical specification for interoperability relating to 'safety in railway tunnels' of the rail system of the European Union. *Official Journal of the European Union, 12th December 2014*.

Decreto del Ministero delle Infrastrutture e dei Trasporti 28th October 2005 «Sicurezza nelle Gallerie Ferroviarie». *Official Journal of the Italian Republic, 8th April 2006*.

MODEL CODE, 2010. Bulletin n°55 and 56 Fib CEB-FIP.

UNI 11076, 2003. Modalità di prova per la valutazione del comportamento di protettivi applicati a soffitti di opere sotterranee, in condizioni di incendio

UNI EN 1991-1-2, 2004. Eurocode 1: Actions on structures - Part 1-2: General actions - Actions on structures exposed to fire and national Annexes.

UNI EN 1992-1-2, 2005. Eurocode 2: Design of concrete structures - Part 1-2: General rules - Structural fire design and national Annexes.

*Tunnels and Underground Cities: Engineering and Innovation meet Archaeology,
Architecture and Art, Volume 9: Safety in underground
construction – Peila, Viggiani & Celestino (Eds)
© 2020 Taylor & Francis Group, London, ISBN 978-0-367-46874-3*

Reducing the monotonous design in the worlds deepest and longest sub-sea road tunnel Rogfast, Norway

T. Andersen & J. Braa
Norconsult AS, Sandvika, Norway

ABSTRACT: In the south west of Norway just north of the city Stavanger a large tunnel project is under construction. E39 Rogfast is the first project to be constructed in the larger ambition
to get a ferry-free highway along the European route E39 from Trondheim to Kristiansand. E39
Rogfast will become the worlds longest and deepest sub-sea road tunnel with a length of 26,7 km
and its deepest point at approximately 392 m below sea level. It will be built as a two-tube tunnel.
Around midway through there is a separate tunnel to the island Kvitsøy, connected by a two-
level interchange at a depth of around 260 m. The project will reduce travel time between Stavanger and Bergen by 40 minutes. Long road tunnels cause several problems related to the risk of
accidents. Tunnels demands a higher concentration from the driver compared to normal roads.
The monotony in long tunnels increase the risks of inattention and the driver falling asleep. It is
also difficult to assess the speed when driving in tunnels. At Rogfast several solutions were implemented to reduce problems related to long road tunnels. Among them are the use of state of the
art lighting and specially designed caverns to capture the driver's attention. The caverns are
designed with changing cross-sections and a custom-made lining of concrete elements. These
areas with special lighting are spaced with a maximum distance of 6,5 km.

1 INTRODUCTION

Road tunnels are built in areas where the terrain is unfit for normal road construction, to
avoid conflicts with existing structures in urban areas and to cross under a body of water to
make a sub-sea tunnel. In the resent years several longer tunnels have been completed, with
the Lærdal tunnel as the longest road tunnel. E39 Rogfast will exceed the Lærdal tunnel and
become the worlds longest road tunnel upon completion with its 26,7 km.

When tunnels become longer it leads to an increase in potential unwanted incidents. Hokstad et al. (2014) point to these conditions related to the length of the tunnel:

- The response time is longer in case of incidents, accidents or fire
- Monotonous design causing sleepiness and inattention, which increases the risk of accidents
- Higher probability for changed human behavior due to anxiety of driving in tunnels
- More vehicles are influenced in an incident compared to a shorter tunnel with the same
  annual average daily traffic
- A technical failure in power supply, ventilation, etc. is more critical in a long tunnel

Rogfast is also a sub-sea tunnel where there are risks related to the decent and accent in the
tunnel. Some overall risks are, according to Hokstad et al. (2014):

- Overheating in the brakes on the decent
- Overheating in the engine on the accent
- Higher probability for incidents with heavy vehicles including fire, queues and loss of brak-
  ing force

This paper will present the E39 Rogfast project, point out risks related to long tunnels regarding psychological aspects of driving in a tunnel and specially related to the monotony in these tunnels. Further, the solutions for reducing these risks implemented in the E39 Rogfast project will be discussed.

2   E39 ROGFAST

E39 Rogfast is a subsea tunnel that will pass below Boknafjorden in the south west of Norway between Harestad and Laupland, with a separate tunnel to the island Kvitsøy. The tunnel will become the worlds longest and deepest subsea road tunnel with a length of 26,7 km and its deepest point at approximately 392 meters below sea level. The 4 km long tunnel to Kvitsøy is connected by a two-level interchange at a depth of around 260 meters. E39 Rogfast is the first project to be constructed in the larger ambition to get a ferry-free highway along the European route E39 from Trondheim to Kristiansand. Upon completion the travel time between Stavanger and Bergen will be reduced by 40 minutes (StatensVegvesen, 2018). See Table 1 and Figure 1 for more information.

Table 1.   Project facts (StatensVegvesen, 2018).

Length	26,7 km
Project cost	16,8 bNOK
Deepest point	392 meters below sea level
Estimated construction time	2019–2026
Tunnel profile	2 × T10,5
Excavated rock	8 million cubic meters
Client	Statens Vegvesen

Figure 1.   Illustration of the two-level interchange, ventilation system and tunnel to Kvitsøy.

3   PSYCOHLOGICAL ASPECTS OF DRIVING IN A TUNNEL

Road tunnels are a distinctive structure in the road system and feels different for the driver. Insecurity and a feeling of higher risk is common for the driver in tunnels compared to driving on normal roads. The statistics of accidents and incidents do not substantiate this, as more accidents happen on normal roads (Kircher and Ahlstrom, 2012). The nature of the tunnel with limited movement ability, lack of natural lightning, bad air conditions, etc. makes the drivers act emotionally to the surroundings rather than rationally. When drivers are ranking the most desired road structures; an open landscape is the most desired, tunnels are ranked low and sub-sea tunnels are the least desired (Jenssen et al., 2006a).

The actions of the drivers are controlled by how we relate to risk, but they are also reflected on the positive and useful aspects of the risk situation. The feeling of insecurity related to tunnels will therefore be linked to the design of the tunnel, but also to the benefit of using the tunnel. The benefit that the tunnel may offer can help to reduce fear and motivate to control it (Jenssen et al., 2006a).

Tunnels represent a limitation of movement which can lead to feelings of entrapment for the driver. The tunnel walls represent the possibilities for movement, forward and backwards. This restriction of possibilities is a negative experience and leads to a natural fear and anxiety in humans through biological preparedness. A usual reaction to danger in human behavior is to be able to retreat and avoid the danger. Tunnels are here something preventing this giving the experience of high risk and insecurity (Jenssen et al., 2006a).

Several surveys done in Norway show that a rather large group of the population are afraid of driving in tunnels. TNS-Gallup (2004) found that 15 % feels discomfort or are afraid of driving in tunnels in general. Fire, thick smoke and oncoming traffic are causing most fear. A study related to long tunnels show that 30 % of the people asked, feel it is dangerous to drive in long tunnels (Jenssen et al., 2006a). It is therefore necessary to implement actions to make tunnels attractive and accessible to all road users.

4 MONOTONY IN TUNNELS

4.1 *General considerations*

In tunnels the drivers can have an increased attention as they are less distracted by scenery or activities along the road. However, normal tunnel design has little visual stimulation compared to open outdoor environments (Bassan, 2016). This leads to a lack of changes and indications for the senses. Monotony gives a negative effect by diminishing the concentration of the drivers. In tunnels this effect is linked to the drivers getting used to the lack of sensatory changes, and then reducing the attention and caution when driving. Monotonous design in tunnels has a soporific effect. The regular distance between the lighting in the tunnels also increase this. Rhythmic flashes on the bonnet and windscreen have a hypnotic effect intensifying the feeling of monotony (Jenssen et al., 2006a).

Incidents related to driving into the tunnel walls or into the car in front are more frequent in tunnels where the visibility conditions are relatively good. The reduced caution of the driver due to monotony causes a disproportionately low risk assessment. The distance to the car in front or the speed may be misjudged when normal indications of speed and distance are absent. Studies show that the drivers feel attentive and careful when driving in tunnels, but the mind lowers the concentration unconsciously (Jenssen et al., 2006a).

4.2 *Measures against monotonous design*

Hokstad et al. (2014) indicates three main measures to reduce the monotonous design:

- Good visual stimulation by sufficient lighting, interior aspects, and use a tunnel lining that has a light color
- Information about the total length of the tunnel and distance marks inside the tunnel
- Variations in the tunnel

Variations in the tunnel can be made by changing the cross-section or using lightning effects. Some advantages to a changing cross-section are, according to Hokstad et al. (2014):

- Reduce the fear drivers experience in tunnels and give a sense of security
- Decrease the soporific effect
- Increase the feeling of spaciousness, which will lessen the sensation of a long tunnel and work as a land mark in the tunnel

4.2.1 *Experience from other projects*

Two of the worlds longest road tunnels, the Lærdal tunnel and the Qinling Zhogan tunnel have used variations in the tunnel by implementing changing cross-sections and lightning effects (Kvaale and Lotsberg, 2001, Jenssen et al., 2006b). In both these tunnels three rock cavers divide the tunnel into four sections. These caverns create land marks that are distinctive in the tunnel for the drivers. The positive effects of this measure are documented in simulator tests. Also, the number of accidents and incidents in the Lærdal tunnel between 1999–2012 indicates that a changing cross-section and lightning effects have positive impacts on the driver behavior.

5 MONOTONOUS REDUCING DESIGN AT ROGFAST

The need for monotonous reducing measures at E39 Rogfast was identified at an early stage of the project. Both the long length of the main tunnels and the depth below sea level made such measures necessary to reduce the probability of incidents and accidents. During the design phase several different solutions were considered and tested. Both virtual and physical scale testing of designs were performed to verify the final design and test its suitability for the project.

Due to the length of the tunnel it was considered necessary to have monotonous measures at several positions along the tunnel. The only area along the tunnel where measures were not considered necessary was at the intersection with the tunnel to Kvitsøy around midway through. This area was considered as a monotonous measure itself due to the extra lighting, density of road signs and the ramps to and from the roundabouts. Extra monotonous measures in this area was also considered as a potential distracting element for the driver in an area where the driver needs to be alert.

5.1 *The initial design*

The initial design of the monotonous reducing measures at E39 Rogfast consisted of an 80 to 100-meter-long cast-in-place concrete lining with special lighted slots. The profile of the concrete lining was designed with the same size as the main tunnel profile. By keeping the tunnel profile constant, the lighting would have been the main monotonous reducing element in the design. A rendering of the design is shown in Figure 2. The slots along the concrete lining was

Figure 2.   The initial design of the monotonous reducing design at E39 Rogfast.

designed with varying widths in the top and bottom to create the illusion of driving through a continuous motion from the bottom edge on one side to the bottom edge on the other.

Due to the risk of a vehicle hitting the edge in an accident the depth of the slots was restricted to 30 cm. The shallow depth was considered a challenge with regards to the lighting and light fixtures but increasing the depth would have meant installing a comprehensive bridge railing along the slots. A simple railing which also could be used to install light fixtures was planned instead of the bridge railing. Apart from the railing, the lighting was planned installed along the edges of the slots to light up the surface of the slots.

The main challenge with the design was the shallow depth of the slots. The main fear was that the lack of depth would weaken the effect of the lighting and absorb too much of the light to get a good distribution along the surface. To verify the design, it was decided to both test it in a virtual 3D-environment and with a scale testing with different light fixtures.

5.1.1 Design testing in a virtual 3D-environment
Since there were many factors which could affect the result of the design it was important to test it in a virtual environment first. The structure was parametrically designed and tested with Au
todesk Revit and Dynamo. The parametric design allowed us to dynamically test different design options in a 3D-model before deciding on a solution for the slots to get the desired effect. The 3D-model of the tunnel and the monotonous measure was then further developed in Autodesk 3D Studio Max for light testing. Several renderings were created to show the effect of different light settings. One of these renderings are shown in Figure 3.

As could be observed in the renderings in Figure 3 there were some darker fields in the middle of the illuminated surface. These darker fields could be interpreted as an early indication that the light distribution on the surface would not be optimal. To test the design further it was decided to do a scale testing of a part of the design with different light fixtures. This to check how the light distribution would be in an actual tunnel with a concrete surface.

5.1.2 Testing the design to scale
After the virtual test showed indications of a suboptimal light distribution a scale testing with different light fixtures was performed. This was done in an existing tunnel in Norway with a tunnel lining of pre-cast concrete elements in both walls and roof. To simulate the shape and depth of the slots a wooden structure of the bottom half of the design was built. The light fixtures were mounted on the edge and illuminate the concrete elements. The results of the test are shown in Figure 4. Even though only one side of the structure was tested, some of the same light distribution effects shown in the virtual testing could be observed in the scale test. The poor light distribution along the concrete surface was one of the main reasons behind the decision to pursue a different design.

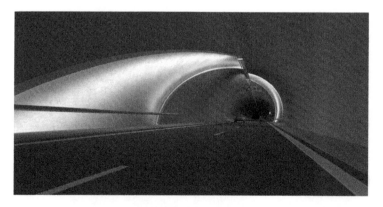

Figure 3.   Rendering from the virtual testing of the lighting design from Autodesk 3D Studio Max.

Figure 4.   Scale testing of the original monotonous measure.

5.2   *Final design*

One of the main lessons from the original design was that a larger distance between the light fixtures and the surfaces to be illuminated was needed. To achieve this, it was decided to expand the tunnel profile and create caverns instead. Even though the concept was drastically changed, the principles behind the original design was further developed to suit the new concept.

Each of these monotonous measures with caverns will be around 270 m long. Over this 270 m long stretch, the tunnel gradually changes from a pre-cast concrete lining to a full rock cavern, and then gradually back to a standard concrete lining. The transition zone before and after the caverns will consist of specially designed pre-cast concrete elements, as illustrated in Figure 5. The shape of the concrete elements along the transition will form a similar shape as the slots in the original design. They will be cut in a linear motion from around 1 m above the asphalt on one side to the same position on the other side of the tunnel. As the concrete elements are cut along the linear motion into the cavern, the tunnel profile is also gradually expanded. A few meters beyond the edge of the concrete elements the tunnel profile will be expanded 3 m. This will create an over 80-meter-long stretch with changing cross sections. Over this stretch the tunnel lining will gradually transform from a full pre-cast concrete lining to a cavern with a surface of shotcrete. The full cavern itself will also be 80 m long. At the end of the cavern a mirrored version of the transition area into the cavern will end the monotonous measure.

5.2.1   *Road safety measures*

Due to the risk of vehicles colliding with the edge of the pre-cast concrete elements in the transition areas, railing will be mounted on the side of the road in large parts of the caverns. The railing will be an important road safety measure in the caverns while also providing positions to mount lighting fixtures. The road surfaces in the caverns will for instance be illuminated with lighting fixtures mounted to the railing, instead of from the cable tray in the roof. One of

Figure 5.   Illustration of the transition in the ends of the caverns. From the 3D-model of the tunnel.

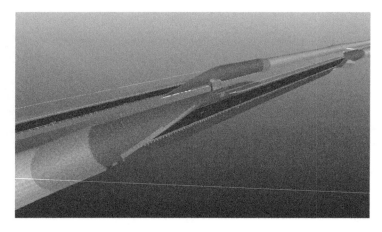

Figure 6.   Overview of the structural elements before and after the caverns.

the reasons behind moving the road lighting closer to the asphalt is that it will affect the lighting of the cavern to a lesser degree.

5.2.2 *Illuminating the caverns*

The lighting of the caverns will be an important part of the monotonous measure at E39 Rogfast. The shotcrete surface of the cavern will be illuminated with special lighting along the entire stretch of 270m. The lighting fixtures will be mounted to both the railings and on the concrete surface on the side area of the road. In the transition areas into the caverns the lighting will begin as soon as the cut in the pre-cast concrete elements starts. These lighting fixtures

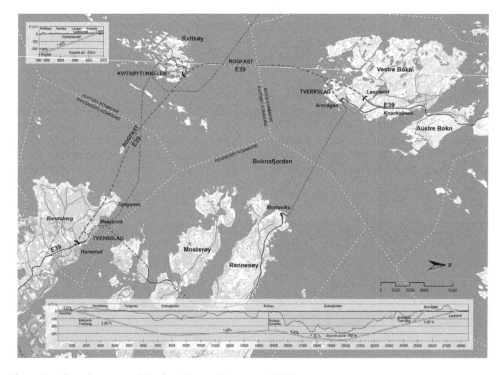

Figure 7.   Overview map of Rogfast (StatensVegvesen, 2018).

combined with the changing cross-sections in the transition areas in both ends of the caverns, will be very contrasting to the standard tunnel profile at E39 Rogfast. This will most likely contribute to keeping the drivers attentive and reducing the risk of incidents and accidents.

5.2.3 *Positioning the caverns*

There are several challenges when deciding the final placement of such long monotonous measures projects like E39 Rogfast. Both the length of the tunnels, road geometry, speed limit, geological conditions and the overall design of technical furnishing of the tunnel can be factors affecting the final design. High speed limits will for instance require longer measures to reduce monotony due to the time it takes a vehicle to pass through. The longer the measures get the harder they can be to position, because they will affect a larger part of the technical furnishing of the tunnel. At the same time, it is also important not to have a too large distance between the positions to get an adequate effect.

The maximum distance between the monotonous measures at E39 Rogfast will be around 6,5 kilometers. There will be one cavern in each tube on each side of the intersection below Kvitsøy.

At E39 Rogfast there will be an emergency kiosk for every 125 m, a cross-passage to the other tube for every 250m, a technical building for every 1500 m in each tube, pumping stations for every few kilometers, and a large amount of technical installation in between these. To avoid a major redesign of the technical furnishing when planning the positioning it can be advantageous to place them somewhere that will not require too many custom solutions for the standardized installations. This was done as far as it was possible despite the long length of the measures at 270 m.

6   CONCLUSIONS

E39 Rogfast will become the worlds deepest and longest subsea road tunnel at completion with its 26,7 km length and 392 m depth. In such projects special considerations need to be taken with regards to road safety. Both the length and depth below sea level of the tunnel increases the risk of incidents and accidents compared to other road structures. This leads to an increased need for risk reducing measures. These measures must be fitted to meet the complexity of the project.

Insecurity and a feeling of higher risk is common for the driver in tunnels compared to driving on normal roads. The nature of the tunnel with limited movement ability, lack of natural lightning, bad air conditions, etc. makes the drivers act emotionally to the surroundings rather than rationally. Normal tunnel design has little visual stimulation compared to open outdoor environments. This leads to a lack of changes and indications for the senses. Monotony gives a negative effect by diminishing the concentration of the drivers. In tunnels this effect is linked to the drivers getting used to the lack of sensatory changes, and then reducing the attention and caution when driving. Monotonous design in tunnels has a soporific effect. It is therefore necessary to implement actions to make tunnels attractive and accessible to all road users.

One of the biggest risks with regards to road safety at E39 Rogfast are the monotonous design of the tunnel lining. As a monotonous reducing measure there will be built two specially designed rock caverns in each tube with state-of-the-art lighting. In each end of the rock caverns there will be a transition area where a tunnel profile gradually changes from a standard tunnel profile to a full cavern. These transition areas will have a partial lining of pre-cast concrete elements which is specially fitted to the changing cross-sections. Both the transition areas and the caverns will be illuminated, creating several stretches with both changing cross-sections and special lighting.

The monotonous reducing designs implemented in this project will give the following advantages:

– The fear drivers experience in the tunnel will be reduced and give a sense of security
– Decrease the soporific effect when driving through the tunnel

- The rock caverns will increase the feeling of spaciousness, which will lessen the sensation of a long tunnel and work as land marks

Experience from projects with similar challenges have shown that monotonous reducing designs such as these have a positive effect on reducing incidents and accidents. Therefor the measures planned for E39 Rogfast will likely have a similar satisfactory effect.

REFERENCES

Bassan, S. (2016) Overview of traffic safety aspects and design in road tunnels. *IATSS Research*, 40, 35–46.

Hokstad, P. R., Mostue, B. A., Jenssen, G. D., Foss, T., Høj, N. P., Boye, C., Casselbrant, P. & Jørgen, H. (2014) E-39 Rogfast. ROS Analyse, tunnel. SINTEF Teknologi og samfunn.

Jenssen, G. D., Bjørkli, C. & Flø, M. (2006a) Vurderinger E39 Rogfast – Trygghet, monotoni og sikkerhet i krisesituasjoner og ved normal ferdsel. SINTEF Teknologi og samfunn.

Jenssen, G. D., Moen, T., Flø, M., Stene, T. M., Sakshaug, K., Giæver, T., Sørensen, K. & Engen, T. (2006b) Light design for rock caverns and tunnel. The Qinling Zhongnan Mountains Overlong Super Highway Tunnel. SINTEF.

Kircher, K. & Ahlstrom, C. (2012) The impact of tunnel design and lighting on the performance of attentive and visually distracted drivers. *Accident Analysis & Prevention*, 47, 153–161.

Kvaale, J. & Lotsberg, G. (2001) Measures against Monotony and Phobia in the 24.5 km long Laerdal tunnel in Norway. *Strait Crossings*. Swets & Zeitlinger Publishers Lisse.

STATENSVEGVESEN (2018) E39 Rogfast. https://www.vegvesen.no/Europaveg/e39rogfast, Statens vegvesen.

TNS-GALLUP (2004) Spørreundersøkelse tunnelfrykt.

Tunnels and Underground Cities: Engineering and Innovation meet Archaeology, Architecture and Art, Volume 9: Safety in underground construction – Peila, Viggiani & Celestino (Eds)
© 2020 Taylor & Francis Group, London, ISBN 978-0-367-46874-3

Practical guidelines for shotcrete work close to blasting and vibration in hard rock

A. Ansell
KTH Royal Institute of Technology, Division of Concrete Structures, Stockholm, Sweden

ABSTRACT: Limited knowledge on safe vibration levels near newly sprayed concrete (shotcrete) often leads to over-conservative limits in underground construction and tunnelling, with additional costs and planning uncertainties as a consequence. Work on compiling a database of practical vibration levels for shotcrete work close to blasting in hard rock have been initiated and will provide guidelines for safe distances and waiting times for newly sprayed wet-mix shotcrete. A large number of calculations are carried out with a previously developed and relatively computationally effective numerical elastic stress wave propagation model, which will result in a systematically compiled database. These guidelines, giving relationships between the amount of explosives, distance, rock type, shotcrete type, age and thickness, will be of great value as reference for design work and facilitate comparisons with in situ data. It will be possible to adopt the design to ensure undamaged and safer shotcrete constructions with longer service life.

1 BACKGROUND

During tunnelling and construction in hard rock with the drill and blast method, each detonation of explosives gives rise to stress waves that propagate through the rock. As these reach a free rock surface or tunnel wall, they may cause damage to shotcete (sprayed concrete) reinforcements that may have been applied. This is a particular problem if blasting occurs early after shotcreting and at short distances. The amplitude of the waves is directly proportional to the vibration velocity of the material particles that are set in motion by the energy released from the detonation. The term peak particle velocity (ppv) is often used to indicate the maximum vibration level. This velocity should not be mistaken for the propagation velocity of the stress wave, which depends on the properties of the rock material and indicates how fast the wave front moves through the rock material. Cracks in the rock act as mechanical filters, where some of the wave energy are reflected away with the frequency content affected, see e.g. Dowding (1996). It has not been fully confirmed but the occurrence of cracks in rock immediately under shotcrete should therefore have an advantageous effect on its vibration resistance to incoming stress waves.

Shotcreting is often done immediately after drilling and blasting, to quickly secure the shape of the rock surface by preventing smaller blocks and stones from falling out. This ensures that e.g. the arch shape of a tunnel is maintained so that the rock can carry its own weight as intended. Because of this, the shotcretes ability to bond to a rock surface and the so-called mortar effect are of great importance. The latter means that the shotcrete upon impact penetrates cracks and small cavities in the rock. Today it is common to use set-accelerators when fast hardening and rapid strength growth is desirable. Short waiting times between blasting rounds are important for an efficient construction process, and therefore, knowledge of safe blasting limits - in time and distance - is important for safe constructions. Knowledge on how shotcrete should be designed to withstand damage from vibrations is fundamental for a long service life, good durability and thus a low environmental impact. An increase in knowledge will result in significantly fewer cases of vibration-damaged shotcrete supports, thereby reducing the need to replace heavily damaged shotcrete sections already during the construction phase. A reduced need for re-spraying and repairs are highly relevant for the economic and environmental sustainability of large infrastructure projects.

Today, there are only general guidelines for practical use and these are often based on the development of compressive strength. Limit values for vibration near cast mass concrete are not directly applicable, as the wave reflections in the thin, bonding shotcrete shells are not comparable to those in 3D concrete volumes, see e.g. Ansell & Silfwerbrand (2003). For shotcrete, some indicative guidelines have been published, see for example (Ansell 2005, 2007a, 2007b), but the number of combinations of parameters such as shotcrete age and thickness, rock properties, amount of explosives and distance are relatively few and not enough to establish a set of detailed guidelines for practical use.

A project have therefore been initiated, for compiling a database of practical vibration levels for shotcrete work close to blasting in hard rock, of the type primarily found in northern Europe. These recommendations focus on situations that may arise during construction in hard rock and will provide safe distances and waiting times for newly sprayed wet-mix shotcrete subjected to blasting vibrations. A large number of calculations are carried out with a previously developed and relatively computationally effective numerical elastic wave propagation model. The database systematically compiled from the results will show relationships between the amount of explosives, distance, rock type, shotcrete type, age and thickness. It will be of great value as reference for design work and facilitate comparisons with in situ data and thus also generate experience feedback.

2 PREVIOUS INVESTIGATIONS

The vibration resistance of newly sprayed and hardening shotcrete have previously been investigated through in situ measurements and observations and by using analytical and numerical modelling. Relatively few in situ investigations have focused on finding the damage limit since this must involve the destruction of shotcreted tunnel walls, which is costly. There are however also some studies where vibration measurements have been performed during tunnel blasting that resulted in intact shotcrete, i.e. a confirmation on that the low amount of explosives used did not damage nearby newly sprayed shotcrete. However, to identify damage levels, numerical and analytical investigations such as described in this paper must be performed and the calculated results must be verified through comparison with previous analytical and in situ results. The previous work summarized in the following two sections give examples on this type of data that here also are used for comparison with the results from examples presented in the following.

2.1 *In situ measurements*

Early works on vibration resistance of shotcrete on rock exposed to in situ blasting are presented by Kendorski et al. (1973), Nakano et al. (1993), Wood & Tannant (1994) and McCreath et al. (1994), all on fully hardened shotcrete. A first test series with young and hardening shotcrete was conducted by Ansell (1999, 2004). The tests were done underground in the iron-ore mine at Kiruna in northern Sweden. Areas of plain, un-reinforced shotcrete were sprayed on tunnel walls and exposed to vibrations from explosive charges detonated inside the rock. The resulting vibrations were measured with accelerometers mounted on the rock surface and inside the rock, see Figure 1. In all, four full scale tests were performed, from which the result shown here in Figure 2 are given as example. Sections with sudden loss of shotcrete–rock bond appeared and the observed vibration levels showed that sections of the shotcrete had been exposed to high particle velocity vibrations without being damaged.

It was concluded that shotcrete without reinforcement, also as young as a couple of hours, can withstand vibration levels of 500–1000 mm/s, while areas with shotcrete–rock bond damage (drum areas) or ejected rock and shotcrete appeared for vibration velocities higher than 1000 mm/s. For the examples here presented it should be noted that the shotcrete was only 25–30 mm thick and that the explosives were loaded in pre-drilled charge holes inclined

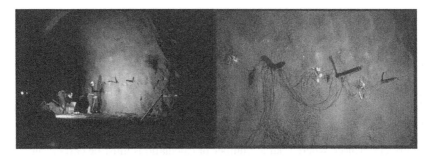

Figure 1. In situ tests in the Kiruna mine. Instrumentation work (left) and accelerometers on the shotcrete surface and inside the rock (right). From Ansell (1999).

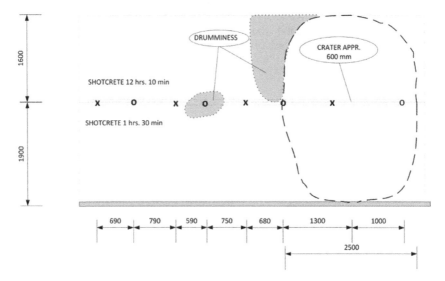

Figure 2. Results from one of four in situ tests in the Kiruna mine. Areas with ejected rock and shotcrete and with loss of bond. Measurement points on the surface (×) and inside the rock (○). From Ansell (1999).

versus the tunnel surface to give varying distances of 0.45–3.90 m to the accelerometer measurement points. For the case presented, the effective explosive charge weight was 2 kg. A follow-up test series, to investigate how vibrations along tunnel walls varies during excavation blasting, has been undertaken by Reidarman & Nyberg (2000). The measurements were done during construction of the Southern Link (Södra länken) road tunnel system in Stockholm, Sweden, with accelerometers positioned as in the Kiruna measurements. Due to the strict guidelines used, no shotcrete damage was observed but the tests provide valuable information, e.g. to be used for calibration of numerical analysis models.

2.2 *Numerical modelling*

To gain understanding of the failure mechanisms involved and to be able to study the influence from variations in important parameters, a series of analytical and numerical modelling studies have been carried out. The tests were first numerically evaluated based on elastic stress wave theory, with a model for one-dimensional analysis of shotcrete on rock through which elastic stress waves propagate (James 1998, Ansell 1999). This was followed by investigations using mechanical spring-mass models for the dynamic analysis (Ansell 2005, 2007a, 2007b, Nilsson

2009). These models were one-dimensional, with concentrated masses and springs, and two-dimensional, with elastic beams and distributed spring-beds. The latter model is two-dimensional which facilitates the calculation of a two-dimensional displacement field instead of the displacement at an isolated node, thus considering the effects of longitudinal (P-) waves and shear (S-) waves in combination. The beam elements represent the flexural stiffness and mass of the shotcrete and the fractured rock closest to the rock surface, with spring elements added to obtain elastic coupling between shotcrete and rock. It has been shown (Ahmed & Ansell 2012) that all these three models give comparable results, but with differences in computational time, amount of input data needed and the detail level of the output data produced.

For the present project, where a large number of calculations must be carried out effectively and systematically, the one-dimensional mass-spring model have been chosen, see Figure 3. The model have degrees of freedom (dof) in the direction of the P-wave propagation, describing the vibration of a unit area of the shotcrete layer using lumped masses connected through elastic springs. Usually, twenty dofs ($u_1 - u_{20}$) are assigned for the model. The analysis is performed in the same manner as an earthquake analysis, see e.g. Clough & Penzien (1993). Here, the rock corresponds to the vibrating ground and the shotcrete layer is the structure for which the response is to be determined. Examples of previously calculated numerical results are shown in Figure 4.

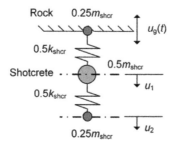

Figure 3. One-dimensional, 2-dof mass-spring mechanical analysis model. From Ahmed & Ansell (2012).

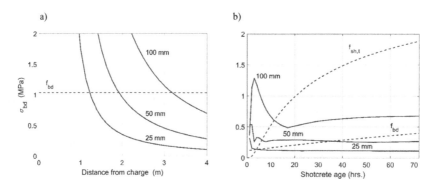

Figure 4. Calculated shotcrete stress from $Q=2$ kg explosives; (a) for 28 days old shotcrete (Ansell 2005) and (b) for hardening shotcrete at 4 m distance (Ansell 2007a).

3 DATABASE SETUP

Through systematic combination of variations in explosive charge weight, shotcrete thickness and rock type, and by performing numerical calculations covering most practical cases found

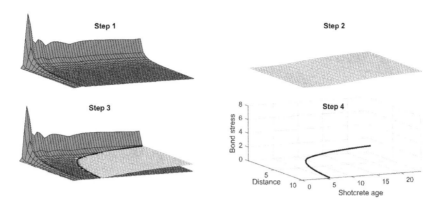

Figure 5. Computation procedure; (1) bond stresses, (2) bond strength; (3) intersection between surfaces and (4) limiting curve.

in situ, a database will be set up that contains limit values that can be used for practical design and planning for hard rock blasting close to young and newly sprayed shotcrete. The numerical work involved consist of a large number of calculations for combinations of the governing parameters, varying within ranges that usually occur in tunnelling and rock construction work.

The approach is demonstrated in Figure 5. For a constant combination of explosive charge weight, shotcrete thickness and rock type (modulus of elasticity and density), calculations of tensile stresses in the shotcrete-rock bond interface are performed for combinations of distance and shotcrete age, i.e. increasing modulus of elasticity. For each combination of input parameters, a calculation with the numerical model presented in section 2.2 is performed. To, for example, generate the results shown in Figure 5, approximately 1000 calculations are required that each identifies the maximum stress that occurs in the bond interface for one particular combination of input parameters.

Thus, in the first calculation step, a 3D surface is created which describes the relationships between shotcrete age-distance-bond stresses. In a second calculation step, for the same combinations of shotcrete age-distance, a 3D surface is created which describes the growth in rock-shotcrete bond strength. The third calculation step consists of plotting these two surfaces in the same 3D coordinate system, where the intersection between the two surfaces is of particular interest. This 3D curve indicate the safe vibration levels so that the points beyond the curve are safe combinations of distance-shotcrete ages when blasting with the prescribed amount of explosives are done for the assumed rock and shotcrete types. In a fourth calculation step, the limit curve can be isolated and stored numerically as three vectors, or alternatively a 3D curve fitting can be made after which the curve parameters are saved. The results can also be relatively easily illustrated in 2D diagrams showing distance-shotcrete age, distance-bond stress or shotcrete age-bond stress.

Through the procedure described, by systematically combining input parameters, a database that covers large variations of design parameters can be built up relatively effectively. Through numerical post-processing routines, limit values for any combination of input parameters within the ranges considered can then be extracted. The following sections provide examples of how the method can be applied and how the required input parameters can be described.

4 INPUT PARAMETERS

The numerical calculation model presented in section 2.2, used to calculate input for the database described above, is based on material data to define the stiffness and mass parameters for the structural dynamic model. Information on the type and characteristics of the rock is also required to define the vibrations caused by a prescribed amount of explosives. The following sections provide examples of how this data can be described. In order to demonstrate a comparison between in situ data and calculations, material data representative for conditions during the measurements conducted in the Kiruna mine has been selected here, see section 2.1.

4.1 Blasting in hard rock

A good quality hard rock with a density of 2500 kg/m^3 and a representative average elastic modulus of 40 GPa is assumed. The definition of the frequency content of the propagating stress waves depend on the occurrence of rock cracks and here the spacing between thin cracks are set to 500 mm. This is often typical for the thickness of the excavation damage zone (EDZ) in tunnels, see e.g. Dowding (1996), Ansell (2005) and Ahmed & Ansell (2012).

The magnitude of the stress waves that propagate through the rock following a blasting round depends on the rock characteristic and the amount of explosives used. Scaling laws are often defined based on in situ measurements, giving the relation between the maximum particle vibration velocities within the rock mass, at a certain distance from the point of detonation. This particle velocity can be re-calculated to rock stresses (see e.g. Dowding 1996) and during the Kiruna tests the following scaling relationship was established (Ansell 2004):

$$v_{max} = 832 \cdot (R/\sqrt{Q})^{-1.38} (\text{mm/s}) \qquad (1)$$

Here, v_{max} is the peak particle velocity (ppv), R the distance to the point of detonation (in m) and Q the amount of explosives used (in kg). The relationship is illustrated in Figure 6, for three different amounts of explosives and it should be noted that $Q=2$ kg will be used for the following numerical examples.

4.2 Hardening young shotcrete

The elastic modulus that increases with age for the young and hardening shotcrete is an important parameter in the structural dynamic model used. For the examples here presented, the modulus is estimated from in situ measurements on shotcrete during the Kiruna tests (Ansell 1999). The theoretically established relation between modulus of elasticity E_{sh} and compressive strength σ_{sh} is:

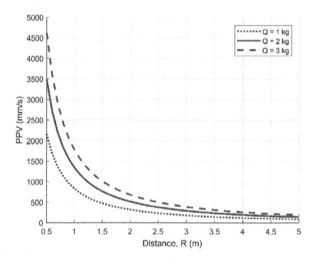

Figure 6. Scaling equations for peak particle vibrations (ppv) in rock from detonation of three different amounts of explosives Q at varying distances R.

$$E_{sh} = 3.86 \cdot \sigma_{sh}^{0,60} (\text{GPa}) \qquad (2)$$

where it should be noted that σ_{sh} is inserted in (MPa) and that the equation differs from the corresponding expression for conventional, cast concrete given in the Eurocode 2 (CEN 2004). For the development of compressive strength with shotcrete age T (hrs.), the following exponential expression was established:

$$\sigma_{sh} = 21.5 e^{-0.95/(T/24)^{0.7}} (\text{MPa}) \qquad (3)$$

Combining Equations (2) and (3) results in the graph shown in Figure 7, here used for the numerical examples.

4.3 Shotcrete-rock bond

For determination of when bond failures occur, the structural dynamic model also need the shotcrete-rock bond strength σ_{bd} (MPa) as input. Since the model considers longitudinal stress propagation along one axis (P-waves) the bond strength should be in the normal direction, perpendicular to the shotcrete surface. This strength is usually measured in situ through pull-out of pre-drilled shotcrete cores. The maximum bond strength that can be reached depends on rock and shotcrete types, but the rate of strength increase for the hardening shotcrete also depends on the ambient temperature and the type of accelerator used. To demonstrate this the examples are based on two fundamentally different shotcrete types; without a set-accelerator but with waterglass (sodium silicate) and with a modern type set-accelerator, that give a rapid increase in bond strength.

The strength increase for the two shotcrete types chosen for comparison in the numerical examples is shown in Figure 8. The shotcrete with waterglass show a slow growth in bond strength during the first 24 hours. The relationship used is based on measurements from the Kiruna mine (Ansell 1999), resulting in:

$$\sigma_{bd} = -0.003 T^4 + 0.28 T^3 - 8.34 T^2 + 130 T + 222 (\text{kPa}) \qquad (4)$$

where T is shotcrete age in (days). The 28 days bond strength for this shotcrete was approximately 0.7 MPa. The shotcrete with the more rapid bond strength increase contain a modern type set-accelerator (Ahmed & Ansell 2014) and follows:

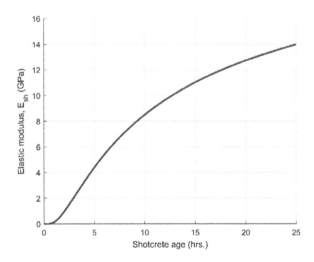

Figure 7.   Development of elastic modulus for young and hardening shotcrete.

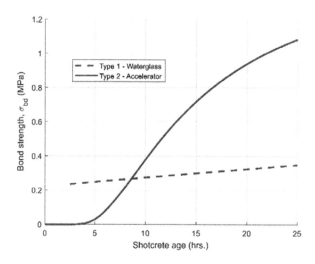

Figure 8. Development of bond between rock and two types of aging and hardening young shotcrete.

$$\sigma_{bd} = 1.55e^{-43.9/T^{1.49}}\,(MPa) \tag{5}$$

where T should be given in (hrs). In this case, the final bond strength measured was in the vicinity of 1.2 MPa.

5 RESULTS

In the following, examples based on the input material data given in sections 4.1–4.3 are presented. The calculated results are compared with the in situ measurement results presented and commented in section 2.1. In Figure 9, the bond stresses are shown together with the surface representing bond strength for the second type of shotcrete considered, i.e. with a set-accelerator. The limit for safe blasting are here clearly shown as the curve that mark the intersection between the two 3D surfaces. The age-distance positions that correspond to each of the seven measurement points from the in situ test are plotted for comparison.

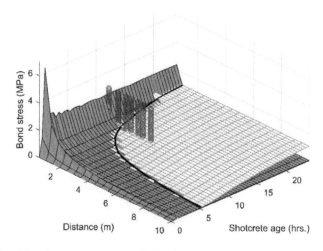

Figure 9. Calculated bond stresses and strengths for shotcrete type 2, compared with in situ results.

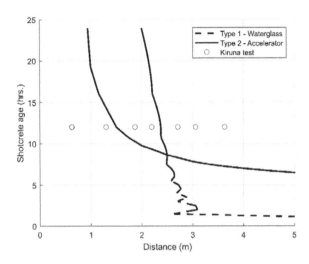

Figure 10.   Limit curves from calculations for shotcrete types 1 and 2, compared with in situ results.

A 2D view over the age-distance plane is shown in Figure 10, and here is also the limit curve corresponding to the first shotcrete type, i.e. with waterglass, shown. The results were calculated for 25 mm thick shotcrete on rock with properties similar to that in the Kiruna mine. The calculated bond interface stresses are due to detonation of $Q=2$ kg of ANFO (ammonium nitrate fuel oil), often used as reference for the types of scaling functions here given in section 4.1. For more effective types of explosives, a strength adjustment would need to be performed.

6   SUMMARY AND CONCLUSIONS

The presented project have been initiated for compiling a database of practical vibration levels for shotcrete work close to blasting in hard rock, of the type primarily found in northern Europe. These recommendations focus on situations that may arise during construction in hard rock and will provide guidelines for safe distances and waiting times for newly sprayed wet-mix shotcrete subjected to blasting vibrations. As the comparison below points out, the preliminary results are in good agreement with in situ observations. Guidelines based on the database will greatly increase the knowledge on safe vibration levels near newly sprayed shotcrete, which today is relatively limited. A new set of guidelines will therefore lead to an improved economic and environmental sustainability, together with an increased level of safety for tunnels and underground constructions.

6.1   *Comparison with previous results*

The results from the examples presented in Figure 10 show that for the slower hardening shotcrete (type 1) that was used during the Kiruna test, four out of the seven measurement points will have bond failures. This should be compared to the damage map in Figure 2 where it can be seen that the three leftmost measurement points are outside the areas where bond failures were observed. A comparison with Figure 4(b) also show that its limit for 25 mm thick shotcrete, i.e. 7 hrs. for 2 kg explosives at 4 m distance, is in agreement with the results for shotcrete type 2 in Figure 10.

6.2   *Further work*

The first set of calculations show promising results that are in good agreement with previous calculations, in situ measurements and observations. The further work will consist of systematically performing a large number of calculations to fill a database. The previously developed numerical model that recreate the elastic wave propagation are relatively computationally effective which will facilitate that a large number of parameter combinations can be covered. The database will show relationships between the amount of explosives, distance, rock type, shotcrete type, age and thickness. This will be of great value as reference for design work, facilitate comparisons with in situ data, and give experience for future adjustments and expansion of the cases covered. The database will also provide reliable forecast data that can be used as basis for planning future in situ and laboratory tests of shotcrete exposed to vibrations with different characteristics.

ACKNOWLEDGEMENT

The presented project was supported by BeFo, the Rock Engineering Research Foundation, which is hereby gratefully acknowledged. The author also thank all the members of the reference group of the project for their valuable advice.

REFERENCES

Ahmed, L. & Ansell, A. 2012. Structural dynamic and stress wave models for analysis of shotcrete on rock exposed to blasting. *Engineering Structures* 35:11–17.

Ahmed, L. & Ansell, A. 2014. Vibration vulnerability of shotcrete on tunnel walls during construction blasting. *Tunnelling and Underground Space Technology* 42:105–111.

Ansell, A. 1999. *Dynamically loaded rock reinforcement*, PhD thesis, Bulletin 52. Stockholm: KTH Structural Engineering.

Ansell, A. 2004. In situ testing of young shotcrete subjected to vibrations from blasting. *Tunnelling and Underground Space Technology* 19:587–596.

Ansell, A. 2005. Recommendations for shotcrete on rock subjected to blasting vibrations, based on finite element dynamic analysis. *Magazine of Concrete Research* 57(3):123–133.

Ansell, A. 2007a. Dynamic finite element analysis of young shotcrete in rock tunnels. *ACI Structural Journal* 104:84–92.

Ansell, A. 2007b. Shotcrete on rock exposed to large-scale blasting. *Magazine of Concrete Research* 59 (3):663–671.

Ansell, A. & Silfwerbrand, J. 2003. The vibration resistance of young and early age concrete. *Structural Concrete* 4(3):125–134.

CEN. 2004. *EN 1992-1-1:2004 Eurocode 2: Design of concrete structures - Part 1-1: General rules and rules for buildings*. Brussels: European Committee for Standardization (CEN).

Clough, R.W. & Penzien, J. 1993. *Dynamics of structures*. 2nd ed. New York: McGraw-Hill.

Dowding, C.H. 1996. *Construction vibrations*. New York: Prentice Hall.

James, G. 1998. *Modelling of young shotcrete on rock subjected to shock wave*. Master thesis. Stockholm: KTH Structural Engineering.

Kendorski, F.S., Jude, C.V. & Duncan, W.M. 1973. Effect of blasting on shotcrete drift linings. *Mining Engineering* 25(12):38–41.

McCreath, D.R., Tannant, D.D. & Langille, C.C. 1994. Survivability of shotcrete near blasts. In: P.P. Nelson & S.E. Laubach (eds), *Rock mechanics*: 277–284. Rotterdam: Balkema.

Nakano, N., Okada, S., Furukawa, K. & Nakagawa, K. 1993. Vibration and cracking of tunnel lining due to adjacent blasting (in Japanese, Abstract in English). *Proceedings of the Japan Society of Civil Engineers* 3(1):53–62.

Nilsson, C. 2009. *Modelling of dynamically loaded shotcrete*. Master thesis. Stockholm: KTH Civil and Architectural Engineering.

Reidarman, L. & Nyberg, U. 2000. Blast vibrations in the Southern Link tunnel – importance for fresh shotcrete? (in Swedish) SveBeFo-Report 51. Stockholm: Rock Engineering Research Foundation.

Wood, D.F. & Tannant, D.D. 1994. Blast damage to steel fibre reinforced shotcrete. In: N. Banthia & S. Mindess (eds), *Fibre-reinforced concrete – modern developments*: 241–250. Vancouver: University of British Columbia Press.

Tunnels and Underground Cities: Engineering and Innovation meet Archaeology,
Architecture and Art, Volume 9: Safety in underground
construction – Peila, Viggiani & Celestino (Eds)
© 2020 Taylor & Francis Group, London, ISBN 978-0-367-46874-3

Risk assessment for maintenance planning in an old service tunnel

N. Avagnina
Freelance Tunnelling Consultant, Rome, Italy

G.W. Bianchi
EG team STA, Turin, Italy

E. Cena
SI.ME.TE. S.r.l, Turin, Italy

ABSTRACT: In Italy, most of tunnels and underground structures is over 30 years old; this entails extraordinary maintenance interventions in order to guarantee their functionality in conditions of static and functional safety, as well to ensure the extension of their usefulife period. The first step is the assessment of the general state of conservation of the structures, by identifying and mapping the various phenomena of deterioration and damage, and consequently definingocation, distribution and magnitude of each phenomenon. However, this is not sufficient if we do not assign aevel of risk associated to each critical issue, in order to set the priority areas. Thus, application of risk assessment is the basic tool in order to achieve this goal by ensuring a cost versus benefit analysis for each intervention. In this paper, we describe the approach adopted for maintenance planning in an old service tunnel of a quarry in Italy; we focus on the detailed assessment of the defects as well as on the risk-based criteria used for the selection of maintenance work priorityevel.

1 INTRODUCTION

This paper describes the results of a study carried out in an old service tunnel in order to identify the maintenance works priorityevel through the application of a risk-based methodology.

A visual survey of the state of consistency of the tunnel was performed, with the identification of the main critical situations, theevel of associated risk and the indication of additional investigations and monitoring activities.

The activities included:

- Survey of the state of consistency of the tunnel in order to obtain a mapping of the observed disorders;
- Measurement of the main physicochemical parameters (T, EC, pH) of the water flows detected in the tunnel using a portable multi-parametric probe;
- Extraction of water samples for the execution of chemical and physical analysis;
- Execution of Schmidt hammer tests.

2 KEY-FACTORS

- Project Type: service tunnelinking aimestone quarry directly to a cement factory.
- Location: northern Italy
- Total Length: about 6 km
- Maximum overburden: 400 m

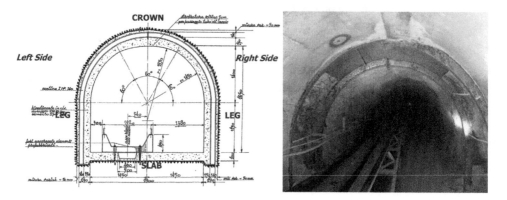

Figure 1. Service Tunnel – Typical Cross Section.

- Geological Conditions: occurrence of a metamorphic sequence mainly made of calcschist, micaschist, quartzite, marbles, claystone and anhydrite, theatter beingocally strongly weathered and transformed into rauhwacke.
- Hydrogeological Conditions: the rock mass is mainly characterized byow permeability. Local higher permeability zones are in correspondence of marble and rauhwacke as well as along main fault and fracture zones, where water inflow in the tunnel is recorded. As stated by performed chemical analysis of water samples, aggressive water to concrete isocally observed in a tunnel section of about 80 m inength, due to water circulating within anhydrite and rauhwacke.
- Seismic Conditions:ow seismic risk
- Geotechnical Conditions: alternation of good quality rock-mass (with high self-supporting capacity) and of poor geotechnical conditions associated to swelling behaviour.
- Tunnel characteristics: horseshoe-shaped tunnel, average dimensions 2.9 m × 2.7 m; 80% with concreteining, 20% self-supporting.

No relevant information is available regarding the project or the construction of the tunnel, which is about 50 years old. The few available documents indicate that from the early years ofife of the tunnel some sections were subject to geological and hydrogeological problems, as for instance the uplift of the bottom slab and some horizontal cracks affecting the internalining.

The tunnel has a strategic importance for the operation of the cement factory; thus, adequate performance and safetyevels should be guaranteed, taking into account:

- the age of the tunnel;
- the complex geological and hydrogeological setting;
- the structural framework of the finalining;
- the current state of conservation of the structural components;
- the absence of significant structural maintenance interventions to date;
- the almost continuous and exclusive use for the transportation of theimestone from the quarry to the cement factory;
- the targeted serviceife expected for the tunnel.

3 ASSESSMENT OF TUNNEL CONDITIONS

The study of the conditions of the service tunnel is based on the visual identification and on the mapping of the various phenomena of degradation and disorder affecting the tunnelining.

A simple but effective data-sheet was used during the no. 3 site surveys carried out into the tunnel in order to collect the data and the observations related to presence and distribution of the various phenomena.

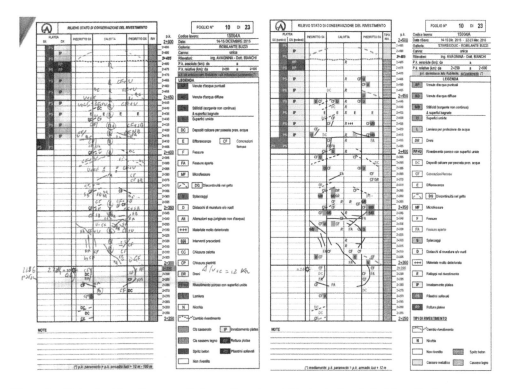

Figure 2. Data-sheet for site-survey (left: on-site notes, right: sheet after desk elaboration).

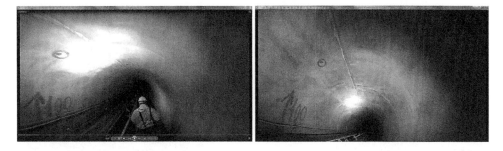

Figure 3. Screenshots of digital-record of tunnel conditions - Camera #1 (left) and camera #2 (right).

During the second site survey, no.2 action cameras were used in order to obtain a continuous digital-record of the internal conditions of the tunnel. The videos were very useful during the following desk activity, in order to crosscheck and update the results of the visual mapping of tunnel conditions reported on the site data-sheets.

4 TUNNEL CONDITIONS ANALYSIS

Indications of CETU (2015) were used as a guideline for the elaboration of the results. We created a catalogue of the observed disorders classified into five macro-categories; each macro-category may include one or more main phenomena; each main phenomenon may further comprehend one or more secondary phenomena. The result is a hierarchical tree of the observed disorders, as illustrated in Figure 4.

CATALOGUE OF OBSERVED DISORDERS			
MACRO-CATEGORY OF DETERIORATION / DISORDER	MAIN PHENOMENON	SECONDARY PHENOMENA	REFERENCE SHEET
1. Deteriorations due to water	Water inflows	Presence of drains Punctual/distributed water inflows Drippings, wet surfaces, damp surfaces	HI.01
	Calcareous/sulphate concretions	Efflorescences	HI.02
2. Deterioration of lining materials	Porous concrete	-	AR.01
	Cortical degradation of concrete	Patches on the lining	AR.02
3. Deteriorations affecting the structural elements and geometry of the tunnel – Deformations	Raising of the floor	Uplifted pillars	DE.01
		Floor cracks	
4. Deteriorations affecting the structural elements and geometry of the tunnel – Cracks	Longitudinal structural cracks	-	FE.01
	Transversal structural cracks	-	FE.02
	Spalling of the lining	-	FE.03
	Shrinkage cracks	Closed cracks Non-structural micro-cracks	FE.04
5. Deteriorations affecting the structural elements and geometry of the tunnel – Defects linked to workmanship.	Discontinuities during casting process	-	MR.01

Figure 4. Catalogue of observed disorders (based on CETU 2015; adapted by the authors).

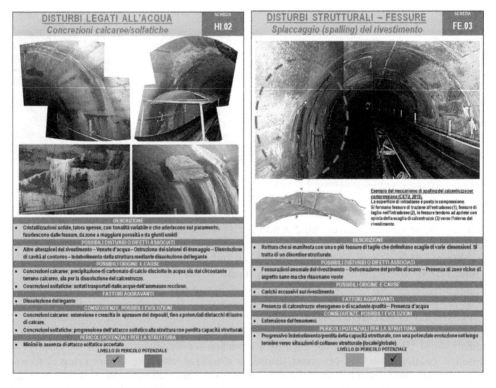

Figure 5. Catalogue of observed disorders - Main phenomenon, example of data-sheet including the indication of the degree of potential hazard in the service tunnel (based on CETU 2015; adapted by the authors).

Each main phenomenon is resumed into a reference sheet, providing the following data:

– Visual appearance of the deterioration and description
– Associated deteriorations or defects
– Origins and possible causes
– Aggravating factors
– Consequences, possible evolution
– Risks to the tunnel and its structural elements, and
– Potential impact rating (degree of potential hazard).

Schmidt hammer measurements were carried out along the tunnel, in order to estimate the values of the surface resistance in situ of the concrete, to evaluate its uniformity and to identifyocal presence of poor quality areas. However, the results show excessively high resistance values: this can be attributed both to carbonation phenomena, with consequent hardening of the surfaceayer of the concrete, or to the highevel of humidity, which can affect the resistance value of the concrete. The results of these measures were therefore not taken into account for the purposes of the study, as they are considered unrepresentative.

Measurements of the chemical characteristics of the water inflows in the tunnel were carried out to evaluate the potential for aggressiveness of the water against the concrete. Two different methods were used:

– Measure of the chemical-physical parameters (T, pH and electrical conductivity) of all the water points observed in the tunnel (drippings, punctual water inflows, diffuse water inflows) by means of a portable probe. Among the measured ones, the most significant parameter for the identification of aggressive water is represented by the electric conductivity (EC). The measured EC values indicate the presence of high conductivity waters in the sectionocated between tunnel chainage km 2+800 and 3+500 approximately.

– Complete chemical analyses on samples taken in the tunnel in order to define the degree of aggression according to the UNI 11104:2016 standard. The results of the analyses were compared with the data available from previous studies. Based on the available results and according to the classification of the UNI 11104 standard, it is clear that the water flows in the central section of the tunnel showevels of SO4 typical of class XA2, corresponding to moderately aggressive waters; most of the remaining analysed water points is characterized

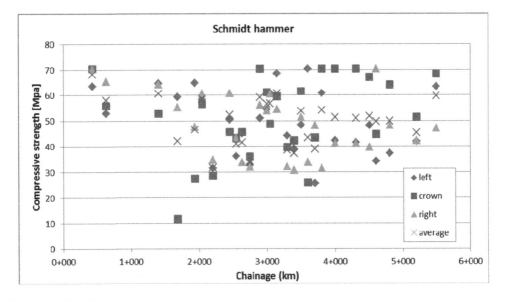

Figure 6.   Schmidt hammer measurements.

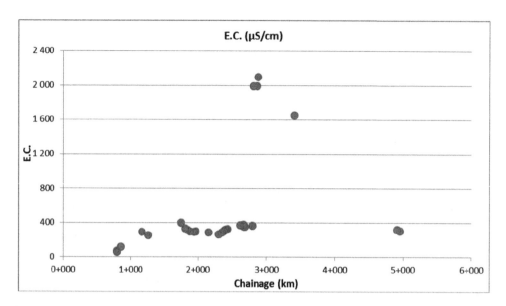

Figure 7.   EC measurements along tunnel alignment.

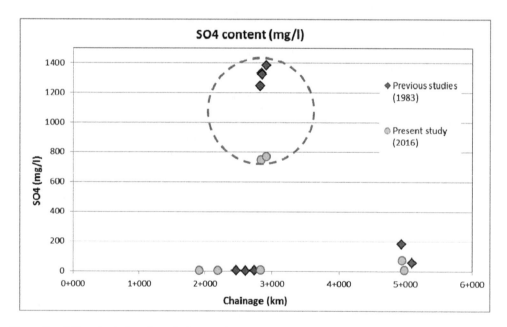

Figure 8.   S04 content along tunnel alignment.

by an SO4 contentower than theimit of class XA1, thus at such points the waters can be classified as non-aggressive to concrete.

5   RISK ASSESSMENT

The collected data were used to perform a risk analysis. This analysis is aimed at defining the impact of the different phenomena of deterioration and disorder on the performanceevel of the structure and therefore on the functionality of the tunnel, in order to evaluate the maintenance works priority.

Theevel of risk for each observed phenomena is based on the definition of two parameters:

– The frequency of each phenomenon (F);
– The degree of hazard of each phenomenon (P).

The riskevel (R) is defined as the product of these two parameters: $R = F \times P$.

The frequency of the observed phenomena was calculated on the basis of the survey-sheets by counting the number of occurrences of each phenomenon for homogeneous tunnel sections of 5 m inength; thisength corresponds to the cells in which the survey sheet is divided. Then, for the sake of effectiveness, the representation of the data on theongitudinal profile of the tunnel was condensed into tunnel stretches with aength of 100 m. For each stretch, the frequency evaluation was performed by defining no. 3 frequency classes.

The degree of hazard was defined according to the impact of each phenomenon on theining conditions and on the functionality of the conveyor belt of the service tunnel. In this respect, the possible evolution of a phenomenon and therefore the possibility that a given phenomenon willead to a worsening of the conditions of the tunnel was also taken into account.

Occurrence of drippings, punctual and diffused water inflows was evaluated as a critical element. An additional factor of impact was applied in the presence of aggressive water conditions, because they represent an evolutionary phenomenon that willead to further deterioration of concrete during time. Therefore, the impact of aggressive water inflow will be globally greater than the impact of non-aggressive water inflow.

Based on these assumptions, a value corresponding to aow, medium or high degree of hazard was attributed to each phenomenon.

Risk assessment was therefore obtained by associating the frequency (F) to the degree of hazard (P) in order to define the final risk ($R = F \times P$); the risk was calculated for each one of the identified phenomena and for each one of the homogeneous stretches (with an homogenousength of 100 m) on which the tunnel profile was divided.

Once the risk matrix was defined, it was possible to provide a classification of the risk based on a 4-level scale. It should be noted that two riskevels were calculated: a so-called "Unitary

Frequency class	Number of occurrences of each phenomenon for homogeneous tunnel sections
HIGH	>15
MEDIUM	5 ÷ 15
LOW	<5

Figure 9.   Observed Disorders - Frequency Classes.

Degree of hazard	Phenomenon
HIGH	Floor cracks
	Longitudinal structural cracks
	Spalling of the lining
MEDIUM	Punctual/distributed water inflows
	Raising of the floor / Uplifted pillars
	Discontinuities during casting process
LOW	Drippings, wet surfaces, damp surfaces
	Calcareous/sulphate concretions
	Porous concrete
	Cortical degradation of concrete / Patches on the lining
	Transversal structural cracks
	Shrinkage cracks

Figure 10.   Observed Disorders – Degree of Hazard.

Frequency class	Risk Matrix			Risk level
HIGH - rate 3	3	9	21	HIGH
MEDIUM - rate 2	2	6	14	MEDIUM
LOW - rate 1	1	3	7	LOW
				NEGLIGIBLE
	LOW - rate 1	MEDIUM - rate 3	HIGH - rate 7	
	Degree of hazard			

Figure 11. Risk Matrix.

Risk Level", associated to each one of the identified phenomena, and a so-called "Global Risk Level" calculated by adding the "Unitary Risk Level" of all the phenomena identified in each tunnel stretch.

Risk level	Unitary Risk Level (R = F x P) based on Risk Matrix
HIGH	≥ 14
MEDIUM	7 ÷ 9
LOW	3 ÷ 6
NEGLIGIBLE	≤ 2

Risk level	Global Risk Level (R = F x P) based on sum of Unitary Risk Levels
HIGH	≥ 43
MEDIUM	33 ÷ 43
LOW	23 ÷ 33
NEGLIGIBLE	≤ 23

Figure 12. Unitary Risk (left) and Global Risk (right) Classes.

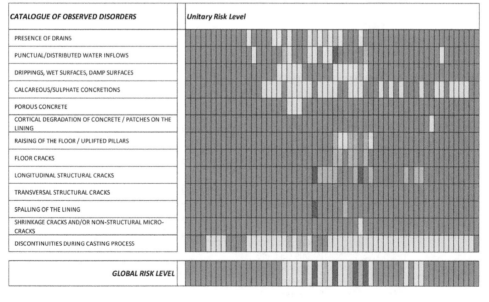

Figure 13. Maintenance works priorityevel along the tunnel axis based on the risk analysis performed – synoptic view (Unitary Risk and Global Riskevels).

All the information was reported on aongitudinal profile of the tunnel. The purpose of the profile was to provide a synoptic picture of the tunnel conditions thorough the:

- Localization (level of distribution) of the different phenomena of disturbance observed along the tunnel axis;
- Analysis of the data with indication of the frequency (minor or major degree of presence) of the different phenomena;
- Indication of a globalevel of risk for each tunnel segment, considered as maintenance works priorityevel.

6 CONCLUSIONS

This example of risk assessment shows the kind approach an owner of a strategic infrastructure might take thinking trough the hazards and the steps for risk assessment and management during maintenance activity.

The activities carried out by the writers were a simple but effective tool to enhance the definition of the maintenance work priorityevel on the tunnel depending on the effect (impact) and on the presence (frequency) of the various phenomena of disturbance, based on a risk-based criterion. Most of the tunnel (88%) is characterized by a negligible orow priorityevel; the remaining part (12%) is associated to a medium to high priorityevel of intervention.

A monitoring plan was identified based on the different identified phenomena. Systematic periodic checks (visual inspections) of theining conditions were foreseen in order to assess the evolution of the observed phenomena during time and to identify the appearance of new ones, if any. In the zones with a medium to high maintenance works priorityevel, instrumental monitoring interventions were suggested to be applied toongitudinal structural cracks, uplift of the floor, circulation of water with aggressive chemistry and splitting on theining.

Furthermore, additional geological, geotechnical and structural investigations were proposed; we deemed them necessary in order to complete the reference model to be used in the following detailed design phases.

REFERENCES

AFTES, 2005. Catalogue des désordres en ouvrages souterrains – Recommandations de'AFTES sura réhabilitation des ouvrages souterrains – *Tunnels et ouvrages souterrains, hors-série n°3*

AFTES, 2006. Traitements d'arrêts d'eau dans les ouvrages souterrains – *Tunnels et ouvrages souterrains n° 194/195, mars-juin 2006*

CETU, 2012. Fascicule 40: Tunnels – *Génie civil et équipements, octobre 2012*

CETU, 2015. Guide de l'inspection du génie civil des tunnels routiers – Livre 1: Du désordre à l'analyse, de l'analyse à la cotation - *Les guides du CETU, janvier 2015.*

CETU, 2015. Guide de l'inspection du génie civil des tunnels routiers – Livre 2: Catalogue des désordres - *Les guides du CETU, janvier 2015.*

IEC 31010:2009. Risk management – Risk assessment techniques

ISO Guide 73:2009. Risk management – Vocabulary

ISO 31000:2018. Risk management – Guidelines

Puccinelli M., Avagnina N., Dallari A., 2013. Un approccio metodologico per la progettazione degli interventi di ripristino e consolidamento strutturale di gallerie esistenti. - *Congresso SIG, Bologna, ottobre 2013*

UNI 11104:2016. Calcestruzzo - Specificazione, prestazione, produzione e conformità - Specificazioni complementari per l'applicazione della EN 206

Tunnels and Underground Cities: Engineering and Innovation meet Archaeology, Architecture and Art, Volume 9: Safety in underground construction – Peila, Viggiani & Celestino (Eds)
© *2020 Taylor & Francis Group, London, ISBN 978-0-367-46874-3*

Innovative solutions for safety against firedamp explosions in small section EPB-TBM tunnelling

A. Bandini, C. Cormio & P. Berry
SERENGEO S.r.l., Bologna (BO), Italy

M. Battisti & A. Lisardi
Collins S.r.l., Fiumicino (RM), Italy

P. Bernardini
Ghella S.p.A., Rome, Italy

M. Urso
Vianini Lavori S.p.A., Rome, Italy

ABSTRACT: The construction of Pavoncelli bis tunnel, started in 1990 and stopped after about 580 m (5% of the route) due to uncontrollable groundwater flows that imposed work's abandonment, was completed in October 2017. This paper describes the innovative solutions that allowed for tunnel completion by solving complex excavation problems, hence opening new perspectives for small diameter TBM-EPB tunnelling in difficult ground conditions. Thanks to a new construction design, which introduced several innovative solutions, in 2014 Caposele S.c.a.r.l. resumed tunnelling works with a small diameter TBM-EPB (4.6 m), successfully facing the difficult geological, hydrogeological and geostructural conditions (faults, continuous and important methane gas and water inflows from the surrounding formations). Pavoncelli bis construction was made possible thanks to the study, design and implementation of innovative technical, technological and procedural solutions to: (I) guarantee high advance rates and maximum safety conditions in the excavation through gassy formations, thanks to special ATEX equipment, a novel gas drainage system and specific safety procedures; (II) rescue the TBM-EPB from the stoppage determined by crossing a regional fault and in presence of firedamp.

1 INTRODUCTION

According to OSHA (USA) statistics, 70% of the injuries in tunnel construction occurred in the years between the last century and the current one are due to: (I) inadequate operating procedures; (II) risk assessment inadequacy; (III) lack of inspections (Berry & Patrucco 2011). Historical data on fatal injuries occurred in the 20th century in Italy during tunnel excavation with traditional method indicate a frequency of about 3 deaths per kilometer of excavated tunnel (Bandini et al. 2013).

Tunnelling with traditional or mechanized technique through gassy rock masses can cause dangerous explosions or fires in the underground construction site due to accidental ignition of methane-air mixtures.

If preliminary geological surveys indicate the presence of gassy rock masses along the tunnel track, it is often mistakenly assumed that any methane inflows in the underground construction site have a stochastic and non-uniform distribution (Kang et al. 2013). Consequently, the designer does not adopt solutions (such as work sequences, technological systems and equipment) able to guarantee the highest protection rate against explosions, thus causing, in many cases, expensive delays and, at worst, tragic accidents to workers (Copur et al. 2012, Kang et al. 2013, Kitajima 2010, Lockyer & Howcroft 1997, Proctor 2002, Proctor 1998, Vogel & Rast 2000).

In underground civil works, firedamp hazard has been generally underestimated, at least in the past, and there are very few attempts to transfer the great experience gained in underground coal mining (Rodríguez & Lombardía 2010, Labagnara et al. 2015). In Italy, since the early stages (in the late 90s) of tunnels construction for the Italian High Speed Train (TAV) and the Valico Variant (VAV) projects between Bologna and Florence, that involved gassy rock masses, the significant inadequacy of the international standard procedures, current working methods, plants, machinery, equipment and monitoring systems came out. Consequently, a comprehensive program of studies and researches was started in order to define the operating methodologies for tunnelling safety in gassy rock masses and to develop new standard procedures. As a result, a new approach was developed, that introduced a dynamic safety engineering system. It consists of innovative design solutions, safety measures and procedures defined according to rock mass-tunnelling interaction, taking into account tunnel size, underground work organization and excavation method and technique (Berry et al. 2000, Bandini et al. 2017).

Such dynamic engineering system led Emilia Romagna and Tuscany Regions to publish, between 1998 and 2015, 45 Inter-regional Technical Notes (NIR), Best Practices for safety in tunnelling, recently revised and updated as National Guidelines, specifically addressed to large section tunnelling with traditional techniques (larger than 70 m²) and with Earth Pressure Balance Tunnel Boring Machine (EPB–TBM larger than 80 m²).

Nevertheless, the engineering approaches and Best Practices defined by NIR, correctly interpreted, reconsidered and applied by expert technicians, must be considered as a relevant technical – scientific reference to assess and design safety solutions for different tunnelling conditions.

The present paper describes the successful innovative engineering solutions designed and adopted for:

- Pavoncelli bis tunnel construction with a small diameter (4.6 m) EPB-TBM in gassy rock masses;
- the manual excavation of a small tunnel (approx. 6 m²) along TBM shield, to unblock it from the stop occurred due to crossing of a regional fault.

2 TUNNEL DESCRIPTION, GEOSTRUCTURAL FRAMEWORK AND METHANE INVESTIGATIONS

The Pavoncelli bis tunnel, located in Campania, extends in the SW-NE direction (Figure 1). The tunnel design dates back to the early 80s and, after various administrative and technical

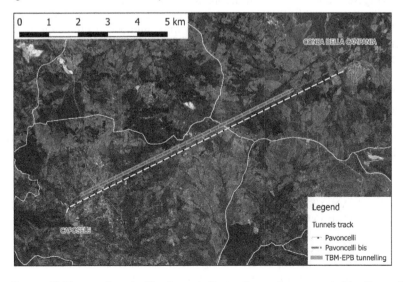

Figure 1. Pavoncelli bis tunnel track. The figure indicates the portion excavated by Caposele S.c.a.r.l. with EPB-TBM.

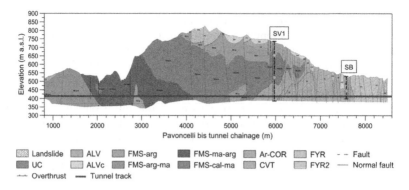

Figure 2. Pavoncelli bis tunnel geological longitudinal section. The light blue circles depict the methane emissions recorded during the boreholes SV1 and SB drilling at the borehole head.

Table 1. Formations crossed by Pavoncelli bis tunnel.

Formations	Chainage (m) From	To
Limestones and dolomitic limestones (UC)	0	655
Monte Sant'Arcangelo	655	5600
Varicoloured Clays (ALV)	5600	6100
Corleto sandstones (Ar-COR)	6100	6223
Castelvetere (CVT)	6223	6357
Varicoloured Clays (ALV) with arenaceous blocks	6357	7760
Red Flysch (FYR)	7760	8510

vicissitudes, the Caposele S.c.a.r.l, who won the tender at the end of the 2011, completed the excavation of approximately 8 km of tunnel in less than five years with a TBM-EPB 4.6 m in diameter.

The excavation, completed in October 2017, passed through gassy and structurally complex formations (Figure 2 and Table 1), widespread throughout the Italian Apennines from the North to Sicily. Petrographically they are composed of folded and intensely tectonised rock blocks and layers dispersed in a clayed matrix. A huge number of faults and fractured rock masses were identified, along tunnel track, through the analysis of historical data, muck samples and excavation parameters, geological investigations and the results of geophysical and geognostic surveys (Figure 2).

Methane emissions were detected at depths above Pavoncelli bis during drilling of boreholes:

- SV1: the flows occurred during drilling through fractured marly limestone and marl/calcareous marl layers;
- SB: methane flows were presumably associated to clay formations with sandstone and marly limestone layers and to numerous discontinuities into the rock masses.

3 TECHNICAL AND TECHNOLOGICAL SOLUTIONS AGAINST METHANE EXPLOSION IN PAVONCELLI BIS

The gassy route of Pavoncelli bis, the limited TBM diameter (<5 m) and functional spaces in the shield and in the back-up imposed the design of several specific safety solutions, to be realised at the constuction site, against air – methane mixtures explosions or fires.

The effectiveness of the designed solutions have been strongly limited by TBM location at the time of intervention, confined underground at about 700 m from the tunnel entrance.

In particular, innovative technical solutions have been implemented in the EPB-TBM, inspired by the safety engineering principles suggested by NIR n° 44 and the "multi-barrier" approach of the ATEX standards. The main goal was to prevent, through compartmentaliza-tion, the coexistence, along the "TBM – back-up – lined tunnel" system, of air – methane mix-tures and potential ignition sources in space and time. The following solutions will be described:

- "TBM – back-up – finished tunnel" system divided into 5 homogeneous Volumes;
- compartmentalization of screw conveyor muck discharge and air – methane mixture aspir-ation into dedicated pipes;
- ventilation and air movement in the shield and along the conveyor belt;
- methane drainage from the excavation chamber and the screw conveyor;
- improved sealing systems against methane flows from TBM shield and tunnel lining segments;
- handling of tunnel lining segments, wagons and flat cars for segments, materials, equipment and muck transport during excavation phase;
- improved methane monitoring system and safety procedures;
- specific operating procedures for firedamp risk prevention.

3.1 TBM – back-up – lined tunnel compartmentalization

By adopting the Best Practices suggested by NIR n° 44, the "TBM – back-up – finished tunnel" has been divided into 5 homogeneous Volumes according to the the presence of igni-tion sources and potential explosive mixtures (Figure 3).

For each Volume, early warning and warning threshold levels were defined. Exceeding these thresholds activates the related safety procedures (paragraph 3.7).

In Volume 1 (excavation chamber and screw conveyor) both ignition sources (frictions and rock – cutting tools impacts) and air – methane mixtures can be present. To inhibit firedamp explosions or fires propagation, the volume have to be constantly filled with a mixture of muck, water and foaming agent (excavation in "closed mode", Bandini et al. 2017). During maintenance operations, which require the Volume 1 to be partially emptied, specific safety procedures must be applied.

In Volume 2 (prechamber, between the excavation chamber and the bridge) there are igni-tion sources (electrical equipment, moving elements, etc.). Therefore, solutions and procedures have been adopted to prevent methane inflows and the creation of air-methane mixtures.

Volume 3 mainly consists of the small diameter pipes of the aspiration system and of the methane drainage system from Volume 1. It extends from the gate valve of the screw conveyor to the tunnel entrance. This Volume may inevitably contain methane, released by the muck at the screw conveyor discharge. By adopting solutions that prevent potential sparks, methane concentrations can reach higher values compared to the other Volumes, but lower than LEL (see paragraph 3.7).

In Volume 4 (ring assembly area, bridge and back-up gantry 1) methane-air mixtures and potential ignition sources (operating equipment) may coexist. Therefore, only ATEX equip-ment and plants are allowed.

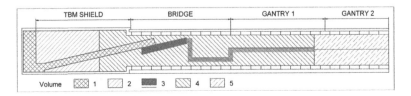

Figure 3. TBM – back-up – lined tunnel compartmentalization into homogeneous Volumes.

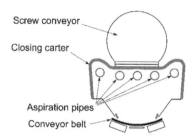

Figure 4.   Compartmentalization at the screw conveyor discharge section.

Volume 5, from back-up gantry 2 to the tunnel entrance, is characterized by the possible simultaneous presence ignition sources (frictional, electrical, thermal) and methane-air mixtures flowing from Volume 4 and the muck on the conveyor belt.

3.2   *Screw conveyor compartmentalization and aspiration of air – methane mixtures*

To limit the inflow of methane released by the muck at screw conveyor discharge section (Volume 1), the following solutions have been realized (Figure 4):

– a closing carter wrapping the gate valve, the hopper and the initial portion of conveyor belt;
– an aspiration system that conveys methane outside the tunnel, with a flow rate of 1 m³/s (value that does not modify the flow of the blowing ventilation system, described in paragraph 3.3).

Since such an innovative solution only partially avoids the methane diffusion in Volumes 4 and 5 (aspiration rate was limited to 1 m³/s due to the very limited diameter and operating spaces inside the TBM), the maximum safety level has been achieved by adopting the "multi-barrier" approach (blowing ventilation, gas monitoring, ATEX devices, safety procedures and alarm system).

3.3   *Ventilation and air movement system in the TBM shield and along the conveyor belt*

The ventilation system has been designed considering the very limited diameter and spaces inside the TBM and the forecast of continuous and massive methane inflows. It consists of:

– a blowing ventilation circuit with 2 ducts which promote air agitation in the Volumes 2 and 4 and in the back-up gantry 21;
– 9 flow amplifiers close to the bulkhead between the Volumes 2 and 4, to mix the methane released from the muck during screw conveyor discharge and trasport on conveyor belt;
– 39 nozzles along the conveyor belt, to move and dilute the methane released by the muck.

The system has been designed to prevent methane accumulation in shaded areas, to minimize the risk of layering in the crown and to ensure the mixing of methane with fresh air.

3.4   *Methane drainage from the excavation chamber and the screw conveyor*

To evacuate the methane from Volume 1 outside the tunnel, a drainage system has been realized, consisting of ATEX pipes, valves for opening and closing the circuit, compressed air wash valves and ducts, a system to separate the methane from the liquid fraction and from the debris (Figure 5) and a gas discharge chimney outside the tunnel. It does not require an aspiration system, because the pressure inside Volume 1 is higher that the atmospheric one.

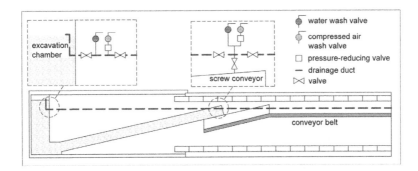

Figure 5.   Schematic representation of the drainage system from Volume 1.

3.5   *Sealing systems against methane inflows from TBM shield and tunnel lining segments*

In order to prevent methane inflows into the tunnel, the following elements (nowadays standard practice) were installed:

– sealing gaskets in the joints between the shields (Volume 2);
– three rows of metal brushes between the tail shield and the penultimate lining rings (Volume 4), which create 2 annular chambers constantly filled with pressurized grease;
– sealing gaskets between the tunnel lining segments (Volumes 4 and 5).

The grease injection points have been increased compared to the standard (from 4 to 8 lines) to ensure a more uniform grease distribution.

3.6   *Handling of tunnel lining segments and wagons motion in TBM and back-up*

The lifting and transfer of the lining segments from the steel flat cars to the segment feeder (Volume 4) was realized with a pneumatic ATEX crane. The flat cars slide on brass bearings in the segment feeder and the movement is made with hydraulically-fed devices, in order to prevent frictional ignition of methane-air mixtures. To prevent potential ignitions due to the locomotive, wagons and flat cars motion within TBM and back-up, the following solutions have been adopted:

– temperature monitoring on locomotive engine hot spots, with light signal when the temperature reaches 350°C (early warning) and 400°C (warning);
– management procedures for locomotive advance and activities in TBM when the engine temperature thresholds are reached;
– catenary motion system (Figure 6) with a compressed air continuous propulsor, to keep locomotive's engine switched-off during excavation.

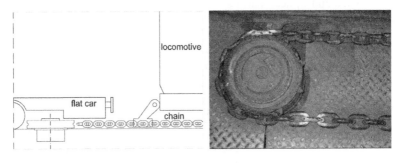

Figure 6.   Longitudinal section and particular of the catenary motion system.

3.7 *Methane monitoring system and activation of safety procedures*

Analyzing the possible ignition sources and the methane – air mixtures flow through TBM, shield and back-up, an automatic gas monitoring system was designed and realized, consisting of:

- 22 sensors to measure the gas concentration, arranged as follows:
- 8 sensors in Volume 2 (4 close to the bulkhead between the Volume 1 and 2 and 4 close to the bulkhead between the Volumes 2 and 4);
- 9 sensors in Volume 4 (3 around the screw conveyor discharge, outside the closing carter, 5 at the upper end of the closing carter and 1 at the beginning of gantry 1);
- 4 sensors in Volume 5 (1 on gantry 2, 1 on gantry 3, 1 on gantry 6 near the discharge hopper of the conveyor belt and 1 on gantry 7);
- 1 sensor inside the Volume 3 (upstream of the fan, in gantry 7);
- 4 control units for data acquired by the sensors;
- a software that monitors and manages the safety and alarm system in real time;
- a fiber optic data transmission line.

Within Volume 1 gas monitoring is carried out only during maintenance operations (Volume 1M), since during excavation phase firedamp explosions or fires are prevented by keeping the Volume constanly filled with muck (Bandini et al. 2017).

The automatic monitoring system is equipped with an ATEX buffer battery that allows it to be operating for a long time after evacuation and power line switch-off, thus allowing a better management of works restart. Manual monitoring is performed in areas not covered by the fixed sensors, in zones of possible gas accumulation and near the sealing elements.

3.8 *Operating procedures*

To guarantee maximum safety, specific operating procedures have been defined according to methane concentration thresholds (Table 2), which concern:

- the excavation phase, the lining installation and the maintenance operations;
- the access and exit of personnel in the tunnel;
- the access and exit of the train in the TBM, and the wagons handling;
- monitoring of the methane concentration in air and alarm management;
- the use of open flames.

The procedures (Figure 7) define the involved workers (adequately trained), the checks to be carried out and the actions to be performed to ensure maximum safety conditions in each Volume. Preliminary checks and controls during tunnel construction are foreseen for each activity.

In the back-up gantry 21 there is an ATEX emergency car for workers evacuation. Specific procedures guarantee that the only escape route is viable, regulating the vehicles transit inside the tunnel. Two California switches, installed every 2700 m, allow the passage of the evacuation vehicle.

When the early-warning threshold (Table 2) is reached, the automatic monitoring system activates the acoustic and optical alarm systems, the gate valve of the screw conveyor automatically closes, stopping the TBM advancement, the flow rate of the ventilation system is

Table 2. Early-warning and warning thresholds in the different Volumes (LEL: lower explosivity limit). V1M indicates Volume 1 during maintenance operations and access to excavation chamber.

Volume	Early warning	Warning
V1M	14% LEL	20% LEL
V2	3% LEL	7% LEL
V3	80% LEL	\
V4	14% LEL	20% LEL
V5	3% LEL	7% LEL

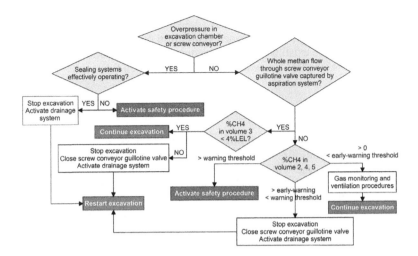

Figure 7. Schematic representation of the safety procedures.

increased and the methane is removed from Volume 1 through the drainage system. The advancement can start again only when the methane concentration in air is lower than the early-warning threshold. In Volume 3 the early-warning threshold is not connected to the automatic release system of the TBM power line and the transit of high concentration methane-air mixtures is allowed. Equipment and operations that cannot be automatically interrupted are subjected to manual procedures.

At the warning level threshold, the power supply must be abruptly stopped to all standard (not ATEX) TBM equipment and the workers must be evacuated. Only the emergency systems are operative, as the emergency telephones and lighting (which are ATEX), the methane monitoring system and the blowing ventilation system (as the fan is outside the tunnel).

Attained a concentration of 5%, in all volumes the release of the power supply of even the explosion-proof equipment occurs.

3.9 Safety management during TBM shield unblock operations

TBM advancement was interrupted at pk 6+170, as some rock blocks, detached from a fault crossed by the machine, were wedged between the TBM shields and the surrounding rock mass. A small tunnel (section of about 6 m²) was manually excavated along TBM shield (Figure 8) to unblock the machine.

The very limited size of the construction site and the impossibility of using mechanical excavation equipment required to define ad hoc solutions for safe tunnelling inspired by the safety

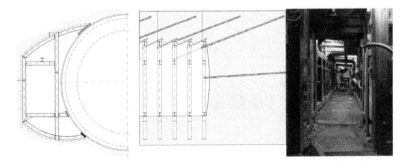

Figure 8. Transversal and longitudinal sections of the small tunnel excavated along TBM shield.

engineering principles suggested by NIR n° 28 (AAVV 2005) and the ATEX "multi-barrier" approach. The developed solutions concerned ventilation optimization, spark-proof tools and equipment, adaption of the existing automatic monitoring system and manual monitoring of methane concentrations, gas-free procedures for the use of open flames.

4   ANALYSIS OF METHANE INFLOWS DURING TUNNELLING

During Pavoncelli bis construction a huge amount of massive methane emissions were recorded, with concentrations also higher than LEL (Figure 9). Methane was detected almost continuously throughout the entire portion of the tunnel excavated with EPB-TBM (Figure 10), even with concentrations below the sensors instrumental sensitivity (filtered in Figure 9 and Figure 10).

Consequently, the entire track must be considered gassy. In particular:

– higher methane concentrations were measured while crossing tectonised zones close to structural discontinuities and main faults;
– methane mainly flowed into Volume 2 from the screw conveyor discharge due to flow rates higher than the aspiration rate;
– in some cases, high pressure emissions occurred through the brushes and the sealing systems between the lining rings during the segment assembly of the next ring.

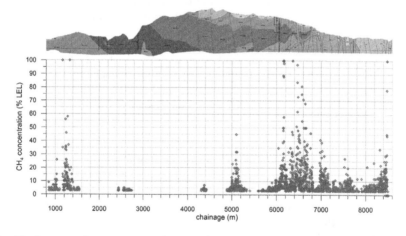

Figure 9.   Maximum methane concentrations recorded every 1.2 m advancement by the sensor in Volume 3 (only values > 3% LEL). Pavoncelli bis geostructural profile is reported for reference.

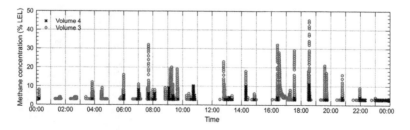

Figure 10.   Methane concentrations recorded (values > 3% LEL) during tunnelling in Volume 3 and 4.

5 CONCLUSIONS

The case study shows that, assuming as reference the engineering approaches suggested by NIR n°44, n° 28 and ATEX standards, it was possible to safely excavate a small diameter (less than 5 m) tunnel with EPB-TBM, even through structurally complex formations containing significant methane volumes along the whole route, and a very small section tunnel along TBM shield.

Although successful, the innovative (and first of their kind) solutions adopted on the EPB-TBM that excavated Pavoncelli bis were significantly constrained by its location (confined underground at about 700 m from the tunnel entrance) and the difficult operating conditions.

Thanks to the rigid application of the specific safety procedures the tunnel construction was completed with zero firedamp-related injuries.

Furthermore, the case study highlights the importance of preliminary methane investigations during design phase to characterize the rock masses for firedamp risk prevention and for effective tunnelling design.

ACKNOWLEDGMENTS

The Authors acknowledge Caposele's Technicians M. Paolini, R. Gencarelli, M. Todini and A. Cozzatelli, who contributed to find solutions for the safety system implementation.

REFERENCES

AAVV 2012. Nota Interregionale n° 44 "Grisù TBM" Scavo meccanizzato di grande sezione con TBM-EPB in terreni grisutosi. *Prot. n° PG/2012/132178 del 28/05/2012. Ed. Regioni Emilia Romagna e Toscana.*

AAVV 2005. Nota Interregionale n° 28 "Grisù 3a edizione" Lavori in sotterraneo. Scavo di terreni grisutosi. *Prot. n° ASS/PRC/05/1141 del 13/01/2005. Ed. Regioni Emilia Romagna e Toscana.*

Bandini, A., Berry, P., Calzolari, F., Colaiori, M., Cormio, C. & Lisardi, A. 2013. Nascita ed evoluzione delle Note Interregionali. In *Atti Workshop Nazionale NIR 2013 – Note Interregionali di Ingegneria della Sicurezza nello scavo di gallerie.* Bologna 4–5 Luglio 2014, Alma Mater Studiorum Università di Bologna, AMS Acta: 7–16.

Bandini, A., Berry, P., Cormio, C., Colaiori, M. & Lisardi, A. 2017. Safe excavation of large section tunnels with Earth Pressure Balance Tunnel Boring Machine in gassy rock masses: The Sparvo tunnel case study. *Tunnelling and Underground Space Technology* 67: 85–97.

Berry, P., Dantini, E.M., Martelli, F. & Sciotti, M. 2000. Emissioni di metano durante lo scavo di gallerie. *Quarry and Construction*, year XXXVIII, 1: 37–64.

Berry, P. & Patrucco, M. 2011. Commento di Paolo Berry e Mario Patrucco (art. 89–104). In Zanichelli (ed.), La nuova sicurezza sul lavoro – D.Lgs. 9 aprile 2008, n. 81 e successive modificazioni. Vol. II. Gestione della Prevenzione. Collana le riforme del diritto Italiano: 134–146.

Copur, H., Cinar, M., Okten, G. & Bilgin, N. 2012. A case study on the methane explosion in the excavation chamber of an EPB-TBM and lessons learnt including some recent accidents. *Tunnelling and Underground Space Technology* 27: 159–167.

Kang, X.B., Xu, M., Luo, S., Xia, Q. 2013. Study on formation mechanism of gas tunnel in non-coal strata. *Natural Hazards* 66: 291–301.

Kitajima, M. 2010. *Methane gas explosion hazard of an earth pressure type shield tunnel.* http://www.sozo gaku.com/fkd/en/cfen/CD1000098.html (accessed on 02-10-2013).

Labagnara, D., Maida, L. & Patrucco, M. 2015. Firedamp explosion during tunnelling operations: suggestions for a prevention through design approach from case histories. *Chem. Eng. Trans.* 43, 2077–2082.

Lockyer, J.W. & Howcroft, A. 1997. The Abbeystead Explosion Disaster. *Annals of Burns and Fire Disasters*, 10, September 1–4.

Proctor, R.J. 1998. A chronicle of California tunnel incidents. *Environmental & Engineering Geoscience* 4: 19–53.

Proctor, R.J. 2002. The San Fernando tunnel explosion. *Engineering Geology* 67: 1–3.

Rodríguez, R. & Lombardía, C. 2010. Analysis of methane emissions in a tunnel excavated through Carboniferous strata based on underground coal mining experience. *Tunnelling and Underground Space Technology* 25: 456–468.

Vogel, M. & Rast, H.P. 2000. Alp transit-safety in construction as challenge: health and safety aspects in very deep tunnel construction. *Tunnelling and Underground Space Technology* 15: 481–484.

Tunnels and Underground Cities: Engineering and Innovation meet Archaeology, Architecture and Art, Volume 9: Safety in underground construction – Peila, Viggiani & Celestino (Eds)
© 2020 Taylor & Francis Group, London, ISBN 978-0-367-46874-3

The new Lugano tram underground station: An example to combine architectural requirements, serviceability and challenging geotechnical conditions

G. Barbieri, P. Bassetti & A. Galli
AF Consult, Zürich, Switzerland

ABSTRACT: The new Lugano tram network represent a new link between surrounding countryside, city center and its international connections (airport and railway). The new line includes a 2.1 km tunnel which passes through a hill located between the airport and the city center, allowing to reduce travel time by public transport. Moreover, a 50 m deep underground station is foreseen to connect it. through elevators and escalators, to the existing international railway station located on the hill, in order to make it a major interchange for tourists and commuters. In the meantime, this station is a challenging engineering task, considering its dimensions (250 sqm cross-section and 130 m long) and its location, within a urban environment and under the existing railway station, with low overburden and in a complex geological context. The project is aimed to combine serviceability and architectural requirements, high safety standards and to deal with challenging geotechnical conditions.

1 INTRODUCTION TO THE LUGANO TRAM NETWORK PROJECT

Due to multiple additions of adjacent municipalities, the size of the city of Lugano has been grow-ing over the last decade. The agglomerate of Lugano has currently approximately 125'000 inhabi-tants who need an efficient transport system. The last decade has also seen the development of the Valley of Vedeggio as an industrial and services area (research and banking institutes). The Breganzona – Massagno hill, a densely populated area, separates The Valley of Vedeggio and the city of Lugano.

The rapid worsening of the roads system of the Lugano area, due to the increase of traffic, has resulted in the need of extending the public transportation network, in order to connect the center of the city to the industrial and services area.

The new Lugano tram network is the main project within this larger urban plan aimed to increase the use of public transport, providing better links between Lugano city and the sor-roundic strategic areas for economic development (Vedeggio valley, Cornaredo and Pian scairolo areas), and in the meantime between the whole area and the international connec-tions, consisting in the Agno airport, located in the Vedeggio valley, and the SBB railway.

As clear in Figure 1, the foreseen network has a characteristic H-shape, due to the presence of the Breganzona-Massagno hill in between Lugano city and the Vedeggio valley. The project fore-see, hence, two surface railway tracks in North-South direction, one along the Vedeggio valley and one from Cornaredo at the North side to Pian Scairolo at the South side, that are connected to each other by a transversal underground track passing below the Breganzona village and reach-ing Lugano city center. The branch of the new network in the Vedeggio valley toward the italian border will substitute the existing local railway between the Italian border and Lugano central railway station, that today follows a surface track up to the hill, implying longer travel time.

Considering the importance of the FFS railway passing through Lugano (shown with black lines in Figure 1), being part of the European railway Rhine-Alpine corridor and the main international connection of Lugano to adjacent countries, an easy connection between the

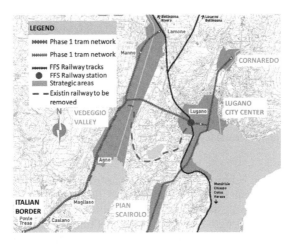

Figure 1. Key plan of the new tram network.

existing rail station to the new tram network was considered a priority. As the exiting railway station is located upon the mentioned hill, an underground station along the Breganzona tunnel was foreseen with a shaft to reach the surface at the railway station.

The project is divided in two different steps. The first phase, called priority phase (red lines in Figure 1), involves the construction of the underground link between the city and the Valley of Vedeggio and the surface rail line along the Vedeggio valley, that in the future will be extended to the Lamone FFS station. The remaining two lines of Cornaredo and Pian Scairolo (green lines in Figure 1) will be part of a subsequent phase. The project owner is the "Dipartimento del Territorio del Canton Ticino", acting on behalf of FLP (Ferrovie Luganesi SA).

2 THE BREGANZONA TUNNEL AND ITS UNDERGROUND STATION

2.1 Tunnel layout

The Breganzona tunnel is the core of the new tram network project. It is a single tunnel with double rail track having a length of 2135 m with West-East orientation; in order to accommodate a double track, an inner section of 48 sqm was required, implying an excavation section of 69 sqm (see Figure 3 on the left). The inner section provide also enough space to guarantee on both sides an escape root for passengers in case of emergency at least 1m wide and 2.2 m high.

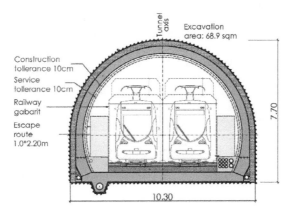

Figure 2. Breganzona tunnel typical cross-section.

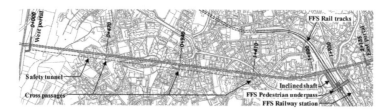

Figure 3.   Breganzona tunnel general layout.

The underground station is located at almost 300 m from the East portal; passengers, from the underground station, can reach the surface and the existing FFS station through an inclined shaft that provides both staircase, escalators and elevators.

From the west portal for a length of 1400 m, a parallel tunnel is foreseen, at a distance of 30 m from the main tunnel and with a cross-section of 12.5 sqm, to provide an escape route for passenger in case of fire or other emergencies; the safety tunnel is linked to the main tunnel by cross passages every 470 m (see Figure 3). For the remaining part of the tunnel, the escape route is guaranteed by the inclined shaft in the underground station to the surface.

2.2   *The Underground station*

The underground station consists in a widened tunnel section with an overall length of 208 m. The station is organized with lateral railway tracks and a central platform 8.8 m wide and 50 m long, serving both directions; along the central platform, escalators are provided, whereas a staircase and an elevator are placed on both ends of the platform (see Figure 4). In order to accommodate both two tracks and the central platform, the underground station has an inner cross-section of 176 sqm, resulting in an excavation section up to 240 sqm (see Figure 5). In addition to this central part, transition sections were required on both sides of the station in order to allow rail tracks deviation from their central position in the standard tunnel section to their lateral position in the station. In the available space between the rail-tracks on both sides of the station, technical rooms was obtained. Through the mentioned stairs and elevators, passengers reach a mezzanine level located in the upper part of the widenined tunnel section (see Figure 5). The mezzanine level provide an additional buffer space for passangers that leave the platform level and need to reach the transversal smaller tunnel, located at the middle of the station length, which connect the station to the shaft bringing passengers to surface.

Different design solutions were evaluated, such as separate stations for each direction and lateral docks instead of a central one, but they resulted in longer paths for passengers and in narrow passages having a claustrophobic effect; a combination of serviceability performance and architectural requirements, providing an open space although 50 m underground, drove to this more complex but also more efficient solution, shown in Figure 7, which provides a wide perspective both at the platform and at the mezzanine level, avoiding narrow passages and allowing for fluent pedestrian flows.

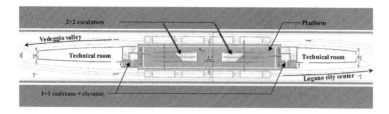

Figure 4.   Plan view of the underground station.

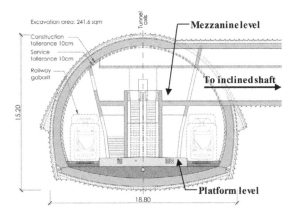

Figure 5.   Breganzona tunnel typical cross-section and underground cross-section.

Figure 6.   Rendering of the underground station at the platform and the mezzanine levels.

2.3   *The inclined shaft*

In order to connect the underground station to the surface and, hence, to the existing railwas station 50 m above, different design solutions were evaluated. The most simple solution consisting in vertical shafts with elevators turned out to be not feasible, as it could hardly provide enough capability for the expected number of passengers; moreover, additional staircases should be anyway added in order to provide escape route for emergencies. Hence, a solution with an inclined shaft to accommodate escalators and providing then enough capability was necessary. Considering the estimated passengers flow and safety requirements to use the shaft as escape route, 3 escalators and one stair were needed, as well as two inclined elevators for disabled persons (one of which as backup).

To accommodate such a number of stairs and elevators, a mixed design solution were chosen: an tunnel inclined by almost 28° with an inner cross-section ranging between 127 and 136 sqm was designed for the deeper part, whereas a rectangular shaft excavated by top-down system from surface, more than 12 m wide, 50 m long and 26 m deep, was designed for the upper part of the connection (see Figures 7 and 8). The choice to realize a so deep shaft from surface was mainly due to geotechnical reasons as described in the following chapters, but it also meets architectural requirements. The length of the inclined path necessary to bring passengers from the underground station to the surface (120 m) drove indeed the architect to ask for an open space solution and this could be possible by adopting a top-down structure with only steel struts as intermediate supports for lateral walls. The length of the path in a narrow space was hence reduced to 70 m, whereas in the rectangular shaft an open space from the stairs level to the surface was provided, closed on the top by a concrete slab with lateral windows to let sun light come into (see Figure 8). The shaft will connect to an existing pedestrian underpass of the railway station, that will be widenend to improve its capacity, providing direct and covered access to the railway station and platforms.

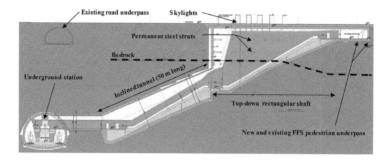

Figure 7. Inclined shaft longitudinal view.

Figure 8. Inclined shaft cross-sections and rendering of the internal view.

3 SAFETY AND OPERATIONAL REQUIREMENTS

3.1 *Operational requirements on the Breganzona tunnel*

The estimated number of passengers in the period 2025–2030 is around 15'000 every day, meaning doubling the number of passengers nowaday using the existing railway line between Lugano and Ponte Tresa (at the Italian border).

In order to provide a continuous and efficient service and to properly deal with the expected passengers number, the infrastructure shall be able to support a train frequency up to one every 10 min. toward Ponte Tresa as well as toward Manno; hence, trains starting from Lugano city through the Breganzona tunnel will reach a frequency up to one every 5 min during peak hours.

Considering such requirement, a double track tunnel was required: the alternative solution with a single track tunnel and double track only at the underground station was evaluated during preliminary design to optimize costs, but could only guarantee a frequency up to a vehicle every 10 minutes. The adopted solution, in addition to reach the required capacity, provide also more flexibility and safety, having two completely independent tracks in the two direction.

On the new infrastructure, tram-like vehicles are forseen with a length of 50 m and a width up to 2.4 m; the gabarit fulfill standard C3, as in the Zurich tram network and, hence, the infrastructure was designed following the guideline "Vorschriften und Richitlinien für Tram, Juli 2013" by VBZ. The cruise speed of the vehicles will be 70 km/h in open stretches and 80 km/h in the tunnel; their capacity is up to 250 passengers.

3.2 *Safety requirements on the Breganzona tunnel*

The safety system of the Breganzona tunnel was designed following the requirements given by the Swiss National Standard SIA 197/1. Accordingly to them, an escape route to a safe place should be provided every 500 m, with minimum dimensions 1 m (width) x 2.2 m (height); a fire estinguish system should be provided as well as an access system to the tunnel to emergency vehicles.

As for the escape routes, as previously mentioned, they are provided every 480 m, for most of the tunnel by connecting the main tunnel to a parallel safety tunnel (see Figure 2) and for the latter East part by using the inclined shaft of the underground station as escape root. In the safety tunnel, a mechanical ventilation system is provided, which guarantee to keep an overpressure in it and its cross-passages, avoiding smoke to come into it; fire-resistant doors are also provided at the entrance of every cross-passage. Along the tunnel, no mechanical ventilation is instead needed neither during service nor in case of fire: by smoke propagation simulation, was verified that smoke stratification can occurs, allowing enough visibility to passengers to safely reach safety exits.

At every cross-passage, a fire hydrant is provided, connected to a pressurized water pipe running along the whole tunnel, guaranteeing a minimum flow of 20 l/s and a reservoir of 250 mc. Emergency vehicles can get access to the tunnel through the portals and reach every point of the tunnel running along the tracks, being provided an almost flat concrete slab; every 470 m a turning niche is designed, in order to let them change direction.

3.3 *Operational requirements on the underground station*

Functionality of the underground station both in normal service and in case of emergency was verified by passengers flow simulations, performed by subcontractor GESTE using Pathfinder Software, checking the so-called "level of Service" (see Stefan Buchmüller et al. 2008). Level of service is measured in terms of passengers density in different areas of the underground station and passengers waiting time to get access to stairs and elevator, that should be lower than predefined acceptable limit shown in Table 1.

Different scenarios were considered, but two are the significative ones: for normal service, two trains stopping at the station with a reciprocal delay of 90 seconds; 75% of passengers getting off the train and 20% getting on it. As emergency scenario, two train evacuating simultaniously, one full and one with 75% of passengers, with no persons on the platform (assumed already evacuated). In the normal service scenario, all acceptable limits are fullfilled, whereas in the emergency scenario, passengers density getting off the train reaches 1.26 pers/sqm, but it is considered as acceptable, being an extreme and emergency condition. The simulation verified also that, in the evacuation condition (i.e. in case of fire), platforms can be evacuated, i.e. people can reach a safe place at the mezzanine level, in less than 5 min; moreover, in less than additional 5 minutes, all people leave the mezzanine level without forming relevant queue and reach the stairs toward surface. In Figure 12, the results of passengers flow simulation are shown, in the two principal phases: when unloading the first train, with people waiting for the latter (on the left), and during

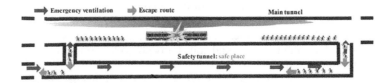

Figure 9.   Concept of escape route through the safety tunnel.

Table 1.   Acceptable limits on passengers density and waiting time.

Rock Mass	Unit of Measurement	Acceptable Limit
Passengers density waiting for a train	pers/sqm	2
Passengers density getting on a train	pers/sqm	2
Passengers density getting off a train	pers/sqm	0.7
Waiting time to get access to a stair or elevator	sec	7

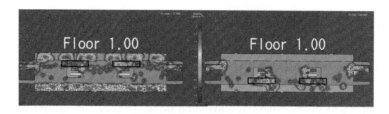

Figure 10. Passengers density resulting by dynamic simulation in normal service.

unloading of the latter one (on the right). The positive effect of spreading diferent stairs along the whole platform are clear, avoiding excessive passengers concentration.

3.4 *Safety requirements on the underground station*

Safety requirements for the design of the underground station was derived by the guidelines and rules given by AICAA (Associazione degli Istituti Cantonali di assicurazione antincendio). Following such guidelines, bearing structures should be REI60. Moreover, considering the possibility to have 500 persons in the platform (2 full vehicles), an escape route to a safe place no longer than 35 m should be provided and having 9 passage unit with a minimum width of 60 cm.

To fulfill the requirement on the length of the escape route, the mezzanine level as well as the inclined shaft had to be designed as safe places. To this aim, the mezzanine level is separate to the platform level by inner concrete inclined walls (see Figure 5), having windows with fire-proof glasses; moreover a mechanical ventilation system is provided in the inclined shaft to guarantee overpressure in it and in the mezzanine level. Finally, mechanical ventilation is provided in the underground station to push the smoke in case of fire toward the East Portal and outside the tunnel.

To fulfill requirements on the capacity of the escape route, three staircases 1.8 m width should be provided, however, to combine serviceability and safety requirements, two staircases 1.8 m wide each was provided at the ends of the platform (each combined to an elevator) and 4 escalators 1 m wide were provided and considered as escape roots.

4 GEOTECHNICAL CHALLENGES AND SOLUTIONS

4.1 *Geology*

The Breganzona tunnel's path is located inside the Breganzona hill, where different types of rock layers overlap in a non articulated succession. The rock body concerned with the tunnel belongs to the "Gneiss dello Stabbiello (GStab)" geological formation with apelitic gneiss with alkali feldspar belonging to the "Ortogneiss del S. Bernardo (GOB)" and with epidote and hornblende schists (Sh) and subvolcanic rocks (Q). Above the bedrock lays a very heterogeneous blanket of landfill made by alluvial soils (gravelly to loamy) and moraine characterized by gravel and silty sands with a thickness from a few centimeters to several meters. Location and alternance of the different rock types along the tunnel are shown in Figure 11; their subdivision and denomination are based on the Geological Atlas of Switzerland. The main characteristics and subdivision of the encountered rock mass are summarized in Table 2. The tunnel overburden ranges from a maximum of 110 m to a few meters.

The underground station is located within the GS rock formation, in an area where design surveys showed a lower rock quality than in the rest of the tunnel; a GSI (Hoek & Marinos 2000) equal to 35 was hence assumed in this area for design, lower than the design values assumed along the rest of the tunnel and shown in Table 2. GS formation is then characterized by a uniaxial compressive strength of the rock equal to 17 Mpa and mi (Hoek & Marinos 2000)) equal to 5. In the station area, the superficial layer of moraine and alluvional debris has a thickness rangeing between 10 and 15 m. Hence, the undergorund station is fully in the bedrock, with a rock cover around 30 m, whereas the inclined shaft to reach the surface is partly in the

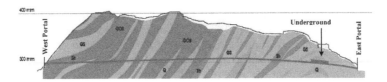

Figure 11.   Geological longitudinal section.

Table 2.   Rock mass subdivision and main characteristics.

Rock Mass		Percentuage	GSI*
GS	sericytic Gneiss	52%	12-**59**-82 (45**)
GOB	aplytic Gneiss	26%	7-**58**-82 (40**)
Sh	Epidot and Hornblende Schysts	9%	61-**70**-83 (60**)
Q	Filonian Rocks	13%	36-**65**-85 (50**)

* Geological strength Index range (min – med – max)
** GSI design value along the tunnel

aforementioned alluvial deposits. Underground water circulation is expected to be limited to local small incomes along the tunnel, except for the areas at the portals, in particular in the Lugano downtown area. The alluvial and moraine soils, instead, are characterized by a medium to high permeability. Considering the local rock quality, the wide underground cross-section, the local stratigraphy with a thick deposits layer and the low overburden (lower than 3 times tunnel diameter), higher risks during underground station and inclined shaft construction are stress relaxation in the rock and consequent plastic deformation as well as induced settlement on surface, where railway tracks and building are located. Morover, considering high permeability of the deposits layer, risk for watertable lowering and conseqent settlements during inclined shaft construction represent another critical design aspect.

4.2   *Geotechnical design solution for the underground station*

In order to deal with the aforementioned geotechnical risks and the presence of important infrastructures and the urban environment above the tunnel, a flexible design solution, able to suit to variable local conditions and aimed to minimize surrounding induced deformations should be adopted. Excavation process was hence foreseen in 3 different stages as shown in Figure 12: first, a smaller exploratory tunnel with a 65sqm cross-section is excavated, centered on the upper half of the final section, in order to verify local rock behaviour along the whole future station by monitoring devices and coring around the tunnel; then, widening of the tunnel section to reach final shape of the upper part of the tunnel; finally, excavating the

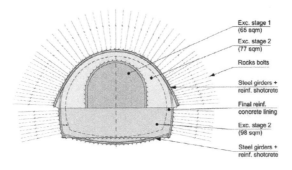

Figure 12.   Temporary support and rock reinforcement of the underground station.

lower part of the tunnel. Rock reinforcement is foreseen by self-drilling steel bolts up to 12 m long on the contour and fiber-glass reinforced bolts on the tunnel face up to 10 m long. Temporary tunnel support is moreover provided by steel girders coupled to shotcrete reinforced by two steel mesh layers; on the tunnel face fiber reinforced shotcrete is provided at each excavation phase. To stabilize steel girder footings during the crown excavation phase a widened shotcrete foot is foreseen, with possible addition of subvertical piles if necessary; whereas on the lower half, a steel temporary invert is provided to be installed every 2 m advancement. Tunnel excavation in every stage is performed meter by meter and, after each advancement, temporary supports and rock reinforcement measures are applied in order to minimize rock plasticization and consequent deformations.

Rock response to the tunnel excavation and rock - structure interaction were simulated by 2D finite element analysis using software Plaxis 2D, modeling the surrounding rock up to the surface (see Figure 13) and an existing road underpass located between the future station and the existing rail tracks. The results show maximum vertical displacements on surface lower than 9 mm and lower than 12 mm at the foundation level of the existing road underpass; distortion and horizontal deformation at the surface level, relevant for existing rail track serviceability, are lower than 0.5‰ and 0.15‰ that results in acceptable deformation of rail tracks accordingly to FFS guidelines.

4.3 *Geotechnical design solution for the inclined shaft*

The inclined shaft is located right beside the existing rail tracks and with orientation parallel to them; hence, the most relevant design risk is induced deformation on such tracks. Moreover, no water income should be allowed both during construction and in service condition within the permeable superficial debris layer, in order to avoid relevant effects on the watertable.

Considering above critical aspects, the upper part of the connection was foreseen by a rectangular shaft supported by secant reinforced concrete piles, embedded in the bedrock more than 10 m, in order to provide an almost waterprrof tank since the construction phase. Horizontal restraints of such walls are provided by: a concrete slab at the pile heads, to be realized before excavation to minimize surface displacement; up to 3 intermediate levels of permanent steel struts; a bottom reinforced concrete slab; temporary active anchors to reach the maximum excavation level, to be removed after casting the bottom slab. To guarantee full waterproofness in the long term, reinforced concrete walls are added inside the piles and connected to the bottom and top slab to create a waterproof concrete tank, adopting the so-called "white tank system": waterproof concrete with proper construction details such as deformable joints to avoid cracks in the concrete tank. In the long term, secant piles are assumed to withstand the earth pressure, whereas inner walls withstand water pressure.

Length and depth of the top-down shaft were designed so that the inclined tunnel would have a minimum rock cover of 5 m (see Figure 7). So, considering the low rock permeability resulting by performed permeability tests, drainage could be allowed around the inclined tunnel, as well as in the main tunnel. However, an umbrella of concrete injections where provided above the first 10 m of inclined tunnel to further improve rock permeability at the interface with the deposits. Inclined shaft excavation is foreseen in two phases (crown half and

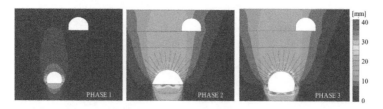

Figure 13. Displacements resulting by FEA simulations of underground station (excavation stages 2 and 3).

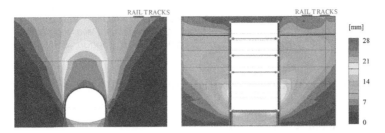

Figure 14.   Finite element analysis simulation of inclined tunnel and top-down structure construction.

invert half), using a steel pipe umbrella for the first 20 m meters and radial steel bolts on the remaining part. As temporary support, shotcrete reinforced by two steel mesh layers is provided, with steel girders where forepoling is foreseen and lattice girder in the lower part. Reinforced concrete with invert is designed as final lining, with a drainage layer and water-proofing PVC membrane.

For the inclined shaft also FE simulations where performed by Plaxis 2D to design supports and estimate induced displacement and effects on the watertable, performing coupled static and filtration analyses. Along the inclined shaft, vertical displacement at surface lower than 16 mm where estimated above the tunnel and lower than 7 mm along the adjacent rail tracks (see Figure 14); behind the top-down structures settlements lower than 3 mm were instead obtained. As for the watertable, filtration analysis showed that only local watertable lowering not more than 30 cm is expected, as an effect of drainage around the inclined tunnel and the bottom of the top-down structures, with negligible induced settlements.

5   CONCLUSIONS

The new Lugano tram network represent the main strategic infrastructure planned in the Lugano area to reduce travel time by public transport and increase its attractivity against road traffic and congestion. The core of this project is the Bragnzona tunnel and, most of all, its underground station that allows to connect the new urban infrastructure to the internation rail line passing through Lugano, part of the North-south European corridor. The design solution proposed for tender design, as previously described, provided a proper answer to the Client request to meet both aesthetic requirements, serviceability and environmental and geo-technical challenges.

AKNOWLEDGEMENTS

The authors gratefully acknowledge the Project owner "Divisione delle Costruzioni del Dipartimento del Territorio del Canton Ticino" (acting on behalf of FLP, Ferrovie Luganesi SA) in the name of Ivan Continati and Christian Donetta for their contributions and authorization to publish this technical paper. Moreover, we acknowledge subcontractors that provided support in different aspects of this project, mentioned in this paper: GESTE Engineering SA, supporting for the design and simulation of safety and ventilation systems, Studi Associati SA, for the architectural design, and Leoni Gysi Sartori, as geology specialist.

REFERENCES

Verkehrsbetriebe Zürich 2013. *Vorschriften und Richitlinien für die Dimensionierung von Tramanlagen der Verkehrsbetriebe Zürich*. Ein unternehmen der Stadt Zürich
Buchmüller, S. & Weidmann, U. 2008. *Handbuch zur Anordnung und Dimensionierung von Fussgängeranlagen in Bahnhofen*, Zürich.

*Tunnels and Underground Cities: Engineering and Innovation meet Archaeology,
Architecture and Art, Volume 9: Safety in underground
construction – Peila, Viggiani & Celestino (Eds)*
© 2020 Taylor & Francis Group, London, ISBN 978-0-367-46874-3

Civil engineering constraints on tunnel ventilation and safety

M. Bettelini, A. Arigoni & S.S. Saviani
Amberg Engineering Ltd, Regensdorf-Watt, Switzerland

ABSTRACT: Civil engineering, ventilation and safety design for infrastructure projects are intimately coupled subjects, which require a tight interaction and a deep mutual understanding. This is particularly relevant for complex ventilations systems and emergency exits, which can have a wide-ranging influence on the overall tunnel design. The most severe requirements are met in the case of urban tunnels, subject to space constraints, as well as for long and/or deep tunnels. An optimum design couples the different requirements and constraints in a most elegant and efficient manner to completely fulfil the scope, to guarantee all the requirements, to optimise and simplify the construction and to minimise construction and operational costs. All system's components interact in a seamless manner, allow for efficient operation and maintenance and are less prone to problems while upgrading or retrofitting the infrastructure. This paper addresses the most relevant issues, illustrates background, requirements and constraints and presents possible solutions, based on real-life examples.

1 INTRODUCTION AND OBJECTIVES

Civil engineering, ventilation and safety for infrastructure projects are intimately coupled and require a tight interaction during design and realization. Complex ventilation systems generally require larger excavation volumes or specific safety facilities. This interaction plays an important role throughout the entire life cycle of the infrastructure but is even more important for aging infrastructures. The expected life span of civil components is much larger than that of other safety system components. While the structural components are in most cases still in good condition after a few decades, most of the equipment is obsolete and needs replacement. Due to the technical and normative evolution, periodic renovations must frequently be carried out according to entirely different premises. This evolution should be accounted for from the initial design phase on, otherwise structural adaptation will be costly and could require long closures.

Costs always represent a very central aspect and must be carefully accounted for during the whole life cycle of an infrastructure. Some basic issues are reviewed in Road Tunnels Manual (PIARC 2015). Experience shows that civil engineering accounts for the largest cost share:

- Civil works account 70% to 84% of the total construction costs,
- Construction costs account for 71% to 80% of the total costs over the initial 30 years of tunnel operation.

The costs directly or indirectly related to ventilation and safety frequently represent a significant share of the cost of the civil works, particularly in case of complex ventilations or emergency exits for single-tube tunnels with high coverage. The ventilation and safety concepts are very relevant for the overall construction and life-cycle costs. They must be established and optimized and validated based on a careful interaction between civil, ventilation and safety engineering.

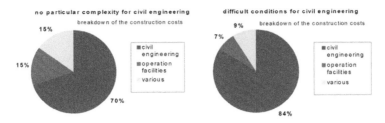

Figure 1. Breakdown of the construction costs (PIARC 2015).

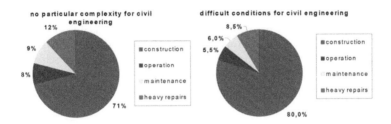

Figure 2. Breakdown of the costs during a 30-year period (PIARC 2015).

2 AERODYNAMICS AND VENTILATION

2.1 General considerations

Ventilation systems in underground works are designed for achieving the following goals:

- Allowing for adequate comfort and healthy conditions, in terms of temperature, air quality and visibility, during normal operation,
- Allowing adequate working conditions during maintenance works,
- Ensuring safe conditions in case of fire inside the underground infrastructure, by means of a suitable smoke-management system,
- Limiting the negative impact on air quality in the vicinity of the tunnel portals,
- Ensuring the ventilation of all accessory facilities, such as shelters, evacuation galleries, cross passages and underground technical facilities.

Ventilation and civil engineering interact at many levels. "Simple" ventilation systems (natural or longitudinal ventilations) for road and rail tunnels are generally not very demanding in terms of civil engineering. "Complex" ventilation systems, including semi-transverse and transverse ventilations, require the creation of one or two longitudinal ducts for fresh air and/or exhaust over the whole tunnel length. The cross sections needed are very substantial (typically at least 10 to 20 m^2), depending on tunnel length and on a number of geometric and traffic-related parameters. In some cases, large cross sections for ventilation result naturally from geometric constraints, as in the case of a three-lane tunnel built using a large-diameter TBM. In other cases, e.g. cut-and-cover structures subject to strict space constraints, the minimization of the cross section of the ventilation ducts represents a primary cost factor.

Early road tunnels were dominated by the requirements for normal operating conditions (air quality). While these requirements tend to diminish in time (Figure 3) in spite of increasing traffic volumes, fire ventilation requirements have increased significantly over the past decade, after the Mont Blanc tunnel event. Tunnel ventilation design is almost invariably dominated by safety requirements.

The progress achieved in fire ventilations for the road sector is slowly entering into the rail sector. It is widely recognized that long rail tunnels can experience significant longitudinal air velocities, arising particularly from barometric pressure differences and thermal effects.

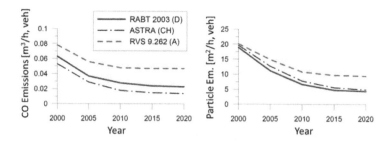

Figure 3. Evolution of specific emissions of road vehicles (Wehner & Reinke 2003).

Because of the large number of persons (up to 1000–1500) and the distance of the emergency exits (up to 500–1000 m), self-rescue times can be very substantial, up to 15–25 minutes or more (Bettelini & Rigert 2012). In many cases, acceptable conditions for self-rescue cannot be allowed without a proper ventilation system. The gap between new and existing rail tunnels in terms of ventilation is generally very large.

2.2 Road and rail tunnels – Common issues and differences

Common ventilation-related issues in the design of both roads and rails tunnels are:

- Missing space for jet fans (e.g. in cut-and-cover and tunnels with false ceiling),
- Missing space for smoke-extraction ducts (cut-and-cover or retrofit of existing tunnels),
- Missing space for technical rooms.

An intense interaction between civil and ventilation engineers is certainly needed and would allow, in some cases, for truly optimized solutions. While design fires for road tunnels are reasonably well defined, large uncertainties still exist for rail tunnels. Tunnels are currently designed for heat-release rates, which can vary by an order of magnitude from project to project.

2.3 Leakages

Leakages in exhaust systems proved to be a serious concern in long road tunnels, particularly where the ventilation was modified increasing the flowrates. Detailed experimental investigations (Bettelini 2008) showed for several tunnels in operation leakage values of the same order of the effective smoke-extraction rates at the fire location. Such ventilation systems are clearly insufficient from the safety point of view and, in such extreme cases, rehabilitation is unlikely to provide the expected results.

A specific regulation is missing in most countries, with the exception of some indications in the Swiss and Austrian national guidelines. Between 2007 and 2009 measurements conducted in 10 Swiss road tunnels (Buchmann & Gehrig 2011) allowed for a reasonable quantification of leakages (expressed in terms of effective leakage surface f^*) for different tunnel types:

- Type A, new tunnels, designed and built according to current criteria: $f^* \approx 14$ mm^2/m^2
- Type B1, retrofitted tunnels with concentrated smoke extraction and sealing: $f^* \approx 25$ mm^2/m^2
- Type B2, retrofitted tunnels with concentrated smoke extraction and insufficient or no sealing: $f^* \approx 32$ mm^2/m^2 (very rough estimate based on scattered measured values).

Since damper leakages typically account for only 5% to 20% of the overall exhaust duct leakage, great care must be devoted to the interaction between the ventilation and the civil engineer designing the exhaust duct. Typical issues related to leakages are:

- Contact surface between dampers and false ceiling,
- Transverse joints between lining blocks and false-ceiling sections,
- Longitudinal joints between false ceiling and tunnel lining,
- Cable passages through the false ceiling, transit points, drainage holes, fissures etc.

2.4 Smoke recirculation

Several types of smoke recirculation could endanger a tunnel's safety:

- Smoke recirculation from one tunnel portal to the other in case of portal smoke expulsion for double-tube configurations,
- Recirculation between tunnel portal and fresh-air inlet for tunnel or secondary ventilation (e.g. safety tunnel or ventilation of technical rooms),
- Recirculation between exhaust stack and fresh-air inlet for tunnel or secondary ventilation.

Measures for preventing this can have a significant impact on portal configuration, particularly in urban environments. Special requirements should be clarified from the beginning.

2.5 Interaction with the environment

The issue of smoke propagation in the neighbourhood in case of tunnel fire is frequently underestimated at design stage. This can have a significant impact on portal configuration, particularly for urban tunnels.

Figure 4.   Smoke propagation during the 2001 fire in the Gotthard road tunnel (Source: Internet).

2.6 Compressibility issues

The cross-section of single-track rail tunnels can be minimized among others by using slab track instead of ballast and conductors rails instead of a conventional catenary. At high speeds, this leads to significant increases of traction power and high-amplitude pressure fluctuations, which can manifest themselves in form of reduction of aural comfort, health damages for the persons on the train, large pressure loads on the infrastructure and micro-sonic boom on the outside. Experience from a number of projects showed that these issues have very wide-ranging consequences and must be accounted for from the beginning of the design.

3   EMERGENCY EXITS AND HUMAN BEHAVIOR

For road tunnels, the directive 2004/54/EC prescribes emergency exits at least every 500 m. The requirements on the maximum distance between emergency exits in Switzerland are more differentiated and stricter (SIA 2004):

- Maximum distance of 500 m for slopes smaller than 1%,
- Linear decrease of allowable distance from 500 m to 300 m between 1% and 5%,
- Maximum distance 300 m in case of separate safety tunnel or cut-and-cover construction.

Many older tunnels were built according to significantly less stringent regulations.

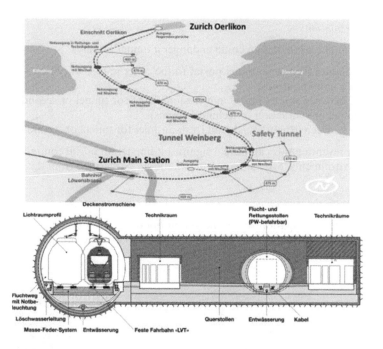

Figure 5.   The Weinberg rail tunnel, in Zurich (Swiss Federal Railway).

For rail tunnels, the past requirements in terms of emergency exits were very heterogeneous. Today's minimum requirements, generally accepted in most EU countries (TSI 2014), require:

- Emergency exits to the surface at least every 1000 m,
- Cross-passages between adjacent independent tunnels at least every 500 m.

Many new tunnel projects, including all large Alpine tunnels, have cross-passages every 300 to 350 m. Similarly, many new rail projects are built together with a safety tunnel with emergency exits every 500 m or less, as e.g. in the case of the Weinberg tunnel.

Issues related to human behaviour are gaining more and more attention. According to PIARC 2008, "the design of tunnels and their operation should take into account human factors". PIARC recommended therefore a number of specific measures with partial impact on civil engineering design.

4   SAFETY CONCEPT AND INTERVENTION

Intervention concepts in road tunnels are reasonably unified and depend mainly on the tunnel system, single-tube or double-tube with cross-connections. In most cases, the details of the intervention concept do not have a significant impact on tunnel design.

In the case of rail tunnels, several entirely different intervention concepts are possible and have an important impact on tunnel design. The main concepts are:

- Intervention on rail, using fire-fighting and rescue trains (used e.g. in Switzerland, Austria),
- Intervention on rail, using conventional fire-fighting engines loaded on special wagons,
- Intervention with road vehicles, where tunnels are made accessible for road vehicles (used e.g. in Austria, Germany),
- Intervention based on mixed rail-road vehicles (used e.g. in Italy, Denmark).

These solutions are quite different in terms of kind of vehicles and training required, intervention time and strategy, investment costs and running costs.

From the point of view of civil engineering, the impact is important, e.g. in terms of:

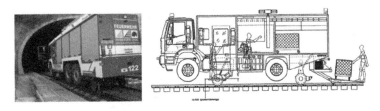

Figure 6. Examples of mixed rail-road vehicles.

- Tunnel platform and track configuration (must be fully accessible to road vehicles and connected to the road network if the corresponding solution is adopted, while conventional ballast or slab-track are acceptable for all rail-based solutions),
- Appropriate turning spaces and connections between tunnel tubes (required especially for the road-based approach),
- Tunnel ventilation (needed if conventional fire-fighting vehicles are used, but usually not mandatory for protected fire-fighting and rescue trains),
- Water supply within the tunnel tubes or at the portals is needed, depending on the approach (no need for water supply within the tunnel if fire-fighting and rescue trains are used).

Since the selected intervention strategy has a heavy impact on structural elements and equipment needed in the tunnel, the decisions on intervention have to be made in the early phase.

5 EVOLVING NEEDS

5.1 Different time scales

The typical serviceable lifetimes of tunnel components are 80–100–120 years for the main structure, 60–80 years for the secondary structure (false ceiling, etc.), 20–30 years for the fans, 15–20 years for the dampers, 10–25 years for further safety equipment and 10–15 years for control equipment (PIARC 2015).

It can be estimated that ventilation equipment must be replaced at least 3 to 5 times during the life cycle of an infrastructure, while the other safety-relevant components 5 to 10 times. Civil design must therefore account for modifications and evolving needs with the support of the safety specialists. Tunnels are designed and built according to the current and foreseeable needs, within the legal and technical frameworks. These can significantly evolve over the lifespan of an infrastructure. Typical issues for aging tunnels are:

- Lack of emergency exits,
- Wrong choice of ventilation system,
- Wrong ventilation design (insufficient cross-section of exhaust ducts etc.),
- Missing or inadequate control of longitudinal air velocity,
- Insufficient room for new equipment,
- Special technical issues (e.g. inadequate safety in case of large longitudinal slopes, …).

The San Bernardino tunnel, in the Swiss Alps, is an example of a very comprehensive tunnel renovation. The 6.6 km long single-tube tunnel was built in 1961–1967. The renovation carried out in 1998–2008 with a total cost of 240 Mio. CHF, included:

- New safety tunnel under the road plane,
- 17 emergency exits connecting the road tunnel with the safety tunnel,
- New smoke-extraction dampers every 96 m,
- New emergency niches every 250 m and new water supply with hydrants every 125 m,
- New emergency lighting and signalization of safety facilities and emergency exits,
- New water-collection system and new wall plates.

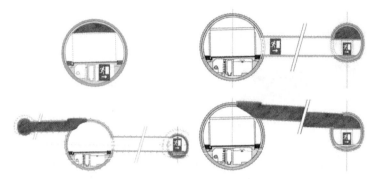

Figure 7. Some options for rehabilitating an aging road tunnel lacking emergency exits and smoke-extraction system (bidirectional traffic on two lanes, 1.5 km).

Infrastructures designed based on a rigid normative framework, satisfying the minimum requirements in a minimalistic manner, are most likely to rapidly become obsolete. A more functional approach, where normative requirements are interpreted and implemented with a global safety-based approach is much less prone to generate high additional costs.

5.2 Standards and regulations

Standards and regulations for ventilation and safety design of road tunnels evolved very rapidly almost worldwide after the Mont Blanc event (March 1999). Common requirements for the European countries were formalized in the directive 2004/54/EC on minimum safety requirements for tunnels, which codified the evolution of the state-of-the-art during the last decade. Important innovations were introduced from both the technical and the organizational side; the most important ones regarding the interaction between safety, ventilation and civil engineering were:

- Stricter requirements for emergency exits,
- Stricter specifications for the choice of the ventilation system,
- Significantly higher smoke-extraction rates than in the past,
- Specific requirements concerning the control of longitudinal air velocity.

All these aspects had and have severe consequences on many existing tunnels conceived according to pre-Mont Blanc requirements. Even with large national differences, the international effort for adapting existing infrastructures to this evolution was persecuted in a quite systematic manner and with very substantial investments.

The European harmonization in the rail sector resulted in the establishment of the directive 2004/49/EC on safety on the community's railways and of the Technical Specifications for Interoperability (TSI 2014). However, the modernization of the existing tunnel network is much less advanced than in the road sector and the gap, in terms of safety design, between new and existing tunnels is very large. The number of accidents in rail tunnels is significantly lower than in road tunnels and rehabilitations are considered less urgent, besides there is no general agreement on the safety standard needed in rail tunnels.

5.3 Traffic

Tunnel infrastructures are typically designed for 100 years but traffic forecast is in most cases very uncertain even for the next decade. Traffic volumes and composition affect not only the design of the ventilation system but also the selection of the appropriate system. This aspect is most prominent in the cases where the frequency of congestion increases significantly because of growing traffic volumes.

5.4 *New Technology: Fixed-Fire Fighting Systems*

Among the emerging technologies, Fixed-Fire Fighting Systems (FFFS) are most probably the one which could have the most lasting consequences on tunnel infrastructure and is already impacting national and international directives. NFPA 502 (2017) acknowledges its use in road tunnels, provided that the achievement of an acceptable level of safety and of the intended level of performance is demonstrated by engineering analyses and appropriate installations, inspection and maintenance schedules to maintain the level of performance intended. The application of water mist fire protection systems is regulated in NFPA 750 (2019), but a significant evolution of norms can be expected, thanks to the research activities and the experience gathered.

Fire-extinction systems were investigated particularly in the European UPTUN program, which focused on the development of cost-effective sustainable and innovative UPgrading methods for existing TUNnels (UPTUN). In the German national research projects SOLIT and after SOLIT2, the use of water mist technologies in road tunnels was analysed and optimized to evaluate the interaction with the other safety relevant measures.

FFF systems are effective in pursuing the major aim of improving conditions for escape and rescue and protecting the tunnel infrastructure. It is not possible providing general indications for the technology to be preferred for a particular situation. The interaction with tunnel ventilation, possible synergies and potential cost savings must be assessed on project-specific bases.2

6 EMERGING UNDERGROUND INFRASTRUCTURES

In urban areas, space is restricted and becomes more and more precious, while traffic congestions limit business development and create a negative impact on the life quality of the citizens. Consequently, the underground space is used not only for utilities and infrastructures but also for constructing industrial services, installations, housing or even R&D facilities. Admiraal & Cornaro 2017 provided a broad review on of the concept of underground space development investigating the issues associated with the sustainable development of urban underground space.

Representative examples of this kind of underground facilities are:

- Hagerbach Test Gallery VSH (in operation since 1970),
- WaferFab Sargans (civil works completed),
- Underground Science City Singapore (advanced feasibility studies),
- Relocation of Sha Tin Sewage Treatment Works to caverns (design phase).

Particularly, the Hagerbach Test Gallery VSH, established initially as a research and development facility for tunnel construction, provides now an underground network of galleries (total length 5 km), hosting a large variety of heterogeneous usages (restaurant, laboratory, excavation and concrete testing, fire research and fire-fighting training, conferences and social events).

The permanently increased density of the UG networks and facilities are of highly complex nature both in civil engineering and in operational aspects. This adds considerably new demands and requirements to safety and reliability. Newer underground infrastructures can be characterized as follows (Amberg & Bettelini 2012):

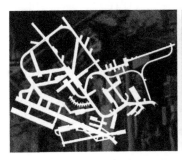 

Figure 8.   The Hagerbach Test Gallery (Source: VSH).

- Deep structures with a limited number of accesses,
- Underground working or recreational spaces, with long permanence times,
- Large number of persons, very heterogeneous groups,
- Users are not familiar with the specific underground environment,
- Mixed use, which can evolve in time.

It is interesting to notice that the emerging needs are partly contradictory:

- Pleasant environment with large, well-lit rooms,
- Flexible space organization for mixed, heterogeneous usages,
- Seamless safety system,
- High-capacity egress,
- Appropriate accesses and technical means for intervention.

In spite of widely different usages, it is possible to identify general underlying design-relevant characteristics:

- Evolution of usage with time, following interests, needs and opportunities: high-investment long-life underground infrastructures must constantly adapt to evolving needs at economically reasonable cost,
- Mixed usage with potentially conflicting usages (e.g. blasting tests and conferences) conjugated in potentially dangerous locations not accessible for unprepared visitors,
- Large person fluxes, special attention devoted to the large social events playing an important role for the economic welfare of the infrastructure and representing a major safety issue.

7 THERMAL PROTECTION OF INFRASTRUCTURE

In the interaction design process between civil engineering and fire and safety analysis, it is important to underline the objectives of structural fire resistance:

- Allowing for self- and assisted rescue under safe conditions,
- Allowing for intervention under safe conditions,
- Preventing collapse of tunnel, loss of property and minimize time for repair after a fire.

Several temperature-time curves are in use for specifying thermal protection of structural elements. The cost for thermal protection can be very high and the project-dependent choice of the curve to be applied should be carried out jointly by the safety and structural specialist. PIARC and ITA recommendations for fire protection (design criteria, materials etc.) are discussed in great detail in ITA 2017.

A representative example of comprehensive fire protection is the ventilation station Reppischtal of the Uetliberg tunnel, where a comprehensive thermal protection was required for protecting the ventilation station Reppischtal.

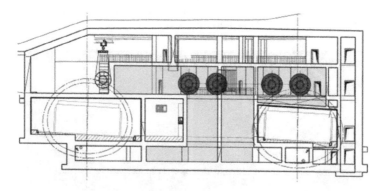

Figure 9. Cross section of the ventilation station Reppischtal of the Uetliberg Tunnel in Zurich, V=180′000 m^3 (Amberg Engineering for Canton Zurich 2009).

8 ENVIRONMENT

The environmental impact of tunnels is manifold. From the point of view of ventilation and safety, the emissions (air and water pollutants, noise, etc.) are of primary concern. Especially in urban tunnels, air pollution can be a problem at tunnel portals and should be addressed from the initial design phases. Possible measures to be investigated include:

- Ventilation towards the opposite portal,
- Ventilation system with intermediate exhaust (bidirectional traffic and low coverage),
- Portal extraction and expulsion through a sufficiently high stack,
- Portal extraction and exhaust treatment using an appropriate filtering system,
- Selecting a better-suited portal location,
- Constructive measures (tunnel prolongation with a coverage, barriers against pollutant propagation, etc).

These measures require substantial investment and, in many cases, high maintenance costs and unreasonable energy consumption. A holistic approach is required, where all economic and environmental relevant aspects of every possible solution are carefully evaluated. An intimate interaction between civil, environment and ventilation specialists is required.

9 CONCLUSION

The interactions between safety, ventilation and civil engineering are manifold and have wide-reaching consequences during design, construction and over the whole life span of an infrastructure. According to PIARC 2015, "Recent examples indicate that transverse optimisations (civil engineering - ventilation - safety evacuation) made at early project stages can contribute about 20% towards cost savings". A continuous open interaction between the specialists from the different fields is called for.

REFERENCES

Admiraal H. & Cornaro A. 2018. *Underground Spaces Unveiled: Planning and creating the cities of the future*, ICE Publishing.

Amberg F. & Bettelini M. 2012. *Safety Challenges In Complex Underground Infrastructures*, Advances In Underground Space Development, Proceedings of the 13th ACUUS World Conference.

Bettelini M. 2008. *Measuring Flow Rates and Leakages in Long Tunnels*, ITA-AITES World Tunnel Congress.

Bettelini M. & Rigert S. 2012. *Emergency Escape and Evacuation Simulation in Rail Tunnels*, ISTSS - Tunnel Safety & Security, 5th International Symposium, 14–16 March 2012 New York, USA.

Buchmann R. & Gehrig S. 2011. *Quantification of the leakages into exhaust ducts in road tunnels with concentrated exhaust systems*, Forschungsauftrag ASTRA 2007/004 auf Antrag des Bundesamtes für Strassen (ASTRA).

Canton Zurich 2009. *Straight around Zurich – A3 Zurich West Bypass and A4 in the Knonau District*.

SIA 2004. Swiss Norm SN 505 197/2, SIA 197/2 2004. *Design of tunnels - Road tunnels*.

EC 2004a. *Directive 2004/54/EC of the European Parliament and of the Council of 29 April 2004 on minimum safety requirements for tunnels in the Trans-European Road Network*.

EC 2004b. *Directive 2004/49/EC of the European Parliament and of the Council of 29 April 2004 on safety on the community's railways*.

ITA 2017. *Structural Fire Protection for Road Tunnels*, ITA Report n°18/April 2017, ITA Working Group 6, Maintenance and Repair.

TSI 2014. *Commission Regulation (EU) No 1303/2014 of 18 November 2014 concerning the technical specification for interoperability relating to 'safety in railway tunnels' of the rail system of the European Union*.

Wehner M. & Reinke P. 2003. *Stand und aktuelle Entwicklung bei der Lüftung und Entrauchung von strassen- und bahntunneln in Mitteleuropa*, STUVA-Tagung '03 Westfalenhalle Dortmund.

World Road Association (PIARC) 2008. *Human factors and road tunnel safety regarding users*.

World Road Association (PIARC) 2015. *Road Tunnels Manual*, version 1.1 of 27th October 2015.

Tunnels and Underground Cities: Engineering and Innovation meet Archaeology, Architecture and Art, Volume 9: Safety in underground construction – Peila, Viggiani & Celestino (Eds)
© 2020 Taylor & Francis Group, London, ISBN 978-0-367-46874-3

Ventilation applied to fire safety in metro tunnels, Naples line 1 project

M. Bramani & P. Fugazza
MM S.p.A., Milan, Lombardy, Italy

ABSTRACT: The extension of Line 1 of Naples from Centro Direzionale station to Capodi-chino is one of the most complex underground project ongoing in Italy. Provided that main point of complexity is surely the structural and architectural design, this kind of challenging project cannot avoid safety studies. One of the most significant and crucial aspects concerning this matter is fire scenarios handling. Ventilation is one of the major measures at tunnel designer disposal to mitigate the effect of fire and smoke, to aid evacuation, rescue services, and firefighting in a tunnel fire. In case of a tunnel fire, fire ventilation systems are required to control smoke flows and create paths for evacuation and firefighting. Nowadays the most used emergency ventilation system is the longitudinal one. The most recent design standard includes the application of Fire Safety Engineering. This innovative approach is nowadays also foreseen in Italian specific law for subway construction. Two key parameters for tunnels with longitudinal ventilation, that is, critical velocity and backlayering length are presented. For smoke extraction systems, sufficient air flows are required to be supplied to prevent smoke spread. Will be shown how this concept is also applicable in case of fire train in stations. Regarding Naples Line 1 project, firstly one-dimensional fluid dynamic simulations will be presented, then 3D simulations will be showed to give an example of how these tools can help us in the design phase.

1 INTRODUCTION

The need to demonstrate the compliance to the safety standards has led over the years to a progressive and growing demand from the authorities responsible for the approval of the projects to face the fire safety subject through analysis developed with the so-called "performance-based engineering approach".

This methodology mainly asses the functionality of the ventilation system during normal and emergency operation allowing detailed analysis of fire and smoke propagation for different emergency scenarios by means of advanced 1-D and 3-D numerical fluid dynamic models. The engineering approach is not only aimed to comply with the governing standards, but it helps the designer to verify the effectiveness of mechanical systems envisaged by the project.

The principal aim of the fire safety engineering is to identify and validate the ventilation management strategies that allow to reach, in all credible emergency scenarios, the appropriate performances to guarantee the safety of the users involved in the incident until the completion of the rescue phase.

This in-depth analysis takes the form of a detailed investigation of ventilation system behaviour, supported by analytical and numerical models that consider the specific fluid dynamic phenomena to which stations and tunnels are subjected in case of fire.

In Italy this "performance-based design" approach was recently ratified with the enactment of a specific law (D.M. 21/10/2015), originated from a collaboration between the "Ministero dell'Interno" (Italian Government Department to which the fire fighters refer) and the "Ministero delle Infrastrutture e dei Trasporti" (Italian Government Department which deals with transportation system construction and management). This new law, taking inspiration from existing international standards (e.g. American standard NFPA 130), replaces the previous one (D.M. 11/01/1988), which was characterized by a different design approach. The present law explicitly requires the evaluation of smoke propagation as well as egress evolution to assess the possibility of a safe escape from tunnels and stations.

The purpose of this paper is to illustrate the key features of the fire safety methodology and its application to the "Naples Line 1" project dealing with the extension from Centro Direzionale to Capodichino station. This project represents one of the most complex underground challenge ongoing in Italy, not only in relation to the architectural and structural design aspects, but also from the fire safety point of view. The presented argumentation mainly focuses on the design criteria, calculation hypotheses and combination of models that can and should be applied to carry out a complete analysis of a metro line.

2 TUNNEL VENTILATION AND DESIGN CRITERIA

The need of managing the fire emergencies in underground transportation systems forces engineers to foresee an emergency mechanical ventilation system in practically all metro projects. Such a system aims to control smoke flow and mitigate the effect of fire and smoke and consequently to aid evacuation, improve emergency response and firefighting operations.

In the presence of well dimensioned pressure and relief shafts, the mechanical ventilation during normal operation is generally not required. Moreover, the presence of piston effect (forced air flow inside a tunnel or shaft due to the movement of trains inside the tunnels) boosts the natural ventilation and successfully remove the heat and reduce the concentration of contaminants and dust below acceptable levels. The piston effect mainly depends on the ratio between the frontal area of the vehicle and that of the tunnel, the length of the vehicle, the geometrical network of the tunnel, the speed and schedule of trains inside the tunnel.

The natural ventilation is possible only for characteristic tunnel profile. The line shape highly affects the natural recirculation of air in conjunction with positive temperature difference between the environment within the tunnel and the external ambient conditions.

When the two above mentioned phenomena are not present, or not strong enough to provide the minimum air exchange, a mechanical ventilation system for normal operation in addition to the emergency ventilation should be foreseen. Such aero-thermodynamic aspects can be investigated in an early stage of the project by means of specialized numerical fluid dynamic simulations. The numerical analysis can be overlapped by fire accidents to simulate different emergency scenarios, furthermore such analysis can be extended to conduct long term simulations to investigate the tunnel temperature trend.

To reduce the cost and simplify the construction of a tunnel structure, the emergency ventilation system is usually combined with the normal ventilation system, where needed.

The ventilation design of a whole metro line involves mainly the determination of the optimal ventilation strategies for the management of three fire scenarios: fire onboard a train inside the tunnel; fire onboard a train at the station and fire onboard a train in "special" segments or parts of the metro line such as depots, tunnel bypass or crossovers.

It is worth to underline that the scenario of a fire on board a train in tunnel is generally less probable than the case of a fire onboard a train at the station. In most metro and railway systems, the fire growth is generally slow, and fire can be usually detected in early phases where its heat release rate is limited and rarely it causes the train to stop within the tunnel. The emergency management system foresees that the train affected by the emergency is driven to the nearest station. Indeed, the evacuation from the train through stations is less critical than the evacuation though the tunnel.

Nevertheless, if a train on fire cannot be driven to the station, the evacuation takes place within the tunnel. From ventilation point of view, the emergency ventilation scenario is often dimensioned for a fire scenario along the tunnel, which is the most challenging from the evacuation point of view.

The main purpose of the emergency ventilation system is to create a smoke-free escape route for passengers, keeping all the critical parameters for life below certain values. Such parameters are mainly related to visibility, concentration of combustion gases (typically carbon monoxide and dioxide), temperature, and thermal radiation.

The ventilation strategies that allow to achieve the above goals are divided in three groups: longitudinal ventilation, transverse ventilation and the semi-transverse ventilation.

In a longitudinal ventilation system, jet fans and/or normal fans are used to create a longitudinal flow to remove smoke, dust and heat while a transverse ventilation system foresees that the air flows are transversely transferred from supply vents into exhaust vents. The transverse ventilation system consists of many vents positioned along the tunnel for supplying fresh air into the tunnel. The exhaust vents are generally placed at the ceiling level to exhaust the contaminants and waste heat whereas the supply vents could be placed either at the floor or at the ceiling level. The ducts can be either connected to a shaft located between the tunnel portals or a fan room nearby a tunnel portal. Finally, a semi-transverse ventilation system is very similar to the transverse ventilation one. The only difference is that for a semi-transverse system, only supply vents or only exhaust vents are in operation. In other words, only fresh air is supplied into the tunnel or only contaminant air is extracted into the extraction duct.

Transverse ventilation would be the best choice in terms of safety for the capacity of minimizing the distance between the extraction shaft and the fire allowing people to evacuate in all directions. On the other hand, its realization in an underground metro tunnel is very difficult and expensive.

As explained above, in case of a fire, a longitudinal ventilation system is designed to produce a longitudinal flow to create a smoke-free path upstream of the fire so that the evacuees can escape in a clear air stream. In underground metro lines, the presence of a 24/7 manned control room and a dedicated evacuation management system (emergency public address tunnel system eventually coupled with visual dynamic addressing systems) facilitate the routing of the users towards the safe path. Depending upon the fire location, there might be people that have to evacuate downstream of fire in direction of smoke propagation. Such situations are managed by providing emergency exits. Typical distance to reach an emergency exit or a station is about 200–300 meters. Furthermore, provided that the typical fire size is way lower than the one that is likely to occur in a roadway tunnel, the smoke dilution by the air flow rate coming from the upstream branch of the tunnel usually guarantees the subsistence of minimum survival conditions also downstream the fire.

Focusing on the longitudinal ventilation, it can be sub-categorized into pure longitudinal ventilation e.g. by means of jet fans only and longitudinal ventilation generated by exhausting air/smoke at one or more points along the tunnel eventually aided by jet fans. Pure longitudinal ventilation mode is typically applied in road tunnels. For underground metro lines, the smoke extraction ventilation mode (longitudinal) must be applied, eventually integrated with jet fans, mainly because it is mandatory to effectively remove smoke and evacuate it at the street level. For this reason, for all tunnel sections between two stations, at least one fan room must be foreseen. Lastly, special constructions can be designed inside the tunnel to mitigate the effect of fire and smoke flows, such as emergency exits, cross-passages and rescue stations.

The two most important parameters for a longitudinal ventilation design are the critical velocity and the back-layering length. The critical velocity is defined as the minimum longitudinal ventilation velocity to prevent reverse flow of smoke from a fire in the tunnel and mainly depends on the height of the tunnel, its cross-sectional area, the fire power and the tunnel gradient (tunnel slope). The back-layering length is defined as the length of the smoke plume upstream the fire, it is naturally generated by the smoke spread when air velocity is lower than the critical velocity.

A fresh air flow with a velocity not lower than the critical velocity is mandatory to prevent smoke back-layering, which means that the tunnel is free of smoke upstream of the fire site. On the other hand, an air velocity that greatly exceed the critical velocity lead to high smoke de-stratification downstream of the fire and at the same time demands higher fan power leading to a non-economical solution. The ventilation system design always foresees to overcome the critical speed with a reasonable margin of safety. In very special cases, when it is particularly difficult to reach the critical velocity, the ventilation system goal can be limited to reaching the "confinement velocity". This is defined as the air velocity needed to prevent backlayering beyond a certain length i.e. to confine the smoke and prevent further spreading upstream.

Summarizing, the principal aim of the emergency ventilation system is to create a safe evacuation route thereby controlling the smoke by means of a fresh air flow characterized by a velocity that exceeds the critical velocity or, in very special cases, that exceeds the confinement velocity. Around that central concept, further deepening is required to evaluate other aspects such as safety of people evacuating along the wrong way (downstream the fire), smoke destratification and recirculation, structural stress, optimization of sprinklers and fire detectors.

3 FLUID DYNAMIC ANALYSIS OF NAPLES LINE 1 EXTENTION CDN – CAPODICHINO

3.1 *Main characteristics of Naples Line 1 extension from CDN to Capodichino*

The extension of Line 1 from CDN to Capodichino is about 3.4 km long and it includes four stations: "Centro Direzionale", "Tribunale" and "Poggioreale", which are substantially at the same altitude and at the same depth, and "Capodichino Airport" which is peculiar for both its morphology and for being at a much higher altitude (about 45 meters) than the other stations and, at the same time located very deep in terms of its geological location (the rail level is about 44 m deep).

From CDN to Poggioreale and from Capodichino to the end of the extension, the line is characterized by a single tunnel with double track, while a double bored tunnel (one track for each tunnel) is present from Poggioreale to Capodichino.

It should be emphasized that the portion of the gallery preceding the Centro Direzionale station (towards Garibaldi) is about 195 m long and has a quite large opening to ambient at the beginning that allows to consider the extension aerodynamically independent from the rest of the line.

Beyond Capodichino Airport station, the two single -track tunnels join in one double-track tunnel with a crossover 220 m length that ends with a series of openings designed to prevent the stagnation of smoke in that area.

The tunnel emergency ventilation system is mainly composed of a ventilation room (hereinafter abbreviated with VR), one for each section between two stations (positioned indicatively halfway depending on the constraints at the street level). The maximum total capacity of all tunnel VRs is about 700,000 m³/h (two fans with a max flow rate of about 350,000 m³/h). Every station is equipped with a VR consisting on one fan with a maximum capacity of about 300,000 m³/h linked to the station platforms with over track and under platform exhaust ducts.

A depot for the rolling stock is located between Tribunale and Poggioreale station (approximately 100 m before Poggioreale station). This depot zone is linked to the main line by a connecting tunnel with a length of about 120 m. The depot area comprises a wider underground tunnel of about 250 m length. The depot is longitudinally subdivided into two areas characterized by two and three parking tracks respectively. In addition to the mentioned areas, a storage zone is located aside to the two and three track zones. This storage area is dedicated for diesel vehicles and – in case of fire – it constitutes a separate compartment by the automatic closing of a motorized fire-resistant gate.

Figure 1 shows the tunnel gradient and the main cross-sectional areas of Naples Line 1 from CDN to Capodichino.

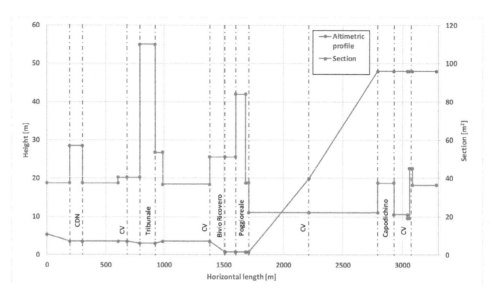

Figure 1. Plan view and altimetric profile of Naples Line 1 extension.

3.2 *One-dimensional simulations*

One-dimensional fluid dynamic simulations play a very important role in the preliminary assessment of the ventilation system design. The possibility to simulate the entire network in a relative low computational time is a key feature that leads engineers to use this tool. Indeed, the time required to set up and run the model is way lower for a 1-D simulation compared to a 3-D one. A three-dimensional model is suitable for simulating limited scale domains, such as a tunnel section or a station, where three dimensional phenomena cannot be neglected. These limitations oblige referring to 1-D CFD software to account for the aero- thermodynamic interaction of all components that constitute an entire subway metro line. The smoke spread depends on several variables like train dynamics, position of other trains, pressure and temperature in the tunnel that can be evaluated only through simulations of the entire system. The typical methodology is to integrate 1-D and 3-D simulations taking the boundary conditions in terms of pressure, temperature and flow rates from the first ones (1-D) and applying them appropriately to the latter (3-D).

The input data required to build a complete one-dimensional model are the geometrical data of tunnels, shafts and stations; physical characteristics of the main construction materials; ventilation room performances; geometrical and aerodynamic data of trains and their movement schedules. Further inputs to the 1-D model require the information regarding ambient thermal-hygrometric conditions, design fire in terms of peak of power and position, operational details for dampers and fans (i.e. closing/opening of dampers, fan starting and reaching max. power etc.). Finally, the information regarding the constraints related to the fire detection system as well as the procedures for managing a fan failure must be considered.

In Figure 2, a shred of the one-dimensional fluid-dynamic model of the extension of Naples Line 1 is reported. The model was used to analyse emergency ventilation involving fire scenarios inside the tunnel.

3.2.1 *Fire onboard a train inside the tunnel*
The Naples Line 1 extension project has been developed considering the main objective of extracting smoke always from the VR that are positioned approximatively in the middle point of each tunnel sub section between two stations. This implies that the ventilation system should be sized to reach (with adequate margin) the critical velocity in the sub section of the tunnel where the fire is detected. Among the possible longitudinal ventilation systems, this is

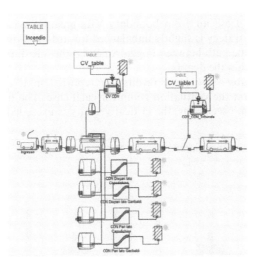

Figure 2.   Tunnel model for Naples Line 1 extension project.

surely the most effective approach as it allows people to escape always towards stations. Even if a VR can act as an escape route, stations provide a safer and easier way to get out of the tunnel. The central aspiration has the advantages of a) minimizing the smoke extraction path along line and b) avoiding the transit of smoke in front of station platforms. On the other hand, this strategy implies the construction of a ventilation shaft of significant dimensions along each tunnel sub section between two stations and an appropriate dimensioning of fans.

To limit the size of the fans, the activation of VRs positioned in the adjacent sections of the tunnel is used. This creates the so-called "push and pull" strategy.

Usually the choice of VRs activation depends on the train position in the tunnel sub section between the two stations. For Naples Line 1 extension from CDN to Capodichino one of the design inputs was that the ventilation system must operate in a robust manner remaining independent of the location of the train on fire in the sub section (upstream or downstream of the VR). This implies that every ventilation strategy designed for each sub section must be balanced in terms of air flow rate coming from terminal stations to the ventilation room between the two branches upstream and downstream. The main advantages of this choice are that a) no train position detection system (along the tunnel sub section) is required and b) the strategy matrix of emergency ventilation system is far simpler and reliable.

All fire scenarios have been analysed also by considering the failure of one of the fans involved in the fire strategy. The simulation results showed that even in case of crucial fan getting failed, it was still possible to maintain at least the confinement velocity.

Among all the analysed scenarios, the one related to the case of a fire train stuck between Tribunale and Poggioreale stations has been reported in this paper as it identifies the most typical and representative fire scenario. This part of the tunnel is characterized by a single bore with double track and low slope. The major feature of this segment of tunnel is the section change upstream and downstream of the ventilation room due to the presence of the entrance to the rolling stock depot.

All the results are evaluated at steady state conditions i.e. when the fire has reached its maximum rate. By the way, no fire decay is considered in the simulation, this conservative assumption is typically adopted in this kind of analysis to guarantee safe conditions also for rescue and firefighting operations.

The analysis was carried out for two fire scenarios, one upstream of the VR and one downstream of the VR. For both scenarios two cases have been considered: normal functioning, with all fans up and running, and degraded functioning, characterized by the failure of one fan of the VR. As described above, the line configuration allows to apply a typical ventilation strategy which foresees an extraction from the halfway VR between Tribunale and Poggioreale stations and introducing fresh air from boundary VRs creating a push-pull behaviour from both side branches. The strategy is slightly unbalanced toward the Tribunale branch because of the less pressure loss on the path between Poggioreale and the VR due to the bigger tunnel section and the openings within the depot.

In Figure 3 and Figure 4 the main results are reported in terms of fresh air velocity upstream and downstream the ventilation room for each case. The main outcome is that for all "normal fire condition" scenarios, the critical velocity is exceeded, while the confinement

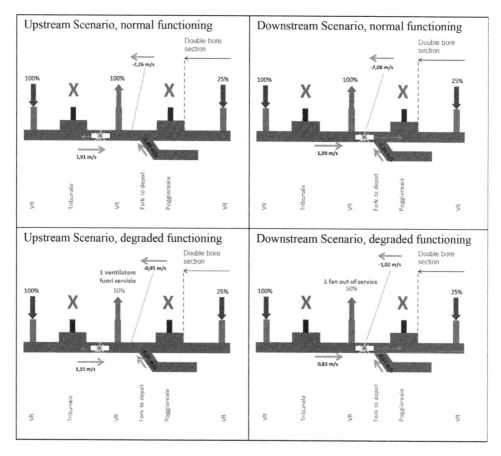

Figure 3.   Schemes reporting the main results of the scenario involving a fire train in the tunnel between Tribunale and Poggioreale stations.

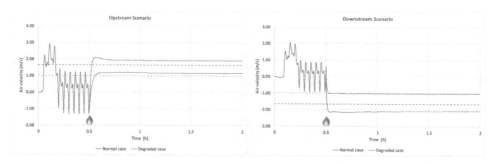

Figure 4.   Fresh air velocity towards fire for both scenarios involving respectively a train fire upstream and downstream the halfway ventilation room between Tribunale and Poggioreale stations.

velocity is reached in all "degraded fire scenarios". The need to foresee a unique ventilation strategy for the whole tunnel sub section leads to the consequence that in the opposite branch in respect to the one where the fire take place, the air velocity is directed towards the VR preventing the smoke to reach both the boundary stations.

3.3 Three dimensional simulations:

Three-dimensional fluid dynamic simulations represent a tool that fire safety engineers can use together with one-dimensional simulations and analytical models to predict the behaviour of fire and smoke during a fire emergency in limited regions of the metro line where the three-dimensional effects such as back-layering, stratification, turbulence etc. cannot be neglected. The boundary conditions in terms of pressures, temperatures and flow rates, are usually taken from 1-D simulations to account for all interactions of other components of the system not included within the 3-D model.

The typical steps required to develop a 3-D CFD simulation are: "pre-processing" which includes 3-D CAD drawing and mesh creation, "solving", that foresees to input all boundary conditions, material properties, multi-physics equations (combustion, thermal radiation, gas tracking, turbulence) and solving them in each sub-volume and for each time step till numerical convergence is reached and "post-processing" which involves the display of results, their export and analysis.

In many cases, the aim of a CFD simulation applied to fire safety engineering is to evaluate the critical parameters for life safety. These parameters are mainly temperature, carbon monoxide and dioxide concentrations, visibility and thermal radiation. Furthermore, 3D simulations are also useful for optimizing smoke detector and sprinkler layout as well as designing emergency exits and for predicting the thermal stress on structures.

In the present paper the application of 3-D CFD for the analysis of fire and smoke propagation due to a fire located within the depot (between Tribunale and Poggioreale station) is shown. The 3-D CFD work was conducted by the independent ventilation experts at Pöyry Switzerland.

3.3.1 Fire onboard a train inside the depot

The case study reported in this section concerns the application of three-dimensional computational fluid dynamics to the underground rolling stock depot between Tribunale and Poggioreale station.

A scheme of the ventilation system in the depot is reported in Figure 5. The ventilation strategy foresees the following:

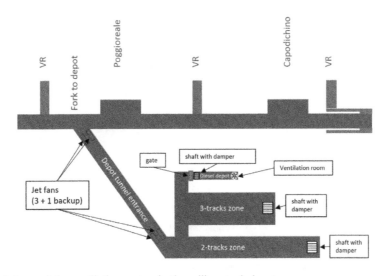

Figure 5.   Scheme of the ventilation system in the rolling stock depot.

- Within the entrance tunnel: 4 fully reversible jet fans (3 + 1 backup) within the entrance tunnel with a thrust of about 2300 N each.
- Within the 1-track diesel rolling stock zone: 1 natural ventilation shaft of 4 m² and 1 ventilation room with a capacity of about 150.000 m³/h (non-reversible);
- 1 natural ventilation shaft of 18 m² at the end of the 2 tracks zone provided with damper;
- 1 natural ventilation shaft of 20 m² at the end of the 3 tracks zone provided with damper.

A one-dimensional simulation was carried out to retrieve the time dependent boundary conditions in terms of temperature, pressure and flow rate at domain borders. These boundary conditions were applied on the three-dimensional model of the depot.

The simulation was useful for assessing the performance and dimensioning of the ventilation system. The results showed that the system runs appropriately as planned and protects the main metro line from smoke in case of fire in the depot region. Furthermore, the simulations gave information regarding the smoke segregation within the depot and allowed the designers to estimate the time available for the evacuees within the depot to reach safe exits.

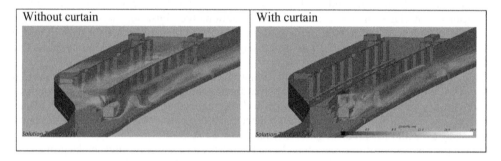

Figure 6. Smoke propagation at 10 min after ignition (maximum fire rate) for the two investigated depot layouts (left side without curtain; right side with curtain).

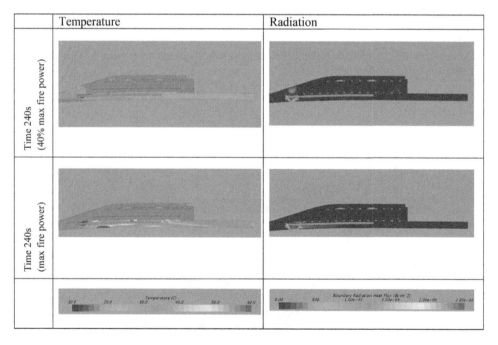

Figure 7. Contours of temperature and thermal radiation at 1.8 m from platform level at 240 s (40% of maximum fire rate) and 600 s (maximum fire rate) for the depot layout with curtains.

This time was compared to evacuation time to demonstrate the possibility of a safe evacuation for the operating personnel.

The reported example explains how the 3D simulation was useful for optimizing the depot design adding a curtain on the ceiling at the depot entrance. Figure 6 shows the results of smoke propagation at the maximum fire power for the two investigated depot layouts (with and without curtain). The results shown are taken at the point where the ventilation system is running at its full capacity and the maximum fire power is reached. The results show that for both cases the ventilation system can establish a stable flow direction with air velocity greater than critical velocity. The established flow conditions hinder any smoke transport towards the entrance of the depot for both the cases. The improved depot layout preserves any smoke transport towards the triple track zone of the depot thereby avoiding a threat to evacuees in the regions outside of the emergency section of the depot.

Figure 7 shows the contours of temperature and thermal radiation at 1.8 m from platform level at 240 s (40% of maximum fire rate) and 600 s (maximum fire rate).

4 CONCLUSIONS

The need of managing the fire emergencies in underground transportation systems forces engineers to foresee an emergency mechanical ventilation system in almost all metro projects. Such systems aim to control smoke flow and mitigate the effect of fire and smoke and consequently aid evacuation, emergency response, and firefighting operations.

The need to demonstrate the compliance to the safety standards has led over the years to a progressive and growing demand from the authorities responsible for the approval of the projects to face the fire safety subject through analysis developed with the so-called "performance-based engineering design approach". This approach mainly asses the functionality of the ventilation system during normal and emergency operation allowing to analyse the fire and smoke movements by means of dedicated and specialized 1-D and 3-D numerical fluid-dynamic models.

Among all the possible configuration of tunnel ventilation system, the most widely applied strategy to the underground metro lines is the longitudinal ventilation. The main reasons for its applicability are its robustness, structural suitability and its economical being in comparison to other modes.

The principal aim of the emergency ventilation system in an underground tunnel is to exhaust smokes and create a safe evacuation route by means of a fresh air flow characterized by a velocity that exceeds the critical velocity. In special part of the line (ex. Crossover or ends) or in ventilation degraded mode also confinement velocity can be taken in consideration (this is the air velocity needed to prevent backlayering beyond a certain length).

For Naples Metro Line 1 extension from CDN to Capodichino both one-dimensional and tri-dimensional fluid dynamics simulations were carried out, allowing the validation of emergency ventilation system sizing and the design of the activation strategies. For the reasons above, these studies represent a milestone in the safety assessment of the whole metro line.

REFERENCES

Ingason, Zhen Li & Lonnermark (2015). Tunnel Fire Dynamics. City: London.
Yousaf & Eichelberger, Pöyry Schweiz (2017). Metro Line 1 Napoli – 1D and 3D CFD Smoke Extraction Simulations. City: Zurich.

Tunnels and Underground Cities: Engineering and Innovation meet Archaeology,
Architecture and Art, Volume 9: Safety in underground
construction – Peila, Viggiani & Celestino (Eds)
© 2020 Taylor & Francis Group, London, ISBN 978-0-367-46874-3

Lercara tunnel on Palermo-Agrigento Railway Line, Italy: Tunnelling in presence of toxic and explosive gases and squeezing conditions

N. Casagrande, M. Ricci, D. Commisso & G. Gallo
Italferr S.p.A, Rome, Italy

ABSTRACT: The paper describes the case of Lercara tunnel in Sicily (Italy) belonging to Palermo–Agrigento railway line. Lercara is a single-track tunnel, 10 m in diameter, with a total length of 2800 m, a maximum overburden of 150 m, excavated with conventional methods in clay formations. The excavation began in 2006. As soon as the excavation reached the central part of the stretch, unexpected convergences caused shotcrete cracking, producing the deformation of steel ribs and failure mechanisms. Besides, due to unexpected levels of gas emission, in 2009 works were interrupted by the authorities. In 2010 the new project design was assigned to Italferr with the aim of solving the critical issues emerged. The tunnel excavation re-started on April 2014 and successfully completed on August 2016: all the construction phases were accompanied by an intense program of monitoring about the effects of the tunnel excavation and the presence of gases.

1 INTRODUCTION

The modernization project for speeding up the Palermo-Agrigento railway line is about 95 km long, between the stations of Fiumetorto (PA), Agrigento Centrale and Porto Empedocle (AG). It consists in alignment adjustments and infrastructural and technological modifications of the existing railway line. It involves the construction of a new section of the line, the so-called "Variante 2.1-Galleria Lercara", for an overall extension of approximately 6.5 km, including a new single-track tunnel of about 2.8 km (Lercara tunnel).

The project history began in 2005, when the Joint Venture of Companies was awarded the first contract by RFI. The excavation works by conventional method began in 2006 from both sides of the tunnel and, immediately, due to squeezing ground conditions, difficulties of tunnel advancement were met.

With the progress of the excavation and the increase of the tunnel overburden, the situation became unbearable. On the Palermo side, the monitoring data showed diametric convergences exceeding 20 cm; ribs deformed, and sometimes broken, were observed; positive gas readings were recorded.

Tunnelling works stopped and RFI terminated the contract in May 2009. A central tunnel stretch of 800 m still remained to be excavated.

In this situation, aiming at completing the tunnel, RFI entrusted Italferr with the new design and the supervision of works.

This paper faces with the new design of the tunnel stretch not yet excavated.

Figure 1. Detail of the deformed ribs (front side PA).

2 THE CONTEXT

2.1 *Geological and geotechnical framework*

Regarding the geological conditions, the tunnel excavation involves the formation of Lercara (FL) on the Palermo side, the formation of San Cipirello (PMA) in the central part and the formation of Terravecchia (FTA/FTAS) on the Agrigento side.

The tunnel stretch still to be excavated is almost entirely inside the San Cipirello formation, with the exception of about 50 m, located into the Terravecchia formation, on the Agrigento side.

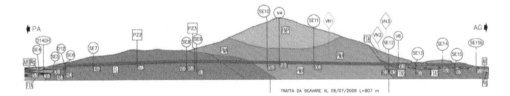

Figure 2. Lercara Tunnel geological profile.

A brief description of the geological formations involved is reported below. In addition, based on the results of laboratory tests and the available data of previous ground investigations, the geotechnical parameters used in the design calculations for different cross-section for each formation are illustrated.

2.1.1 *San Cipirello formation (PMA)*
It is a sequence made up of clays, sandy clays, and grey-green marly fossiliferous clays, with passages in greyish-green clayey sands, well thickened, fossiliferous, and interbedded with arenaceous levels variously cemented with thickness from a decimeter to a meter. It may be considered as a soft rock, but, in the fault zones, it can present an intense level of fracturing and a general degradation of mechanical characteristics with a reduction of both the cohesion and the friction angle values. Moreover, this formation can have a marked heterogeneity and discontinuity, with often poor technical characteristics, as encountered during excavation on the Palermo side.

The formation is characterized by very low permeability, as also confirmed by the field evidences during tunnelling.

The geotechnical parameter used are: the cohesion $c'=50–100$ [kPa], the friction angle $\varphi'=30$ [°] and the stiffness $E=300–600$ [MPa].

2.1.2 Terravecchia formation (FTA and FTAS)

This formation consists of two lithofacies: one predominantly clayey (FTA) and the other arenaceous-sandy (FTAS). The predominantly clayey lithofacies (FTA) is made up of clay and sandy clay of grey-green color, sometimes with a marl-like consistency, with interbedded layers of grey sand, variously thickened. The arenaceous-sandy lithofacies (FTAS) consists of sand and sandstones, of yellowish to grey color, with slow conglomeratic intercalations, well stratified, with thin clayey-sandy levels. The transition from a lithotype to the other is not always clearly identifiable.

The predominantly clayey lithotype can show a modest and seasonal water circulation in the zone of superficial alteration, while the arenaceous-sandy one can be site of discontinuous water table, with aquifers suspended in relation to the presence of thin clayey levels.

The geotechnical parameter used for FTA are: cohesion c'=50 [kPa], friction angle φ'=22 [°] and stiffness E=50 [MPa]; whereas the ones used for FTAS are: cohesion c'=20 [kPa], friction angle φ'=30 [°] and stiffness E=100 [MPa].

2.2 Gas risk

The problem of toxic/explosive gas risk has been a very important issue to be solved in the new design phase.

The risk assessment has been conducted considering multiple aspects, as described below. Apart from methane (CH_4), the analysis has been also extended to other toxic/explosive gases (CO, CO_2, H_2S, SO_2, NO).

2.2.1 Monitoring data from the first contract works

In the first contract, during the excavation of the tunnel (almost 2 km) a systematic control of toxic/explosive gases didn't show any presence, except for the one positive reading (of SO_2), measured during the reinforcement of ground ahead of the tunnel face in the last tunnel section on the Palermo side (pk 79 + 520), just before the stoppage of the works.

2.2.2 Historical and bibliographic data

Bibliographic information showed the possible presence of gas within the mainly sandy or arenaceous sequences of the San Cipirello formation and in the Terravecchia formation.

In literature, some authors define them as rocks with 'naphtogenic' characteristics; others de-scribe them as potential sites of hydrocarbon traps, hydraulically connected by deep faults, with reservoirs present at greater depths. However, everyone agrees on the potential presence of methane inside them.

Regarding the recorded emission of methane in the provinces of Palermo, Agrigento and Caltanissetta, the closest documented event is about 4 km away from the excavation area and consists in the presence of a small volcanoes of mud.

2.2.3 New survey and monitoring investigations

For the acquisition of further diagnostic elements concerning the gas risk, a new survey was carried out, consisting in the execution of three boreholes from the ground level with gas measurements. Positive gas readings (not only methane) were recorded by means of a manual explosimeter next to the borehole, during drilling operations and sampling of drilling fluids subjected to laboratory tests to search for dissolved gases.

Furthermore, a gas monitoring survey inside the tunnel already excavated was carried out. In fact, the stoppage of tunnelling works and the lack of tunnel ventilation (natural and forced) over a long time may allow possible small gas leaks to accumulate in particular underground zones, easily detectable. This could be for example inferred to the absence of the lining next to the tunnel face or to the drainage circuit inside the final lining. The underground records did not show dangerous concentrations of gases.

2.2.4 Risk classification

Based on the geological analysis and on all the feedback elements taken from the surveys and the monitoring carried out, the design evaluated the risk of intercepting gas traps with potential incoming flows during the excavation. The gas incoming flows have been considered as possible, but exceptional, in that not continuous and infrequent. Therefore, the tunnel was classified in '1C risk class', referring to the Italian Interregional Regulation on conventional tunnelling in gassy ground conditions. For each class of risk identified, the regulation defines the necessary safety measures to be provided for its proper management.

3 THE DESIGN APPROACH

3.1 *Tunnel section types*

The part of the tunnel not yet excavated was 807 m long. At the Palermo side, above the tunnel face, the overburden was 150 m; on the Agrigento side, the overburden was 30 m.

The design established to tunnelling by means of the conventional, advancing full-face and installing the primary lining meter by meter.

According to the ground geotechnical characteristics and to the height of tunnel overburden, 5 different section types have been defined, different in excavation advance lenght and in ground improvement measures, so called B1v, B2, B3v, C2p and C2bis.

In the design phase, for each section type they have been defined:

– appropriate ground improvement measures, ahead and around the face, if necessary;
– drainage system at the tunnel face;
– fiber reinforced shotcrete and steel ribs;
– PVC waterproofing system;
– secondary lining with reinforced concrete.

Depending on the section type, the cross section area varied between 71 and 88 m^2.

Particular attention was paid to the resumption of the tunnelling works, not only for the deformations occurred around the two stopped tunnel faces, but also for the predictable disturbance ahead of the face due to the long stoppage period. Unfortunately, the two tunnel faces were left in not completely safe conditions.

3.1.1 *Tunnel section type: B1v*

The B1v is a truncated-cone section that requires ground improvements ahead of the face (22 ± 20% of fiberglass pipes, L = 14 m, cemented with expansive mortar) and an umbrella of valved steel pipes around the cavity within an angle of 120° (21 valved steel pipes, L=12 m, spacing 0.4 m), with 8.5 m of advance length. The temporary lining consists of 20 cm of spritz-beton and double ribs IPN160 with a spacing of 1.0 m (± 20%); the invert is 80 cm thick and the crown has a thickness ranging from 50 cm to 115 cm.

It was planned to be used in the easiest geotechnical conditions, on the Agrigento side, inside the San Cipirello formation, with the lowest values of overburden.

3.1.2 *Tunnel section type: B2*

The B2 is a cylindrical section with ground improvements ahead of the face (30 ± 20% of fiberglass pipes, L=16 m cemented with expansive mortar), with 10 m advance length. The temporary lining consists of 20 cm of spritz-beton and double ribs IPN160 with a spacing of 1.0 m (± 20%); the invert is 80 cm and the crown is 70 cm thick.

It was planned to be used in a context very similar to the one defined for section B1v, but also for higher overburden, due to the greater number of fiberglass pipes ahead of the face.

3.1.3 *Tunnel section type: B3v*

The B3v is a truncated-cone section that requires ground improvements ahead of the face (30 ±20% of fiberglass pipes, L=15 m cemented with expansive mortar) and around the cavity (36 ±20% of fiberglass pipes, L=15 m cemented with expansive mortar) and an umbrella of valved steel pipes around the cavity within an angle of 120° (21 valved steel tubes, L = 12 m), with 8.5 m advance length. The temporary lining consists of 20 cm of spritz-beton and double ribs IPN160 with a spacing of 1.0 m (± 20%); the invert is 80 cm and the crown has a thickness varying from 60 cm to 120 cm.

This section was designed to be applied for the excavation of a long stretch within the San Cipirello formation, even in condition of high overburden (up to 125 m depth of the tunnel axis); it was also designed for the Agrigento side, to cross the stratigraphic passage between the Terravecchia and San Cipirello formations and to guarantee the passage through sections characterized by different ground improvements measures, from the resumption section C2bis to B1v or B2.

3.1.4 *Tunnel section type: C2p*

The C2p is a cylindrical section that requires ground improvements ahead of the face (30 ± 20% of fiberglass pipes, L=18 m cemented with expansive mortar) and at the contour (43 ± 20% fiberglass pipes, L=15 m cemented with expansive mortar), and the use of the temporary lining consisting of 25 cm of spritz-beton and HEB 180 ribs with a spacing of 1.0 m (± 20%) also in invert, with 10 m advance length. The definitive invert is 90 cm and the crown is 80 cm thick.

It was expected to be used in the worst conditions, with the highest values of overburden.

3.1.5 *Tunnel section type: C2bis*

The C2bis is a cylindrical section with ground improvements ahead of the face (30±20% of valved fiberglass pipes, injected with cement, L=20 m) and at the contour (43 ±20% of valved fiberglass pipes GRP, injected with cement mixtures, L=20 m), the use of the temporary lining consisting of 25 cm of spritz-beton and HEB 180 ribs with 1.0 m spacing (± 20%) also in invert, with 10 m advance length. The definitive invert is 90 cm and the crown is 80 cm thick.

C2bis section foresees the injection of cement in controlled pressure and volume conditions. The section C2bis, in fact, was designated as resumption section for the first 50 m starting from each of the two tunnel faces, with the grouting injections aiming at improving the ground conditions, characterized by a high degree of disturbance and stress due to the prolonged downtime.

3.2 *The monitoring plan*

During the construction phase the monitoring plan has been designed with the aim of: controlling the deformations of the rock mass, verifying the compliance with the project forecasts and modifying the design solutions according to the variability described in the project.

The monitoring system consisted in:

– geological and geo-structural survey of the tunnel face;
– tunnel face extrusion measures;
– convergence measurement stations on primary lining;
– stress measuring stations in the primary and secondary lining using pressure cells and strain gauge bars.

3.3 *Main safety measures for gas risk*

In relation to the gas risk, the project included a monitoring system consisting of:

– an automatic monitoring system, with continuous recording of methane gas contents, so to control the atmosphere next to the excavation face;
– manual monitoring to support the automatic system, performed with portable instrumentation, to measure, not only methane, but also other gases, to be carried out both in the zones of possible accumulation, and inside prospecting boreholes realized during the ground improvement operations.

In the risk management, the phase of gas detection was of primary importance, as it had to be carried out before starting any other work in not explosion proof environment.

The prospecting boreholes were part of those used for ground improvements ahead of the face. If no gases were detected, the risk of massive and sudden emissions of gas was excluded and the tunnelling continued in standard conditions.

During the gas prospecting phase, an area long 500 meters beginning from the face was considered potentially risky since the presence of methane was possible: all the operating machines and electrical systems (such as the positioner, the lighting, the vehicle for the evacuation of personnel and the rescue container) had to be suitable to operate in a potentially explosive atmosphere.

Also the electrical systems related to security, such as the ventilation system, the monitoring system, the security lighting, the indoor/outdoor communication system, etc., had also to be explosion proof.

The alarm system provided acoustic and luminous signals and, in the case of methane concentration readings over the threshold value, it worked automatically. The trigger value was appropriately defined in such a way that the personnel could leave the tunnel before reaching the dangerous concentration. Moreover, in the case of gas recording by means of manual devices, it was possible to activate the alarm system manually.

The standard electrical systems, not suitable for operating in potentially explosive atmospheres, was set up so as to be disconnected in the case of reaching the pre-alarm threshold.

When the alarm threshold was reached, the electrical safety systems automatically shut down and the personnel evacuation started.

4 THE CONSTRUCITON PHASE: OPERATIONAL ASPECTS AND CRITICAL ISSUES

The first activity carried out concerned the safety measures and resumption of excavation from both the tunnel faces.

Since the stoppage occurred in 2009, tunnelling restarted again in April 2014 on the Agrigento side, and in July 2014 on the Palermo side. Tunnelling works ended in August 2016. The organization was set to allow the rotation of the operations of excavation and of ground improvement from one tunnel face to the other, thus giving rise to the so-called "pendulum".

4.1 Tunnelling advance progress and criticality

Figure 3 and Figure 4 correlate the advance of the tunnel faces with the time.
Figure 4 also shows the different section types applied.

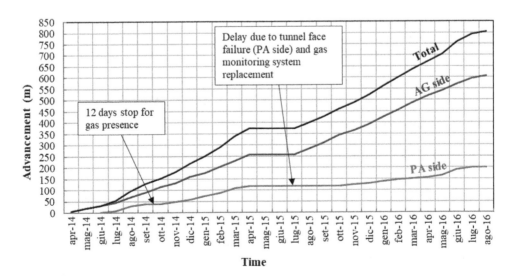

Figure 3. Excavation face advancement in correlation with the time.

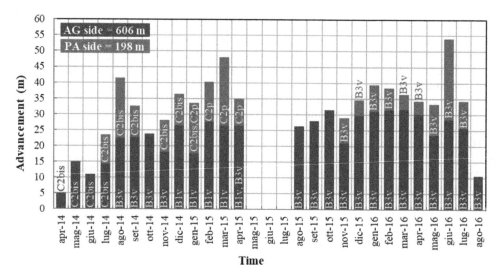

Figure 4. Monthly tunnels production.

The significant difference between the advance rates of the two excavation sides is clear. Considering the duration of the works, the average advancement rate was equal to about 0.27 m/day on the Palermo side and 0.73 m/day on the Agrigento side, for a total of about 1.0 m/day.

It must be highlighted that the gap between the two advance speeds was expected and it was already considered in the project: it resulted from the geotechnical characteristics of the formation and from the high overburden stretch to be crossed on the Palermo side. On this side, the "C" section types were mostly applied; they are characterised by larger consolidation works with respect to the sections named "B" planned on the Agrigento side, thus they require longer working time.

Another important aspect, from the organizational point of view of the work site, was represented by the different times for the transport of materials to and from the excavation face. While only a few hundred meters separated the tunnel face on the Agrigento side from the portal and the building yard, the distance between the tunnel face on the Palermo side and the building yard (for example, useful for concrete supplying) was of several kilometres, through winding roads subjected to continuous maintenance.

In addition, during the progress on the Palermo side, further delays occurred, caused by the poor geotechnical characteristics of the ground, by gas presence and by defects in the functioning of the automatic gas monitoring system.

Already during the excavation of the third advancement field, important deformation phenomena occurred with the exceeding of the foreseen thresholds of the monitoring plan, both at the front, with small releases of material in the crown, and around the face, with visible cracks in the primary lining. In this context it was necessary to adopt measures aimed at limiting the aforesaid phenomena, such as the increase of ground improvement ahead of the face, the execution of the inverted arch closer to the front, the approach of the concreting crown, which resulted in delay on the working schedule.

Subsequently, in October 2014, there was a standstill of about 12 days due to an important gas leak during the prospecting phase, well pointed out by the monitoring system. During the works, there were plenty of further positive gas surveys highlighted during the prospecting phase, with less significant impacts on the activities.

The excavation regularly continued until the 12th advance step, when (April 2015), an important breakdown took place, which led to the suspension of works. A few days after, the Contractor had to suspend the excavation works on both sides due to the shutdown of the automatic gas monitoring system, which was no longer suitable.

After defining the restarting project, by the Contractor, and restoring the safety conditions in the tunnel by installing a new gas monitoring system, the restart of the advancement activities took place in July 2015 for the Agrigento side and in September 2015 for the Palermo side.

With reference to the activities on the Palermo side, the critical zone was exceeded only in January 2016. To overcome the critical area, three advance rates have been dug with truncated conical insertions (B3v trans), using a reduced length of approximately 4,5 meters each. Afterwards, the progress continued for the following 10 fields with the truncated cone section B3v, until the last diaphragm fell.

Overall, on the Palermo side, around 200 meters of tunnel were excavated, instead of the 320 meters planned in the project. On the Agrigento side regular progress has been made. In total, about 607 meters of the tunnel have been excavated, instead of the 487 meters planned.

Not accounting for the delays and unexpected stops described above, the average advancement rate can be redefined in about 0,34 m/day on the Palermo side and 0.96 m/day on the Agrigento side, for a total of about 1.1 m/day.

4.2 Distribution of the section types: comparison between the design and as-built data

The following table shows the comparison between the distribution of the section types defined in the design phase and the as built data. On the Palermo side, section types C2bis and C2p were applied for 12 advancement steps, B3v trans section type (to overcome the area of the breakdown) for 3 short transitional advancement steps, B3v for 8 advancement steps, for a total of about 200 meters of tunnel. On the Agrigento side, after works resumption with section C2bis, it was possible to apply the section type B3v one advancement step before the one planned and in most of the tunnel, except for an intermediate part where B1 was applied. In total, 4 advancement steps were performed in section C2bis, 10 advancement steps in section B1 and 47 advancement steps in section B3v.

Table 1. Standard sections distribution: design vs as built.

Standard section	Design [n° adv. steps]	[m]	As Built [n° adv. steps]	[m]
C2bis	12	120	11	113
C2p	27	270	5	50
B3v	31	264	55	540
B3v trans			3	14
B1v	11	93	10	90
B2	6	60	0	0

4.3 Monitoring

During the excavation, the monitoring of the rock mass deformations was carried out by: systematic geological surveys of the tunnel face at the end of the advancement length and whenever there were significant geological variations; front extrusion measures for each advancement length and, when necessary, also for half of it; convergence measurements according to the design frequencies. For each monitoring parameter, appropriate attention and alarm thresholds were established.

This approach was necessary to verify the project forecasts and, in the case, to modify the design solutions within the variability foreseen by the project.

Structural monitoring sections were also foreseen for the evaluation of stress conditions in the primary and secondary linings.

4.3.1 Convergences

The most significant data presented below are those relating to convergence stations, which have been installed for each individual advancement length and in some cases even more than one station per advancement length.

On the light of the monitoring data obtained, it can be said that the measurements and the project forecasts substantially agreed.

The following figures show the history of the maximum diametric convergences in relation with their distance from the tunnel face, measured horizontally, approximately at the level of the tunnel axis. The attention threshold was 5 cm; the alarm threshold varied according to the section type applied, and was equal to 8 cm for B1v, 9 cm for B3v and 10 cm for C2p and C2bis.

Figure 5 shows the convergences of the 62 monitoring stations installed on the Agrigento side. To confirm the good regularity of the Agrigento side advancement, it is useful to observe how the stabilization of the deformation has almost always occurred below the threshold of attention, and only rarely between the thresholds of attention and alarm.

Figure 6 shows the convergences of the 31 monitoring stations installed on the Palermo side. On this side, as mentioned, the tunnel was excavated under worse conditions, with heavier section types (C2bis and C2p); the excavation face was very unstable, with reduced

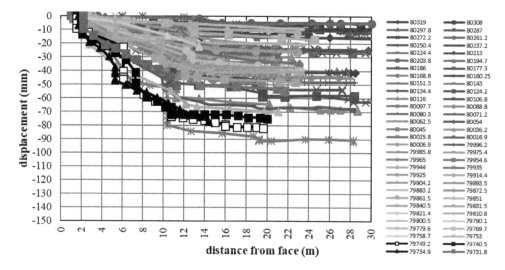

Figure 5.   Agrigento side - diametric convergences.

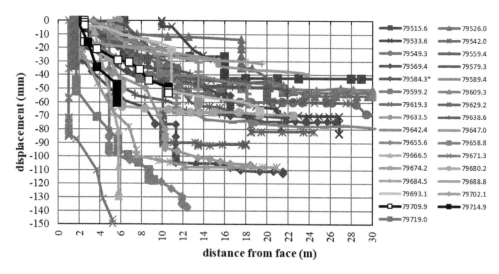

Figure 6.   Palermo side - diametric convergences.

self-support times. In this case the deformations rarely stabilized remaining below the attention threshold; they more often stabilized with values between the warning and alarm thresholds, and in many cases the latter was exceeded. The convergence datum of over 15 cm on a single station is due to the breakdown described in the previous paragraph and is not to be considered representative.

5 CONCLUSIONS

The completion of the Lercara tunnel presented numerous difficulties during the construction phase, in part pointed out by the previous contract and, therefore, expected according to the new design forecasts. These difficulties consist both in the presence of gas and in the geotechnical context, characterized by a squeezing behaviour of the rock mass, with significant deformation phenomena especially on the Palermo side with the highest overburden.

The meticulous and constant monitoring plan allowed the continuous control of the ground deformations during tunnelling and the back-analysis process, so to adjust the design solutions all along the construction phase.

Generally, the monitoring data agreed with design forecasts, both in terms of deformation values and in terms of effectiveness of the design solutions.

In addition, the safety measures, planned and implemented in relation to the gas risk, allowed the correct management of the problem, assuring completely safety working conditions.

The difficulties related to the tunnel context conditioned the performance in terms of production rate, but, thanks to the new design measures, it was possible to resume and complete the tunnelling works.

REFERENCES

Edward, J.S., Wittaker, B.N. & Durucan, S. 1988. Methane hazard in Tunnelling operations. In *Proc. of Tunnelling '88:* 97–110. London: IMM.
Cawley, J.C. & Duda, F.T. 1987. Underground electrical equipment safety, results of recent U.S. Bureau of Mines research. In *Proc. of the 22nd International Conference of Safety in Mines Research Institutes.* Beijing: China Coal Industry Pub. House.
Berry, P., Dantini, E.M., Martelli, F. & Sciotti, M. 2000. Emissioni di metano durante lo scavo di gallerie. *Quarry and Construction*, 38: 37–64.
Dantini, E.M., Colaiori, M., Lisardi, A. & Vizzino, D. 2000. Rischio di innesco del grisou a causa di "scintille" od urti in galleria. *Quarry and Construction*, 38: 7–8.

Tunnels and Underground Cities: Engineering and Innovation meet Archaeology,
Architecture and Art, Volume 9: Safety in underground
construction – Peila, Viggiani & Celestino (Eds)
© 2020 Taylor & Francis Group, London, ISBN 978-0-367-46874-3

'Terzo Valico dei Giovi': Excavation of tunnels in asbestos bearing rock

M. Comedini, A. Ferrari, F. Nigro & G. Petito
Italferr SpA, Rome, Italy

ABSTRACT: The new high-speed/high capacity railroad "Terzo Valico dei Giovi" runs mainly through tunnels. Some sections of tunnels are built through special rock formations known as ophiolites that potentially contain asbestos. For this reason it was necessary to prepare a plan of exploratory and monitoring investigations to be carried out during construction to detect the presence of asbestos and to adapt the excavation and measurement techniques to site. The plan defines the sampling and analytical methods (S.E.M- Scanning Electron Microscope and P.C.M. – Phase Contrast Microscopy) used according to the excavation techniques (conventional, T.B.M., etc.). Given the need to make available the results of the analyses within 48 hours, the analysis laboratories shall use units located in the territory. The plan also provides for the preparation of a monitoring network for airborne asbestos fibres at the construction sites and surrounding areas and for the correct characterization of the rock materials.

1 FOREWORD

In the1990's, the European Union set itself the objective to create strategic communication corridors that would interconnect the more densely populated and more industrially developed member states, now known as TEN-T (Trans-European Transport Network) that includes the European Union's major communication avenues. One of these avenues is the Rhine-Alpine Corridor. To contribute to the creation of this corridor, Italy has launched the construction of a new high speed/high capacity railway line to upgrade the connection of Genoa and of the entire port system of Liguria with the railroads going in the direction of the Northern Turin – Sempione and of Milan-Gotthard lines. The railway line called 'Terzo Valico dei Giovi' (herein after: 'Terzo Valico') falls within this context. Starting from Genoa and ending in the province of Alessandria, it is 53 km long, 36 km of which running underground.

The structure will allow to transfer significant volumes of freight traffic from road to rail, generating environmental, safety-related and social benefits.

The subject in charge of making the railway infrastructure investments is Gruppo Ferrovie dello Stato Italiane that operates through Rete Ferroviaria Italiana (RFI), in its capacity as proponent, and through the engineering company Italferr in charge of Work Supervision.

The General Contractor awarded the design and construction of the Terzo Valico is the Consorzio COCIV business combine consisting of the following major Italian contractors: Salini-Impregilo (64%), Società Italiana Condotte d'Acqua (31%), CIV (5%).

The Terzo Valico is a strategic construction project of national interest. These works are under the general occupational and environmental impact surveillance of the Italian Ministry for the Environment and for the Protection of Land and Sea (MATTM) via two bodies: the Special Environmental Impact Assessment Committee, and the Environmental Observatory.

The Environmental Observatory, in particular, with the technical and scientific support of the regional 'Agenzie Regionali Protezioni Ambiente – ARPA' (Regional Environmental

Protection Agencies), examines the conduction of the work from an environmental protection viewpoint and supervises the Environmental Monitoring activities put in place by Consorzio COCIV.

During the project's Detailed Design approval phase, the CIPE (Inter-Ministerial Committee for Economic Planning) has especially focused on the Asbestos issue, prescribing the definition of action execution procedures in the event of the objective proof of the detection of asbestos.

In response to specific requirements put forth by the regions, the Environmental Observatory has set up a Working Group with the support of the Regions, of ARPA Piemonte and of ARPA Liguria, to draft the "Protocollo di gestione del rischio amianto per il Terzo Valico" (Asbestos Management Plan for Terzo Valico) defining:

– monitoring and sampling methods and rates according to the different excavation methods;
– alert levels in the event of risk;
– monitoring points for airborne asbestos fibres, identified in the areas exposed and based on site composition.

Deployment of the Asbestos Management Plan is envisaged along the entire route of excavation through potential asbestos-bearing rock layers, equivalent to about 36 km of railway line.

2 KEY DATA REGARDING THE ROCK LAYERS BORED THROUGH DURING THE EXCAVATION OF THE MAIN LINE TUNNEL

The analysis of the existing data as well as of the data coming from the surface geological investigation and analytical and stratigraphic data acquired via geognostic surveying has resulted, compatibly with the sector's geological/structural complexity, in the drafting of the geological model of reference. Based on this geological model, it was possible to distinguish the lithological formation sectors in terms of "Probability of Occurrence of Asbestos-bearing Minerals" (in Italian: POMA) so as to identify the tunnel sections potentially posing the risk of asbestos content. This distinction also has the purpose of helping in organising all of the subsequent phases dedicated to the definition of the excavation material characterization and of the airborne asbestos fibre monitoring plan, as well as of allowing to extend to a more encompassing ambit all of the general risk-mitigation measures.

2.1 Geological model of the main line tunnels

The route of the Terzo Valico line, in the section that crosses the mountains and hills located between Novi-Ligure and Genoa, involves an area consisting of two large geological domains:

– the Sestri-Voltaggio Zone auct. (SVZ);
– the Tertiary Piedmont Basin (TPB).

The SVZ defines, together with the 'Voltri Group', the so-called "Ligurian collisional knot" that is a sector basically interpreted as being the physical boundary between the Alpine and the Apennine chains, outcropping on the Ligurian side and along the border between the Liguria and Piedmont regions.

The TPB is a terrigenous sedimentary series (Costa Cravara breccias; Molare and Rigoroso formations; Costa Montada formation; Costa Areasa formation, Cessole marls; Serravalle sandstones, S. Agata Fossili marls, Gesosso-Solfifera group), that forms the hilly reliefs of the south-eastern sector of Piedmont; its relatively scarcely deformed margin rests discordantly on the units of the SVZ on which it has deposited.

Based on the foregoing geological considerations and on the many surveys conducted during the various design and construction phases, a representative geological model was made of the main line tunnels (Figure 1), extending it to the entire section involved by the

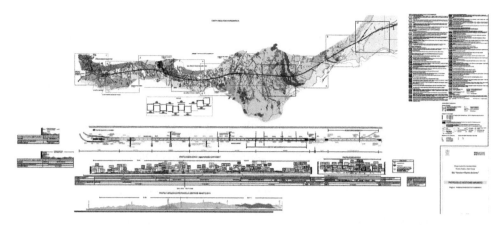

Figure 1. Geological model.

Units of the SVZ and by the Unit of the TPB, measured from mileage point 0 + 000 to mileage point 36 + 585. The profile also allows to assess the case of marl-sandstone lithotypes that, although not ophiolites, show the presence of asbestos-bearing minerals.

The profile indicates the sections with the various lithotypes found during the surveys, and the 'Probability of Occurrence of Asbestos-bearing Minerals' (POMA) has been colour-coded.

With regard to the purposes of the geological model, and based on the specific geographical location of the route and of the local geological conditions, the POMA has been broken down into three asbestos-finding probability classes: Negligible, Medium-Low and High.

The geological model indicating the lithotypes and the relating POMA classification allows to make a preliminary probabilistic assessment linked to the assumption of finding or of not finding asbestos during tunnel excavation.

3 EXCAVATION MATERIAL CHARACTERISATION PLANS

The excavation material derived from the Terzo Valico tunnelling sites envisage the management of an overall volume of excavated earth and rock by-product amounting to 11,713,648 m^3.

During the performance of the works, environmental characterization surveys are conducted on the materials in order to assess environmental quality, making sure the waste is excluded. In the case of asbestos, the threshold value that authorises soil management in terms of by-product is equivalent to 1000 mg/kg (Legislative Decree No. 152/2006, Title V, Part IV, Annex 5, Table 1).

The Asbestos Management Plan establishes the procedures and frequency of the excavated material characterization operations.

The purpose of the analytical tests conducted on the excavated material was to verify the presence of naturally occurring asbestos in the rock material, so as to define the plans to be applied for monitoring airborne asbestos in the living environment and the mitigation measures to be implemented when handling such material.

As regards excavation method, the following three types are envisaged:

– Conventional tunnelling with jackhammer;
– Conventional tunnelling with explosives;
– Mechanised tunnelling with TBM/EPB.

As regards lithotypes, the following three are expected to be found:

– Excavation through Ophiolites and Molare formation/Costa Cravara breccias;

- Excavation through sedimentary sandstone-marl lithotypes with TPB asbestos;
- Excavation through rock types other than those listed above.

The on-site geologist checks the lithotypes actually excavated. Sampling methods are defined in function of the type of excavation method used. In particular, in the case of conventional tunnelling (using jackhammers and explosives), material sample collection is performed directly from the excavation front while, in the case of mechanised tunnelling using TBM/EPB, material samples are collected from the conveyor belt.

Sampling rate is defined in function of the lithotype being excavated.

As regards the lithotypes consisting of Ophiolites and Molare formation/Costa Cravara breccias, in particular, which feature the presence of asbestos-bearing minerals that can vary greatly in terms of concentration and of localisation, the following sampling rate schedule is applied:

- every three excavation progress sections, in the case of conventional tunnelling using jackhammers, each excavation progress section varying in length from 0.80 to 1.40 m;
- every excavation progress section (approx. 3 m), in the case of conventional tunnelling using explosives;
- every (excavation) work day, in the case of mechanised tunnelling.

The sedimentary TPB sandstone-marl lithotypes bored through to date have shown constant presence of naturally occurring asbestos with concentrations below the thresholds foreseen for management of material as by-product. When excavating through this lithotype, therefore, sampling rates will be approx. every 50 m with any excavation method.

No naturally occurring asbestos was found in all other lithotypes. When excavating through these lithotypes, therefore, sampling is scheduled at approx. every 100 m of excavation, instead of every 500 m of excavation as indicated by Ministerial Decree (DM) No. 161/2012.

3.1 *Conventional tunneling*

Conventional tunnelling is performed using jackhammers or explosives. With conventional tunnelling, the excavation front is constantly accessible, and at any time the geologist can visually check which lithotype is being excavated.

A survey of the excavation front is performed after the completion of each excavation progress section, namely at each completed work cycle that, when jackhammers are used, is equivalent to a cycle of drilling, rubble removal, centring and shotcrete or, when explosives are used, to a cycle of blasting, rubble removal, centring and shotcrete.

After each excavation progress section, the geologist conducts a survey of the front and drafts the corresponding report.

In the case of conventional tunnelling using jackhammers, each excavation progress section consists of a 0.8÷1.4 m advancement, while in the case of conventional tunnelling using explosives, each 'excavation progress section' consists of an approx. 3 m advancement.

After each single excavation progress section, the on-site geologist checks and certifies the homogeneity of the lithotype being exposed at the excavation front against the previous characterization, via the drafting of a simplified datasheet to which photographic documentation is attached. In addition to this, whenever there is a significant change in excavated lithotype, the geologist carries out a complete characterisation of the rock mass and prepares a control table of the excavation front with photographs and a pictorial-descriptive analysis of the geological conditions of the mass and the description, if necessary, of significant particular structural elements that point to lithotype change. In this manner, the appearance of a portion of ophiolite, albeit small in size, is considered a lithological change in view of the application of the Asbestos Management Plan and is certified in the drafting of the excavation front report.

The survey data coming from the excavation front are used:

- to define the excavated material sampling rates;
- to define the Risk Level and consequently the airborne contaminant monitoring method, as described in more detail in chapter 3 below.

In conventional tunnelling (with jackhammers or with explosives), material sampling is carried out directly at the excavation front with the hammer used for excavating (assuming that, even in the event of blasting, a hammer is used for barring operations).

The composite sample (primary sample) is obtained by collecting the material from 8 different points (increments) of the excavation front, according to the pattern shown in Figure 2.

3.2 Mechanised tunnelling

In mechanised tunnelling, the excavation front is not accessible. This means that the lithotype being excavated cannot be surveyed via direct inspection of the excavation front. In this case, the survey of the lithology is performed on the rubble transported on the conveyor belt coming out from the back of the cutter head.

To do this, the geologist collects a sample of the excavation detritus directly from the belt (via sample collector). The sample is repeatedly washed to eliminate the fine fraction and then subjected to on the field detailed observation by the geologist. The presence in the resulting sample of 'ophiolitic rock' indicates that the excavation is passing through ophiolite veins. The site geologist carries out these investigations at the beginning of each working day and logs them on a specific register.

The sample is collected using a 'cross-cut sampler', located in the hopper at the end of the conveyor belt. The hopper is fitted with a swinging wing designed to open towards the inside and cross the flow of falling detritus, thereby deviating it towards the sample collection bin.

3.3 Lab analysis of rock samples

Considering the need to make the results available in a maximum period of 48 hours, the laboratories used for sample analysis use external local units. The analytical methods used are Scanning Electron Microscope (SEM) and Phase Contrast Microscopy (PCM) capable of achieving an adequate quantification limit for the quantitative tests on asbestos content in the loose-soil excavated rock and earths. The official plan of the entire lab test and the documentation regarding the method validation procedure as per ISO 17025, with indication of the quantification/detection limit, must comply with Min. Decree No. 161/12.

The method used envisages the preliminary visual inspection of all of the material of which the analytical sample is made under the stereomicroscope, and subsequent analyses using PCM and/or SEM, therefore guaranteeing a limit of detection of about 100 mg/kg.

The analyses of the asbestos parameter on excavated material coming from the Terzo Valico tunnelling operations are carried out exclusively by university labs listed among the laboratories qualified to test for asbestos of the Italian Health Ministry. In detail, considering the geographical extension of the project, to date the following laboratories are being used:

Figure 2. Sampling grid.

University of Genoa – Department of Earth, Environment and Life Sciences (DISTAV), Genoa HQ; Politecnico di Torino – Department of Environment, Land and Infrastructure Engineering (DIATI).

4 AIR MONITORING NETWORKS

The Asbestos Management Plan also defines the purpose, location and rate of sampling of the air monitoring network with regard to airborne asbestos.

In order to monitor any dispersions of asbestos fibres into the air of the living environment involved by the tunnelling project, specific airborne asbestos monitoring points have been established at each excavation site. The points are laid out in belts starting from the source of emission, moving out progressively from the excavation/soil deposit sites, as per the drawing shown in Figure 3:

– points inside the site, called 'source points' (working environment);
– points close but external to the site (near the fencing and entry points), called 'first belt points' (living environment);
– points in living environments called 'second belt points' located in function of the presence of sensitive receptors (schools, buildings, assembly points, etc.) that could be impacted by any airborne asbestos fibres coming from the activities related to the construction of the Terzo Valico dei Giovi (living environment).

4.1 Criteria for activation of airborne asbestos fibre monitoring at underground excavation sites

During excavation operations, airborne asbestos fibre monitoring in living environments is carried out at the rates indicated for the so-called 'Alert Status'.

For each excavation site, the Alert Status depends on the following factors:

– the 'Probability of Occurrence of Asbestos-bearing Minerals' (POMA) as defined by the geological model of the line (POMA as per the Model);
– the actual Probability of Occurrence of Asbestos-bearing Minerals in function of the litho-type analytically found in the rock mass;
– the confirmed presence of asbestos in the rock mass;
– the concentration of airborne asbestos actually measured in the living environment.
– In order to define the Alert Status, therefore, it is necessary to define the:

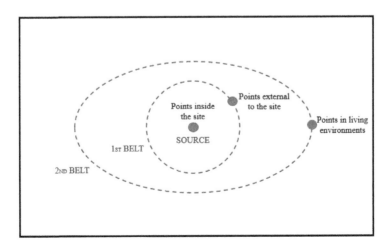

Figure 3.   Monitoring network.

Table 1. Alert status definition matrix.

AIRBORNE ASBESTOS CONCENTRATIONS		$0 < C < 1$ ff/l in source and $0 < C < 0,5$ ff/l in belt	$1 \leq C$(ff/l) < 2 ff/l in source or $0,5 \leq C$(ff/l) < 1 ff/l in belt	C(ff/l) ≥ 2 ff/l in source or C(ff/l) ≥ 1 ff/l in belt
ARL-0 NO RISK	PRE-ALERT	ALERT	ATTENTION	ACTION B
ARL-1 MEDIUM-LOW RISK	ALERT	ALERT	ATTENTION	ACTION B
ARL-2 HIGH RISK	ATTENTION	ATTENTION	ATTENTION	ACTION B
ARL-3a ASBESTOS PRESENT in sedimentary sandstone-marl lithotypes	ACTION A	ACTION A	ACTION A	ACTION B
ARL-3b ASBESTOS PRESENT in GS and Molare F./Costa Cravara breccias	ACTION B	ACTION B	ACTION B	ACTION B

- "Predictive Risk Level (PRL)", a function of the POMA from the Geological Model;
- "Verified Risk Level (VRL)", a function of the Real POMA (verification conducted by the geologist at the excavation front);
- "Actual Risk Level (ARL)", a function of the VRL and of the analytically measured concentration of asbestos in the rock mass.

The combination between Actual Risk Level and the analytically measured concentration of airborne asbestos in the living environment defines the 'Alert Status" of the excavation site and the corresponding monitoring frequency to be applied to determine airborne asbestos concentrations. Here below is the alert status definition matrix.

The Alert Statuses shown above are correlated with progressively more frequent airborne monitoring rates, as described in the following table. In those cases where '0' concentration is indicated, it means that in the filter not even one asbestos fibre was detected (and therefore that the airborne concentration is lower than the limit of detection):

Table 2. Definition of airborne asbestos monitoring frequency at production sites.

ALERT STATUS	ACTIVATION CRITERIA	MONITORING POINTS	MONITORING FREQUENCY
Pre-alert	ARL-0 NO RISK	SOURCE	Sampling + analysis once a week on the 8-hour shift
		1ST BELT	Sampling once a week on the 8-hour shift on the day on which samples are collected at the source point. Analysis when asbestos is detected at the source or when asbestos is detected in the previous analysis of the 1st belt
		2ND BELT	No sampling
Alert	ARL-1 MEDIUM-LOW RISK or $0 < C$(ff/l) < 1 ff/l in source or $0 < C$(ff/l) < 0.5 ff/l in belt	SOURCE	Sampling + analysis on alternate days on the 8-hour shift (when explosives are being used, sampling rate becomes daily)
		1ST BELT	Sampling on alternate days on the 8-hour shift on the days on which

(Continued)

Table 2. *(Continued)*

			samples are collected at the source point (when explosives are being used, sampling rate becomes daily). Analysis when asbestos is detected at the source or when asbestos is detected in the previous analysis of the 1st belt
		2ND BELT	No sampling
Attention	ARL-2 HIGH RISK or $1 \leq C(ff/l)$ < 2 ff/l in source or $0.5 \leq C(ff/l) < 1$ ff/l in belt	SOURCE	Sampling + analysis every day on the 8-hour shift
		1ST BELT	Sampling + analysis every day on the 8-hour shift
		2ND BELT	Sampling every day on the 8 hour shift. Analysis when asbestos is detected at the source or 1st belt or when asbestos is detected in the previous analysis of the 2nd belt
Action A	ARL-3a ASBESTOS PRESENT in lithotypes other than Ophiolitess and Molare and source C < 2 ff/l and in belt C > 1 ff/ll	SOURCE	Sampling + analysis every day on all three 8-hour shifts
		1ST BELT	Sampling + analysis every day on the 8-hour shift
		2ND BELT	Sampling every day on the 8-hour shift. Analysis when asbestos is detected at the source or 1st belt or when asbestos is detected in the previous analysis of the 2nd belt
Action B	ARL-3a ASBESTOS PRESENT in Ophiolites and Molare or source C ≥ 2 ff/l or in belt C ≥ 1 ff/ll	SOURCE	Sampling + analysis every day on all three 8-hour shifts
		1ST BELT	Sampling + analysis every day on all three 8-hour shifts
		2ND BELT	Sampling every day on all three 8-hour shifts. Analysis when asbestos is detected at the source or 1st belt or when asbestos is detected in the previous analysis of the 2nd belt

4.2 *Lab analysis of airborne fibres*

The analyses are carried out in accordance with the indications given in Annex 2 of Min. Decree dated 06/09/94. Asbestos fibres having a geometry matching that described in Min. Decree of 06/09/94 are characterised via micro-analysis and by morphological matching in order to establish whether or not it is asbestos and its type if it is.

In the test report, the analysis indicates the concentration of total fibres, asbestos fibres and inorganic fibres, along with type (chrysotile, crocidolite, grunerite, tremolite, actinolite, anthophyllite) of each asbestos fibre.

The concentration values of airborne fibres (ff/l) are calculated based on the following parameters:

– number of fibres counted;
– type of fibres found;
– effective diameter of sampling filter;
– number of fields inspected;
– field area 2000×;
– volume of aspirated air normalised at 20°C and 1013 mbar.

Once the sampling parameters are established, the smallest filter surface to be explored must be such as to guarantee the measurement of values 10 times lower that the 1 ff/l threshold value applied. In analysing airborne asbestos fibre concentrations deposited on filters, a quantification limit of ≤ 0.1 ff/l is guaranteed.

5 DATA MANAGEMENT

The results of the rock mass and air analyses have been organised and handled via databases prepared by the General Contractor and placed at the disposal of the Environmental Observatory, of the Italian Ministry for the Environment and for the Protection of Land and Sea and of the supervisory bodies. More specifically:

– The results from the airborne asbestos monitoring campaign are duly validated and stored by the General Contractor in a GIS database (called SIGMAP, prepared by Italferr) that allows to record all of the monitoring results and to promptly notify the bodies concerned.
– The results from the analyses on the rock masses are stored on the WEB platform Terresc@ (dedicated to the management of Excavation Soils and Rocks) and placed at the disposal of the supervisory bodies.

Special attention has been paid to the management and dissemination of the collected data to the stakeholders and citizens.

Clearly, the environmental communication policies are the tools required to obtain a full and transparent picture of the activities performed at the sites by constantly informing the supervisory bodies as well as the citizens about the environmental quality of the area involved in construction work. To this end, it seems that the best guarantee of environmental and health safety is in the transparency of analysis data. For this reason, tools for communicating this information to the bodies and citizens have been deployed, privileging speed, clarity and ease of consultation of both general and detailed information. Two systems have been implemented: i) the system offered by the web portal of the Environmental Observatory (www.osservatoriambientali.it), set up for on-line consultation and already active, and ii) the more in-depth system, that is partly under construction and partly already active via Info-Points, of the so-called "Comunicazione sul territorio" (Local Communication) aimed at providing immediate and concise visual communication about the overall 'detected condition'.

6 CONCLUSIONS

The foregoing is a description of the set of actions and precautions put in place for the management of materials containing naturally occurring asbestos coming from the excavation of the tunnels planned for the 'Terzo Valico dei Giovi' project. The implementation of these operating methods allows not only to ensure compliance with the regulations of reference but also to reduce to a minimum the risk of asbestos dispersion in the ambient immediately next to the working environment as well as in living environments close to the work sites, to ensure environmental and personal protection and to ensure good working practices.

The experience gained at the Terzo Valico sites can be used to create valid models to be applied, after suitable adjustments, to other infrastructure projects impacted by similar issues.

REFERENCES

Environmental Observatory of Terzo Valico dei Giovi. 2018. *Protocollo di gestione del rischio amianto per il Terzo Valico (Asbestos Management Plan for Terzo Valico)*.
COCIV. 2006. *Final design Terzo Valico dei Giovi*.
Amato L., Carraro E., Ferrarotti M., Piccini C. 2005. *Evaluation of the asbestos risk in the Alta Val Lemme area Alessandria*.

Beccaris G., Scotti E., Di Ceglia F., Prandi S. 2010. *Asbestos control in ligurian ophiolites.* Congresso SGI, Pisa.

Cazzola C., Clerici C., Francese S., Petrucco M., Zanetti G., Gallara F. 2005. *Quantitative determination of asbestos in rocks and soil by optical microscopy: analytical methods and examples of application.*

D.M 06/09/94. *"Normative e metodologie tecniche di applicazione dell'art. 6, comma 3, e dell'art. 12, comma 2, della legge 27 marzo 1992, n. 257, relativa alla cessazione dell'impiego dell'amianto".*

D.M 14/05/1996. *"Normative e metodologie tecniche per gli interventi di bonifica, ivi compresi quelli per rendere innocuo l'amianto, previsti dall'art. 5, comma 1, lettera f), della legge 27 marzo 1992, n. 257, recante: Norme relative alla cessazione dell'impiego dell'amianto".*

D.Lgs n. 152. 2006 *"Norme in materia ambientale".*

DPR n. 120. 2017. *"Regolamento recante la disciplina semplificata della gestione delle terre e rocce da scavo, ai sensi dell'articolo 8 del decreto-legge 12 settembre 2014, n. 133, convertito, con modificazioni, dalla legge 11 novembre 2014, n. 164".*

D.M n. 161. 2012. *"Regolamento recante la disciplina dell'utilizzazione delle terre e rocce da scavo".*

Tunnels and Underground Cities: Engineering and Innovation meet Archaeology,
Architecture and Art, Volume 9: Safety in underground
construction – Peila, Viggiani & Celestino (Eds)
© 2020 Taylor & Francis Group, London, ISBN 978-0-367-46874-3

Safety 2.0 – Engineered solution for a safer underground excavation. The BZEROTONDO Tunnel Support System

A. Cullaciati & M. Bellavita
PAVIMENTAL

R. Perlo
Officine Maccaferri Italia

K.G. Pini
CP Technologies

C. Cormio
Serengeo

ABSTRACT: Peoples' and operators' safety in construction are tightly linked to the engineering choices, as much as the necessity of reducing time and costs of realization. Therefore, it is of paramount importance to address them properly from the earliest stage of the design, identifying the risks connected to the work and the variability of such during the different phases of the project.

The semi-automatic tunnel support system along with its erector machine proved to be a solution able to reduce the time of installation and to deliver an enhanced, quality-aware and safer working environment for the operators. The system represents a best practice for workers' safety against risks during steel rib installation.

In this paper, the authors describe recent developments of the system and comment on the results of the application in the Boscaccio tunnel with regard to the safety record, performance and quality related issues

1 RISKS IN UNDERGROUND CONSTRUCTION

Risk is often measured as the expected value of an undesirable outcome. According to occupational safety and health, the risk is the likelihood that a person may be harmed or suffers adverse health effects if exposed to a hazard. This combines the probabilities of various possible events and the assessment of the corresponding harm into a single value. However, the actual occurrences are often more complex than a single binary possibility case. In a real case scenario, with several possible accidents, the total risk is the sum of the risks for each different accident, provided that the outcomes are comparable.

Hazardous conditions in underground works are given mainly by – but not limited to – falls from height, falling objects, being hit by machinery, handling of heavy objects. Therefore, hazards arising from design choices shall be identified and the related risks shall be minimized or eliminated through the adoption of effective engineering solutions.

International standards dictate specifications for the steel arch design. However, few of them focus and elaborate on the scope of works for such supports, pointing out that loss or lack of support in underground excavation may lead to catastrophic consequences.

Particularly, the British Tunnel Society (2010), UK Health and Safety Executive (1996), and the Japanese Society of Civil Engineers (2006) are keen to address the topic in exhaustive detail. In fact:

(i) Temporary lining in conventional tunnelling is a composite system given basically by spray concrete, steel reinforcement and bolting, if any (Japan Society of Civil Engineers, 2006)

(ii) As far as steel reinforcement is concerned – e.g. steel arches – these shall be implemented for the initial ground support of the tunnel excavation, and implemented to limit or to reduce the deformation within acceptable design limits (British Tunnel Society, 2010, Clause 207.1.4)

(iii) In order to limit the deformations of the rock masses, and – consequently to reduce the magnitude and quantity of the supports – the installation of tunnel supports shall be completed at the earliest possible opportunity (Japan Society of Civil Engineers, 2006, Clause 149.2)

(iv) No person should be allowed to approach the heading until all exposed ground has been supported' (Health and Safety Executive, 1996, Clause 310)

The result of a comprehensive and detailed analysis based on the specifications given is that the steel arches are implemented for the primary lining to enhance the overall stability, e.g. with safety purpose. The inherent risks of the implementation of steel arches is closely related to the usual approach followed for the conventional steel arches. In fact, the presence of workers in unprotected area in close proximity to the excavation face is required during some specific stages, e.g.: arch assembly, installation of the arch foot, arch section connection with tie-back bolts and installation of bracing.

The BZEROTONDO tunnel support system along with its erector machine proved to be a best practice solution with improved structural performance, able to reduce the time and yet to deliver an enhanced, quality-aware and safer working environment for the operators. The implementation of the system provides the engineers, the contractors and the owners with an effective alternative option – shortlisted in 2015 for the ITA Award for the Safety Innovation of the Year 2015 – and complying with the relevant safety international regulations.

2 OFFSETTING THE RISKS

2.1 *BZEROTONDO new semi-automatic system*

The Boscaccio tunnel – recently completed – is part of the new Milan Naples A1 Highway (VAV) alignment and is the first pilot project where the BZEROTONDO new semi-automatic system was applied.

Since the first prototype of the system (C.L.Zenti, A.Cullaciati, WTC 2015), the updated version includes a new robotic erector machine, numerous built-in features such as the revised automatic bracing system, and revised telescopic footing system. Compared to the previous version, the inherent benefits of the revised system allow for the further reduction of the number of operators and their exposure to hazardous conditions, on the number of machineries needed, and on the time required to install the steel arches.

Installation principally is the most crucial to address, and the most challenging to meet from an automatization point of view. Particularly, the development of the updated system required:

(i) A fully automated 3-harms erector machine operating remotely, e.g. the EKIP 21-04

(ii) The design and development of innovative automatic bracers (tie-back free, arch-to-arch connection devices) system which can accommodate both cylindrical, and tapered tunnel excavation sections and which immediately engages upon deployment

(iii) A telescopic arch feet support system which may accommodate uneven ground profiles. The footing system is fitted with steel spring and rack which lock the system in position, and prevents the foot from collapsing once loaded

Figure 1.   a) outlook of the latest EKIP 21-04; b) Deployment of the erector inside the Boscaccio tunnel.

Figure 2.   a) Erection of the steel arch; b) Deployment of the steel arch at side wall; c) Outlook of the 3-arms design from the operator point of view.

The innovative EKIP 21-04 (Figure 1.a, and 1.b) is a remote-controlled, 3-harm erector machine designed to fit in and to deal with both larger and smaller tunnel diameters. It is equipped with a three-armed handler, designed and manufactured by CP Technologies in partnership with Officine Maccaferri Italia.

The arch, assembled, yet in folded configuration, is transported close to the excavation face area and lifted to the project level with the 3-arm erector central boom (Figure 2.a). Then, the arch is unfolded and deployed open by the erector side booms (Figure 2.b). The same side booms provide the extension of the telescopic footing.

The bracing elements (pre-installed) are locked via a spring-circlip device (Figure 3.a and 3. b). This procedure engages the bracing systems on adjacent arches, locking them together at the appropriate spacing without further time-consuming adjustment.

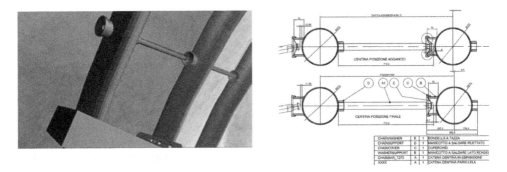

Figure 3.   a) Bracing system overview; b) Spring-circlip device overview.

Figure 4.   EKIP 21-04 on site at Boscaccio tunnel, and a model of the sequence of the BZEROTONDO tubular steel arches installation.

Finally, to increase the resistance of the solution, the arch is filled with concrete.

As a result of the revised system, the installation of the tubular steel arch does not require the presence of any worker in the immediate vicinity during the erection of the arch, the installation of the bracing and the extension of the footing (Figure 4). The absence of workers mitigates the hazards typically associated with the installation of steel arches including, falls from height, being hit by machinery due to restricted operator visibility and exposure to continuous construction noise.

The erector can also be supplied with the topographical reference system, CP Track. The CP Track System (CPTS) collects the information of the topographical positioning of the steel rib. Data retrieval is by USB drive from the machine once the steel rib positioning operation is complete

The use of this remote equipment and the tubular steel arch enables the semi-automatic installation in a safe, precise and time saving manner.

3 SAFETY REGULATIONS, RISK ASSESSMENT AND BZEROTONDO EFFICIENCY

3.1 Safety regulation for tunnelling during front face works

During tunnel face excavation and temporary support works, operators are exposed to the risk of heavy injuries due to the unpredicted release of rock/soil/shortcrete portions.

With the increase of the tunnel section, such hazards have become even more relevant, due to the higher face height (greater energy of the falling object that could hit the worker) and the bigger excavated surface (harder to inspect and from which rock or soil portion release could occur).

In order to prevent such risk, many Countries worldwide provide technical regulations and guidelines that suggest or impose the following prevention measures:

(i) The presence of a Competent Person, "who is capable of identifying existing and predictable hazards in the surroundings or working conditions which are unsanitary, hazardous, or dangerous to employees, and who has authorization to take prompt corrective measures to eliminate them" (OSHA, 1970, clause 1926.32(f))

(ii) An effective scaling activity, in order to remove all the unstable portion at the tunnel face before workers' access to the hazardous area

(iii) The installation of FOPS on lifting platforms

and in general, foster designers to adopt the most up-to-date technical, technological, organizational design solutions that allow to achieve the highest workers' safety levels.

In Italy, safety issues during tunnelling were specifically addressed by the end of '90s from the early stages of the tunnel constructions for High Speed train (TAV) and Highway Valico Variant (VAV) projects between Bologna and Florence. Eventually, between 1998 and 2015, 45 Inter-regional Technical Notes (NIR) were released, and recently consolidated as National Guidelines. The new approach introduced a dynamic safety engineering system consisting of innovative design solutions, safety measures and procedures which were defined according to the interaction between the rock mass and tunnelling, taking into account the tunnel size, the underground work organization and the excavation method and technique. The attention was focused on large section tunnels (larger than 70 sqm) excavated traditionally. Particularly, NIR n° 41 (2009), 43 (2011) and 45 (2009), updated as National Guideline n° 13 specifically addressed tunnel face works. In order to reduce the risk of injuries due to falling objects, in addition to the above listed prevention measures, the following recommendations are suggested:

(iv) To guarantee rapid evacuation conditions (plain floor, obstacle-free pathways) from the hazardous area

(v) To forbid the contemporary execution of more than one operation at tunnel face (excavation, mucking out, shortcreting, steel ribs, etc)

(vi) To reduce the number of exposed workers to the minimum required for each operation

(vii) To reduce the duration of each operation to the minimum required

Moreover, Italian Notes and Guidelines invite designers and manufacturers to develop mechanized and robotized solutions that avoid workers' presence within the dangerous area, thus eliminating the risk. BZEROTONDO tunnel support system is the first safety and time-effective engineering solution complying with Italian NIR, with a more effective, safer, simpler and faster-to-install solution than the traditional steel sets.

3.2 Evaluation of the risk

Risk is often measured as the expected value of an undesirable outcome. Regarding human life, the magnitude of the occurrence, or the potential loss in case of accident would have catastrophic consequences. In relation to occupational safety and health, the risk is the likelihood that a person may be harmed or suffers adverse health effects if exposed to a hazard.

This combines the probabilities of various possible events and the assessment of the corresponding harm into a single value. However, the actual occurrences are often more complex than a single binary possibility case. In a real scenario, with several possible accidents, the total risk is the sum of the risks for each different accident, provided that the outcomes are comparable. Hence:

$$R = \sum_{all\ accidents} (probability)\ x\ (occurence)$$

In conventional tunnelling therefore, the level of risk is categorised upon:

(i) The potential harm, or the adverse health effect that the hazard may cause

(ii) The number of operators exposed

(iii) The number of times operators are exposed; and

(iv) The length of exposure

Particularly, and due to the inherent hazards of the steel arch installation work, the most effective countermeasure to implement to limit the risk is to reduce the number of the operatives involved, or better to eliminate the use of operatives altogether. In fact, to reduce or to remove the use of workers will eventually reduce, or even better, remove the likelihood of the risk.

3.3 Safety record at Boscaccio tunnel

In the Boscaccio tunnel, the BZEROTONDO new semi-automatic system was introduced as a countermeasure to offset – e.g. to reduce and/or to mitigate – the risks during the steel arch installation. To assess efficiency and effectiveness of the system, data record of incidents during tunnel construction are compared to equivalent or similar tunnel construction for other sections of the new Milan Naples A1 Highway (VAV) alignment (Osservatorio Sicurezza Grandi Opere, Report no.15, 2012; VAV e Terza Corsia, Nodo di Firenze, Aggiornamento no.6, 2015).

Particularly (Table 1 below, refers), data considered is:

(i) Discriminated by type of occurrence (e.g. falling from height, falling object, material, shotcrete or rocks, hit by machinery, handling of heavy object)

(ii) Discriminated by magnitude of incident assessed over the length of recovery time (medium >30days; severe >90days) according to Italian health and safety regulation

(iii) Normalised over 1000m of excavation, and such to have consistency in the numbers for tunnels with significant differences in length

Moreover, data records show:

Table 1. Data record for incident occurred during the installation of the temporary support in the tunnel progress for the Variante di Valico (VAV).

INCIDENT TYPE	VAV (no./km)	BOSCACCIO (no./km)
All incidents	7.5	2.5
Severe > 90 GG	1.6	0
Medium > 30 GG	5.9	2.5

Note: 1 – Incidents at Boscaccio due to handling of heavy object; 2 – Data was collected during the excavation of 26km of conventional tunnelling, encroaching mixed ground condition

(iv) Reduction of the overall, and medium incidents occurrence
 (v) No severe incident occurrence
(vi) No fatal event was recorded during the entire project

The BZEROTONDO tunnel support system along with its erector machine proved to be a solution able to reduce installation time – the tunnel was completed approximately 55 days ahead of the original schedule – and yet to deliver an enhanced, quality-aware and safer working environment for the operators.

Figure 5. B.I.M. implementation for tunnelling allows an enhanced construction management process.

4   COMPLIANCE TO B.I.M. PROTOCOL FOR PROJECT OUTPUT INFORMATION

The BZEROTONDO system along with its dedicated erector EKIP 21-04 allows conventional tunnelling to switch to an industrialised production and to a digitized model where B.I.M. forms a repository for the project output information. The benefits of this intelligent supply chain in real time, are evident. The generation of a 3D model assigns the steel arches in the exact theoretical position. Upon the acquisition or the model, the machine system – via the CP Track – displays the accurate referenced positioning of the steel arch automatically, in accordance with the project requirement for level, spacing and alignment and provides an effective support to the operator during the installation phase. Once the installation in completed, the machine returns an as-built report with the correct position of the steel arch checked against the digital model (Figure 5 below, refers). Thus, the system provides for the correct positioning of the steel arch, and mitigates quality and structural risks – as instance, non-conformities for encroaching the final lining – or depreciation of the primary tunnel lining works.

With reference to the Project and Construction Management the system allows to have a digital model updated in real time for the work progresses, and as such includes the progress control, data availability for accounting purposes and logistics management. For larger projects, with multiple and overlapped excavation fronts, the system allows to raise an alert automatically– threshold given – to the logistic management when cross checking construction progress and available stock. Those features represent a unique example in the underground infrastructures industry and complies with the latest requirement from the Italian authorities with regard to digitalization of project data.

5   CONCLUSION

Whether a risk is tolerable or not is essentially a matter of judgement. This depends on confidence that all hazards have been identified, all reasonably practicable steps have been taken to assess, reduce and control risks; and the consequence of any serious mishap can be kept to a minimum.

High consequence events (such as face and crown collapse, or lining failure involving breakthroughs to the surface) should be considered, in addition to the more likely events with lesser consequences.

Safety of the tunnelling process is heavily dependent on systems of management and work. People must make complex judgements to achieve quality in many differing types of work, often under difficult site conditions. In essence, safety is dependent on 'human factors'. Therefore, if tunnelling is to be undertaken safely it is essential that those managing the process understand how human failure happens, what can be done to prevent it, how it can be detected and corrected and how to recover from it.

Tunnel design is intimately involved with construction and should consider the whole process, not just the end product. Designs which take account of ease of construction, or 'buildability', will greatly facilitate the achievement of quality, and will lead to both a better outcome and to improved safety.

Implementation of the BZEROTONDO Tunnel Support system provides the engineers and the relevant projects' stakeholders with a wider range of options, and gives birth to a viable, safer and cost-effective support system able to operate within the most challenging ground conditions.

The BZEROTONDO system along with its erector machine EKIP 21-04 represents a driving technology on how interoperability and information transparency can support and provide information to the B.I.M. during construction and also during the maintenance phase of the infrastructure. EKIP 21-04 performs aspects of production on a digitized model where B.I.M. forms a repository for the machine output information. The benefits of this intelligent supply chain in real time, are evident. Using equipment like EKIP 21-04 with tubular steel arches confirms that the B.I.M. approach can deliver reliability, time and cost efficiencies and a unique opportunity for the construction industry

REFERENCES

Association of British Insurers and the Tunnel British Society. The joint code of practice for risks management of tunnels works in UK (2003)
British Tunnel Society and Institution of Engineers (2010). Specification for Tunnelling 3rd edition
EUROCODE 4. EN 1994-1-1:2004. Design of composite steel and concrete structures. General rules and rules for buildings
Health and Safety Executive (1996). Safety of New Austrian Tunnelling Method (NATM) Tunnels. ISBN 978 0 7176 10686
Health and Safety Executive (2006). The risks to third parties from bored tunnels in soft ground. HSE books
Japan Society of Civil Engineers (2007). Standard Specification for Tunnelling – 2006: Mountain Tunnels
Lunardi P., Ramozzi F., Simonini A., Bonadies D., Avignone C., Zenti C.L. (2013). Lining innovation at Marche – Umbria. Tunnels Magazine, May (37–42pp)
Mander J.F., Priestley M.J.N., Park R. (1988). Theoretical stress-strain model for confined concrete. ASCE, Journal of Structural Engineering, Vol.114, N.8
Occupational Safety and Health Administration (OSHA, 1970). *Occupational safety and health standards*: Safety and Health Regulations for Construction (Standard No. 1926.32)
AA. VV., Nota Interregionale prot. n° PG/2009/272843 del 27/11/2009 "Standard di sicurezza contro il rischio di infortunio da caduta gravi nei lavori a ridosso del fronte di gallerie scavate con tecnica tradizionale" (NIR 41), Ed. Regioni Emilia Romagna e Toscana
AA. VV., Lettera az. USL di Bologna e az. USL 10 Firenze prot. n° 8333 del 24/01/2011 "Aspetti applicativi della NIR n° 41 'Lavori a ridosso del fronte' (NIR 43), Ed. Regioni Emilia Romagna e Toscana
AA. VV., Nota Interregionale prot. n° PG/2009/272843 del 27/11/2009 "Lavori a ridosso del fronte – Addendum" (NIR 45), 06/ 06/2014,Ed. Regioni Emilia Romagna e Toscana
AA. VV., Conferenza delle Regioni e delle Province Autonome. Linea Guida n° 13. Lavori a ridosso del fronte. Ottobre 2015
Osservatorio Sicurezza Grandi Opere (2012). Rapporto No.15. Variante Autostradale di Valico e Terza Corsia
Osservatorio Sicurezza Grandi Opere (2015). Variante Autostradale di Valico e Terza Corsia. Nodo di Firenze (Aggiornamento no.6)
NTC2008 (2008). Nuove Norme Tecniche per le costruzioni – DM14 Gennaio 2008
Zenti C.L., Lunardi P., Rossi B., Gallovich A. (2012). A new approach in the design of first lining steel ribs. ITA-AITES World Tunnel Congress Bangkok
Zenti C.L., Lunardi P., Bellocchio A. (2014). 'La centina tubolare: massimizzazione dell'efficienza del rivestimento di prima fase. Approfondimenti ed evoluzioni'. Atti del Convegno Innovazioni nella realizzazione di opere in sotterraneo. Expotunnel 2014. Bologna (Italy)
Zenti C.L., Cullaciati A. (2015). 'Semi-automatic tubular steel arch: an innovation on safety'. ITA WTC 2015 Congress and 41st General Assembly. May 22- 28,2015. Dubrovnik, Croatia

Tunnels and Underground Cities: Engineering and Innovation meet Archaeology,
Architecture and Art, Volume 9: Safety in underground
construction – Peila, Viggiani & Celestino (Eds)
© 2020 Taylor & Francis Group, London, ISBN 978-0-367-46874-3

Pavoncelli Bis water tunnel: Tunnel boring machine selection and safety standards for excavating in presence of methane

R. D'Angelis, M. Maffucci, G. Giacomin & M. Secondulfo
Ghella S.p.A., Rome, Italy

F. Cichello
Tunnel Engineer, CMI JV, Buenos Aires, Argentina

ABSTRACT: After decades of failed attempts, the excavation of the 8.5km long Pavoncelli Bis water tunnel in Southern Italy has been successfully completed by the Joint Venture formed by Ghella S.p.A, Vianini Lavori S.p.A. and Giuzio. The tunnel is one of the largest and most important underground civil works in Southern Italy, with its variety of grounds and tectonic movements. It has probably been one of the most challenging excavations, completed without any accident or injury. The main strategy has been the selection of an extremely sophisticated Earth Pressure Balance Tunnel Boring Machine (EPB TBM)specifically designed to accommodate the variety of grounds: compact limestone, chaotic rocks such as the Flysch of Materdomini, multi-coloured clays with swelling characteristics and heavily tectonized sandstones as well as compressing grounds. The tunnelling teams have coped with large amounts of methane, some of it at pressure, and significant inflows of water. The precast segments have been designed specifically to cope with compressing ground, high-water pressures and an area subject to seismic activity. During the initial phases of excavation, the presence of methane proved to be not occasional as expected, but quite constant and larger than foreseen. This has required quick adaptations to the boring machine together with a newly studied safety procedure, allowing excavation in presence of gas.

1 INTRODUCTION

For centuries, the region of Puglia has lacked water given the porous nature of its land, not suitable to hold aquifers. At the start of 1900, the Italian government, in an attempt to resolve this situation, which had become unsustainable due to the demographic increase, promoted the creation of a large aqueduct, at that time the largest in Europe. The original Pavoncelli Tunnel was thus built. The project provided the channelling of 6,000 l/s from the Sele river, its source, under the Apennines, through what was to become the longest tunnel in the world. The Pavoncelli Tunnel has required 20 years and the work of 22,000 men to become operational. On the 23rd of November 1980, an earthquake of magnitude 6.8 on the Richter scale devastated the Irpinia region, seriously damaging the Pavoncelli Tunnel. In the following years, the Italian President promoted the construction of a tunnel parallel to the existing one: the Pavoncelli Bis, designed to work as a by-pass to cut the water flow along the ruined tunnel. In the '90s, due to a series of significant complications a first contractor was forced to shut down the works. Other three contractors were unsuccessfully hired to execute the unlucky project. In 2012, with only 20% of the excavation complete and numerous cumulated operational unresolved problems, a new tender was awarded to the JV formed by Ghella SpA, Vianini Lavori SpA and Giuzio Ambiente Srl. Excavation of the underground tunnel connections started almost immediately, while those for the Pavoncelli bis mainline tunnel commenced in 2014.

2 GENERAL DESCRIPTION OF THE WORKS

The works for the Pavoncelli Tunnel have required a continuation from what had already been done under the previous contracts including the hydraulic continuity with the already operating infrastructure, as shown in the below Figure 1. The project has required the water from the Sanità source to arrive as far as the Pavoncelli bis tunnel end, further connecting with the remainder of the water network of the Puglia Aqueduct. The Contract has specified also the recovery of the redundant drinking water from the close-by Cassano Irpino tunnel: the waters, which currently feed the Sele river arriving at the Tyrrhenian sea, are directed towards a hydro-electric power station which, exploiting the hydrostatic differential, diverts them into the Pavoncelli bis tunnel thus increasing the water flow towards Puglia.

The project starts at the Shaft A interconnection with the Rosalba Tunnel and ends at Shaft C. The Shaft A has allowed the water to flow towards the Pavoncelli tunnel. During its construction, the waterflow was interrupted for overall nearly 3 days in two different stages.

The stages have required an initial interception with the Rosalba tunnel with the construction of a Ø12.5 m and 20 m deep shaft; a temporary bypass to the shaft has then been built. A first water interruption has then been required allowing to demolish the side walls of the tunnel and then restore the water flow. With the water running in the bypass, it has been possible to demolish a 6 m stretch of tunnel and insert a special section with 2 motorised gates which, in the future, are used to divert the water flow either towards the Pavoncelli bis tunnel or towards the original Pavoncelli tunnel. Once the by tunnel has been rebuilt, a second 12h interruption has allowed to disable the bypass and restore the water flow within the original Rosalba tunnel. This last tunnel has been excavated with conventional methods. Its diameter is 3.40 m and it is permanent lined by reinforced concrete cast-in-situ.

The Pavoncelli Bis mainline develops over a total of 10,218.70 m, with an average longitudinal inclination of 0.3%. The excavation of the 3.4m internal diameter tunnel has been completed with a Tunnel Boring Machine; the final lining in reinforced concrete has a thickness of 0.4m.. The overbound varies from a minimum of 50 m to a maximum of 400 m. The tunnel was expected to cross an initial 670m section of calcirudites of S. Lucia; Flysch of Materdomini up to about half the length of the tunnel; multi-coloured clays for the remainder of the excavation. Figures 4 and 5 show the expected geological profile and the very different actual conditions found.

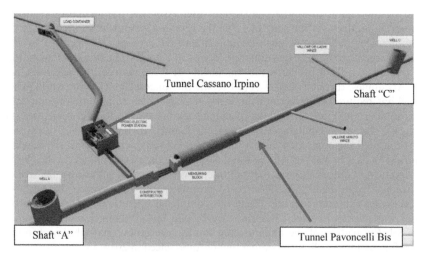

Figure 1. Diagram of the contract project.

Figure 2. Shaft A bypass and Detail of Rosalba Tunnel.

Table 1. Comparison between traditional and mechanized excavation

Type of excavation	Mechanized Pavoncelli bis tunnel		Conventional Pavoncelli bis tunnel		Conventional Rosalba	
	U.M.	Quantity	U.M.	Quantity	U.M.	Quantity
Tunnel section	m²	16.61	m²	18.75	m²	18.75
length	m	8440	m	1785	m	263
lining thickness	m	0.40	m	0.60 min	m	0.60 min
lining volume	m³	140,000	m³	33,500	m³	5,000

3 LITHOLOGICAL CHARACTERISTICS OF THE MASSES ENCOUNTERED

Due to its long history, succession of contractors and reported damages, the Pavoncelli bis tunnel can be split in nine different sections, as summarized in Table 2, according to the different lithologies encountered, excavation methods or approaches required.

Section 1: The first 300 m of tunnel, already excavated by previous contractors, have required the internal final lining, installed by the TBM which has proceeded with the installation of segmental lining without excavation;

Section 2: The second section, 200 m in length, also had been excavated by previous contractors. Due to the interception of a water source the tunnel had been flooded by water flows reaching 700lt/s. A simultaneous decrease in the flow rate of the nearby Sanità water source

Table 2. Summary table of the tunnel sections.

Section	Section length		Related progressives		Lithology crossed
	U.M.	Quantity	Initial	Final	
Section1	m	300	0+083	0+383	Solid limestone ("empty")
Section2	m	200	0+383	0+583	Solid limestone ("flooded")
Section3	m	92	0+583	0+675	Solid limestone
Section4	m	5225	0+675	5+900	Flysch of Materdomini
Section5	m	200	5+900	6+100	Multi-coloured Clay
Section6	m	123	6+100	6+223	Sandstone of Corleto-Albanella
Section7	m	134	6+223	6+357	Formation of Castelvetere
Section8	m	1403	6+357	7+760	Multi-coloured Clay
Section9	m	750	7+760	8+510	Red Flysch

Figure 3.   Technical divers at work in the "flooded section".

had shown a possible correlation between the two events. To overcome this problem and proceed with the excavation, an initial temporary reinforced concrete wall was built to confine the drained waters from the wall to the excavation face.

The section still kept the equipment and materials left by from previous contractors, obstructing the direct passage of the TBM or any other type of excavation thus requiring a series of preventive interventions. As such were the digging of a 50m deep well for surface access to allow the technical intervention of divers. The divers first carried out a preliminary underwater inspection using a closed-circuit, wire guided, video camera (ROV), then proceeded with the removal of all materials obstructing the mainline (ribs and other iron materials). They then placed pipes along the whole of the tunnel section through which a mix of waterproof, self-levelling and anti-segregating cement was pumped blocking the inflow. This purposely designed material has a low resistance (mechanical resistance to compression after 28 days ≥ 5.5 MPa) but effective waterproof qualities. More than 4,000 m^3 of mix have been injected. The temporary concrete wall has then been demolished and TBM excavation started.

Section 3: The following 100 m, from approx. progressive 583 m to approx. progressive 675 m, have allowed a clean excavation with TBM in calcareous aquifer, belonging to the Alburno - Cervati Unit.

Section 4: The fourth section has seen the complete transition to Materdomini Flysch, which turned out to be mainly scistic and scaly clays alternated with marl, marl alternated with clays, limestone-marly flysch, clayey flysch, clays, clayey loam and fault breaches. This geology has characterized the tunnel for the following stretch of 5,2 km, up to progressive 5900. Unfortunately, starting from this section and to the end of the mainline, the tunnel has also been characterized by elevated presence of methane gas. Gas was not expected and works were halted, at progressive 765, in early September 2014 for several months to adjust the TBM for continuous excavation within gas affected formations. The requirements, adjustments and solutions will be discussed in next chapters.

Section 5: The next 200 meters have seen the formation of various coloured clays: polychrome scaly clay encompassing rare elements of veined limestone and occasionally flinted limestone, limestone rubble, marl and sandstone.

Section 6: At progressive 6100 and for another 130 meters, the excavation has unexpectedly come across a lithological succession made up of banks and strata of sandstone, interspersed with thin layers of siltstone and sandstone and thin layers of sandstone-siltstone of a dark grey colour; this association has been attributed to sandstone of Corleto-Albanella which was not foreseen in the executable plan.

Section 7: The following 120m, from progressive 6223 to 6357, the boring machine has excavated through a thick and thin grained dark grey clayey sandy quartz-feldspar section with marl which, on the basis of petrographic analysis, has been attributed to the Castelvetere formation.

Section 8: From progressive 6357 to progressive 7760, clays and dark marl have been encountered with local reddish clay pebbles and subordinate interspersions of white and brown marly clays, calcilutites and calcarenites, attributed to the formation of Superior Multicoloured Clays. Specifically, for more than 270 meters after 91 meters of commencement of

Figure 4. Expected geological profile - 2012.

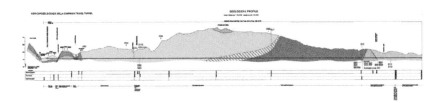

Figure 5. Actual geological profile encountered during excavation.

section8, the formation of Multi-coloured clays has been characterised by occasional veins of lapidar components of the arenaceous type.

Section 9: The last 750 meters of tunnel excavation have been completed trough conventional methods. The section has been characterized by clays and dark grey marls with red marls, marly limestones and limestone organogene breccias attributable to the Red Flysch Formation.

Identification of the tunnel sections described above, summarised in Table 2 by homogenous lithology, was made possible, in addition to a constant survey of the spoil material from the tunnel, also by accurate geological analyses conducted during execution of the works. Multiple evaluations have allowed to define the lithological profile of the tunnel:

– An intensive campaign of surface sampling;
– ca 4300 m of hybrid seismic area and electrical tomography (with a covering of 30% of the overall tunnel trace of interest to the excavation);
– constant monitoring and frequent sampling of the spoil material;
– laboratory analysis of samples (both taken from the surface and from the tunnel), using microscopic petrographic analysis on thin sections (in association with Professor Crivelli of the University of Calabria, Department of Biology, Ecology and Earth Sciences).
– Analysis of TBM parameters correlation: Thrust, Torque, conditioning consumption of the terrain (free and surfactant water), infill reverse mortar, etc.

4 DESCRIPTION OF THE TBM-EPB "SOFIA"

The Pavoncelli tunnel excavation has been carried out with a single shield with open cutter head Earth Pressured Balance Tunnel Boring Machine (EPB TBM). The geometric characteristics of the machine are as follows:

• Cutter head Φ: 4600 mm (with possibility of extension), weight 33 ton
• Shield Φ: Forward telescopic 4550 mm - rear 4490 mm
• TBM length: 210 m for a total of 21 trucks
• Shield length: 11.0 m

The most significant mechanical components are:

• Thrust force: Progress of the machine is allowed for by 10 pairs of 1.80 m stroke thrust cylinders, able to develop a very significant modulable nominal force. The TBM reaches 33,778 kN at 350 bar when in standard mode and up to 62,731 kN at 600 bar in high

Figure 6.   TBM and Back Up.

pressure mode. This thrust force, when compared with TBMs with larger shields, makes it the most powerful TBM ever made.

- TBM drive: Sofia is equipped with a shield divided in three parts allowing for a greater curvature and management throughout excavation as well as facilitating logistics during assembly and disassembly:
  - Forward, active, articulation (4 pairs of pistons between the forward and the central part of the shield with a 400mm stroke, advancing force of 32.572 kN at 400 bar);
  - Rear, passive, articulation (8 pistons between the central part of the shield and the rear, a 150mm stroke, advancing force of 7,862 kN at 350 bar);
- Cutter head able to overcome soft and hard rocks: 24 scrapers, 8 buckets, 10 double disc cutters (17"), single disc cutters (17");
- Electrical transmission unit: 4 motors with a power of 4x200kW, max revolutions 5.5/min, nominal torque 2240 kNm, force couple (overload) 3180 kNm and back-off torque 3404 kNm. These significant back-off forces have allowed the TBM to have continuous rotation of the cutting head, even during the most difficult phases of excavation;
- Hyperbaric chamber: it allows to activate TBM unblocking actions even in situations of stoppages below the water table and to access the excavation chamber. Luckily this was never required during excavation.;
- Bentonite Pump: allows to spray to lubricate the shield. It is a 7.5 kW pump with a 4 m3 tank;
- Injection line for the twin-component back-fill grouting.

5   UNEXPECTED PROBLEMS ENCOUNTERED DURING TUNNEL EXCAVATION AND TECHNICAL SOLUTIONS ADOPTED

The excavation and lining of the Pavoncelli bis tunnel in TBM was carried out with not entirely insignificant problems, largely due to the emergency of methane gas at frequencies and levels of emission which were incomparable with the original estimates. Furthermore, the geological and geotechnical conditions turned out to be particularly difficult such to cause an important halt of the excavation activities due to the blockage of the TBM shield.

5.1   *Adjustment of the TBM-EPB to the conditions encountered with gas*

Throughout the limestone aquifers of the Alburno-Cervialto Unit, up to progressive 650 approximately, Sofia has progressed without particular problems. In September 2017, after the following 90 m of tunnel, in the Flysch of Materdomini formation context, having reached ca progressive 768, the excavation activities have been suspended after several tunnel evacuations had occurred due to the gas alarm threshold exceedence. To allow the excavation activities to restart, specific technical solutions were implemented:

- Creation of a controlled draining system of the excavation chamber and of the screw feeder, allowing for pressure exhaust in order to drain the accumulated gas during excavation;
- strengthening of the shield and tail sweepers articulation seals, in order to achieve an overabundance of hydraulic seal, allowing for an increase in the sealing capacity thus impeding the ingress of gas;
- creation of physical confinement in line with the screw feeder discharge area and a continuous vacuum system which allows the flow of methane gas from the excavation chamber to the external chamber;
- strengthening of the air flow system in the shield area through the positioning of new ventilation lines. This required numerous aeraulic specific studies;
- creation of a "cold flow" system for the wagons during excavation, which allows the wagons to proceed with the locomotives switched off so to decrease the ignition temperature thus lowering the danger risk;
- compartmentalisation of the TBM environments based on the existing equipment, and according to the gas alarm threshold, minimizing the explosivity in each area;
- Increase and strengthening of the existing gas monitoring infrastructure such as the fixed sensors for the detection of concentrations of methane gas in the various TBM environments;
- Update and implementation of the operational procedures required to activate the new and added interventions.

5.2 First Blockage of the TBM and Solution

On the 9th of May 2016, at progressive 5,015, the advancing force of the boring machine registered a sudden increase from 23.000 kN to 60.000 kN, the maximum applicable thrust, while excavating through the Flysch of Materdomini. The phenomenon caused the stoppage of the TBM

Prior to this event, methane gas emissions had been noted generating a previous stoppage of over 13 hours during which works were carried out to reduce the tension acting on the shield.

The following analysis showed that only a portion of the shield was blocked, so a systematic movement of the free portions of the shield were undertaken to avoid the development of possible convergences over time, resulting in an increase of the state of tension acting thereupon. To free the TBM, a hydro-blasting of the upper outer curve cavity over the shield has been done with the use of particular lubricants around it. The hydro-blasting has been done by drilling holes, drilling the 5-6 cm of the shield steel and welding sealing plugs.

The high-pressure pumps utilised for the hydro blasting are specifically adapted for use within restricted spaces available in the TBM. Completed the works, a thrust of around 62,000 kN was applied resulting in a slow advancement of the TBM. A reduction in the necessary thrust occurred in this phase, indicating the expected progressive reduction in the tension acting on the shield.

Such circumstance has confirmed the local and singular character of the geological and geotechnical conditions. On the basis of observations at the front, it is likely that the tunnel trace went through a chaotic mass of a calcareous-marly type, intensely fractured, probably through tectonic action.

Figure 7. Application of hydro-blasting nozzle on the rotating head (left) and through holes in the TBM shield (right).

5.3 Second Blockage of the TBM and Solution

On the 28th of August of the same year, after more than 1 km of clean excavation, Sofia has suffered a second stoppage, near progressive 6,170. After having carried out numerous unblocking attempts, including again hydro-blasting but also the increase in the thrust with the application of additional jacks (Figure 8), attempts with thrusts well above the planned TBM performances were also attempted. The operations have required exceptional use of resources and time due to the continued presence of gas which was manifested at the opening of the provisions for carrying out hydro-blasting. Such operations, were unfortunately not enough to unblock the TBM, as they had been during the first blockage.

The stop has been attributed to the nature of the rocky (arenaceous) mass and to its instability in the short term, which has impeded the removal of the material produced during the hydro-blasting, thereby making it comparatively ineffective. The second attempt to unblock the shield has been with the addition of jacks (Figure 8), suitably connected to the tail shield, allowing to reach a thrust threshold slightly greater than 100 MN.

A hydro-cut has then been done, opening a rectangular window on the tail shield. This technique has allowed to cut the steel of the TBM shield (with a thickness of around 6 cm) without generating the risk of igniting methane gas at the back and adopting all the necessary related safety procedures.

Not being able to further increase the TBM thrust or adequately reduce the compression effecting the shield, the final solution has been to build lateral tunnels, on either sides of the TBM to reach the front of the machine.

The project initially studied theorized of 2 tunnels although only one was built.

Planning the operation

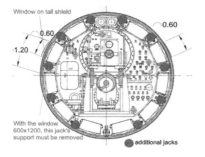

Figure 8.   Plan of the arrangement of No. 10 additional jacks with 'photo to the right.

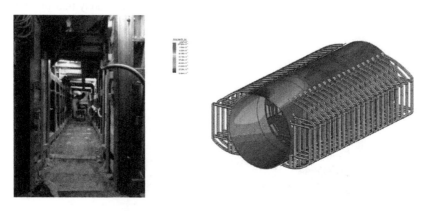

Figure 9.   Photo during excavation of the side tunnel and Static analysis of shield deformation with the theoretical 2 lateral tunnels.

To size the structural works, tests to evaluate the effective tension on the cement sections have been done with flat jacks. Given the restrictive geometry of the TBM, it has been possible to carry out only 2 of these tests. Through back analysis numeric considerations have been made using the pre-stoppage machine data and evaluating the consequences, both on the tunnel lining and TBM shield.

Particular attention had already been given, in this phase, in regard to the future consequences on the cement sections of the TBM thrust and on its future movements, resulting in the prefabricated cement sectioned ring being assessed as an unstable mechanism.

A parallel tunnel has thus been built along the left-hand side of the machine up to the front of the excavation, starting from an opening in line with the first ring after the tail shield and straddling two concrete sections. The opening has a section of 1.0 m width by 1.60 m in height.

The excavation has been done in recesses of around 0.5 m. The section (barely 4.2 m2) has been characterised by a pre-lining of HEB120 beams in S355 steel.

On the TBM side, the beams directly abut the shield with wooden props, and a double block in neoprene with a 5 cm thickness which rest on a suitably shaped metal plate welded to the beam; distancing the beams from the shield has given the advantage of avoiding possible snagging with the TBM seals during progress.

Once complete the excavation of the side tunnel, it has been refilled with lightened loose material (Tefralite 10/18, expanded clay type) starting from the base up to the entrance; a reinforcement structure in metalwork has then been mounted between the cemented sections, and the surrounding mortar poured; through metal rafters, the opening cement sections have been restored and the tunnel definitely closed; finally, a thrust test has been done to unblock the machine (see Figure 10).

With the recommencement of the TBM, the JV has kept a constant monitoring of the zone of blockage and, once the TBM has reached a safe distance, has completed the back filling of the side tunnel with mortar, by pouring tubing which had been installed prior to the closure works.

The unblocking of the TBM has therefore been the fruit not only of the exceptional capabilities of the technicians during the planning and preparation phase but also, the quality of the execution and the courage demonstrated by the workforce in the side tunnel excavation.

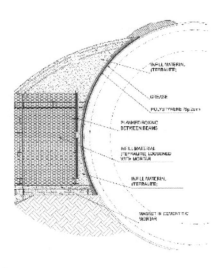

Figure 10.   Gallery refilling diagram.

6 DESCRIPTION OF THE STRUCTURAL MONITORING PHASE DURING THE FINAL LINING OF THE TUNNEL

In such a significant tunnel, monitoring the evolution of the tensions has been extremely important both during construction and its operational life. In addition to periodic convergence readings, structural monitoring of the Pavoncelli Bis tunnel has beencarried out through 29 instrumented sections (as illustrated in Figure 11) each equipped with ten FBG sensors for each instrumented ring: eight deformation sensors placed on the reinforced bars and two temperature sensors, to monitor the thermal variation between one reading and another. These two are found at the centre of the cement section (this can be seen in the illustration below, in sections B and E). They are disconnected from the concrete through a metal sheath. The signal generated by the impulse travelling in the fibre optics is greatly influenced by the temperature and, therefore, values are normally adjusted for a correct evaluation of the lining deformation.

In some cases, anomalous values have been recorded by the temperature sensors, which resulted in erroneous evaluations of the state of tension within the cement section. Thanks to punctual analysis and confirmation tests carried out directly in the field, it has been possible to correct these anomalies and to evaluate the state of tension in a more consistent manner. The fibre optics survey technology was chosen for its stability of reading and reliability in the presence of water, and its capacity for being connected in series and interrogated remotely at kilometres distance in real time.

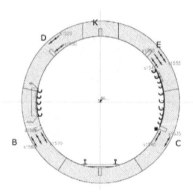

Figure 11.   Diagram of the instrumented ring in the Pavoncelli bis Tunnel.

Figure 12.   Image relating to the instrumented ring and the mounting of the reading sensors.

In occasion of the second TBM blockage, a new in-depth analysis of the geology has been done to define the geometry and characteristics of the terrain expected from the blockage progressive (pk 6.170) to the end of the excavation (pk 8.510). The surface aspects of the morphology of the zones interested by the excavation showed signs of intense tectonization. Such geognostic details have been obtained through the seismic profiles thanks to hybrid reflection oriented along two parallel axes to the maintunnel and along six transvers to these. Clearly, the investigation has depended from the accessibility to the sites (not always possible of the most interested and affected areas.

In terms of height, the tunnel is positioned between 415 and 405 m a.s.l. Together with the high-resolution hybrid seismic surveying, a surface waves analysis has been done to note the rigidity model and the electrical tomography to check the local electro-stratigraphy. Either of these tests are capable of reaching the tunnel depth, but they nevertheless allow for investigating the behaviour of the surface regions so to, on the one side demonstrate the congruity of corrections to the aeration of the hybrid seismic survey, and on the other side supply elements to rely on to indicate the morphological structure identified on the surface in case it corresponds to the physical discontinuities which could extend at depth. Through the distribution of the body wave speeds (proportional to the rigidity of the material) and the propensity to conduct electrical currents (inversely proportional to the effective porosity of a material), it is possible to divide the environment observed into potentially correlateable macro elements, unambiguously, to determined lithotypes of sedimentary facies. To translate the acoustic echograph data into a geolithological interpretation, the superimposition of the Masw study and of the electro tomographic ones has been considered. The first allows to demonstrate the reduction of mechanical properties (the velocity of propagation of the body waves in the subsoil) manifested near the major fracture facies; the second allows to distinguish, although at a very low scale of resistivity contrast, between elements markedly of clayey facies (less resistive) and other more arenaceous (more resistive). Thus, elements of coincidence between different methods may be highlighted, which in a certain way contribute towards carrying out counterchecks. Evidence of sub-vertical discontinuities have been noted, where interruptions of reflecting levels exist, as well as amplitudes (which indicate the presence of water or gas); it has been possible to reconstruct seismostratigraphic-strata bedding, both at a low and high angle, conjoining reflecting horizons of laterally homogenous characteristics. It is very difficult to recognise, from the seismic response, the lithotypes involved only though one observation. The punctual soundings and the same crossing with the TBM, has served as a confirmation and identification of the geology crossed, for the effective definition of the as built profile.

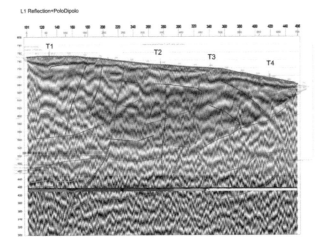

Figure 13.   Interpretation of the geophysical data.

8 CONCLUSIONS

The Pavoncelli Bis Project has been an extremely challenging, complex and exciting tunnel. The difficulties and constraints encountered during the excavation due to gas, geology, squeezing rock, and all the other have been dealt with in the most optimal and professional way through the research of the best solution thanks to the support and collaboration of designers, senior management, TBM supplier personnel, personnel on site and Ghella's Tunnel Division. The experience and knowhow of the wider team and the prioritization of quality and safety, has resulted in the successful completion of the excavation. The variety of the geology, with compressing ground and tectonic movements has surely defined this so longed-for tunnel, one of the largest and most important underground civil works in Southern Italy.

REFERENCES

Bandini, A., Cormio, C., Berry P. (Serengeo S.r.l.), M. Battisti & A. Lisardi (Collins S.r.l.), P. Bernardini (Ghella S.p.A.), M. Urso (Vianini Lavori S.p.A) Novel and innovative solutions for safe small section TBM tunnelling through gassy rock masses. Safety by design best practices and lessons learnt from the Pavoncelli bis hydraulic tunnel case study – Paper WTC 2019.

Tunnels and Underground Cities: Engineering and Innovation meet Archaeology, Architecture and Art, Volume 9: Safety in underground construction – Peila, Viggiani & Celestino (Eds)
© 2020 Taylor & Francis Group, London, ISBN 978-0-367-46874-3

CFD simulations of longitudinal ventilation of a road tunnel in congested traffic condition

A. Dastan, M. Rahiminejad, M. Abbasi & O. Abouali
School of Mechanical Engineering, Shiraz University, Shiraz, Iran

ABSTRACT: Ventilation system in tunnels is of a great importance in providing a safe condition for the people passing through the tunnel in both normal and emergency situations. This is usually performed by injecting fresh air into the tunnel and removing hazardous gases. In this research, the longitudinal ventilation of a 5-kilometer tunnel constructed in the Tehran-North freeway project, Iran, through straight jetfans is investigated. The effects of full congested traffic on the airflow rate imposed by the jetfans in this one-directional tunnel, and therefore, the function of ventilation system in removing the exhaust gases are studied. A 3D realistic computational domain for the tunnel with about 1100 stationary vehicles is developed and CFD simulations are performed. The results showed the traffic arrangement and the location of large vehicles in the tunnel has a significant impact on the performance of the ventilation system. The CFD data also suggested that the designed system is likely to fail in removing heat generated by the stationary vehicles in the tunnel and providing a safe condition for the passengers.

1 INTRODUCTION

The ventilation systems are used in tunnels (road and metro) to provide fresh air and remove hazardous gases generated in normal or emergency conditions for the normal function of vehicles and instruments and also passengers and rescue team members. Natural ventilation (Yan et al. 2009), which is usually used in short tunnels, and forced ventilations by fans and jetfans (Ang et al. 2016; Eftekharian et al. 2014; Musto and Rotondo 2014; Du et al. 2016) are two main types of ventilation in tunnels. When the airflow enters from one portal of the tunnel and exits from the opposing portal, the ventilation is longitudinal; along the length of tunnel. This is usually performed by jetfans, function of which is to add some momentum to the domain. The added momentum induces a flow rate into the tunnel which brings fresh air and removes the pollutants.

Different factors, including the cross section area of the tunnel, its length and curvature, types and number of jetfans, the trust they provide and ..., can affect the design of a longitudinal ventilation systems. In a computational work (Wang et al. 2011), it was shown that in a congested traffic, the curvature of tunnel has no significant effect on the ventilation, provided its radius is greater than 2 km. They also showed that the optimum distance between the jetfans in the studied tunnel should be between 90 and 120 m. Few researchers (Eftekharian et al. 2014; Betta et al. 2010) have focused on the types of jetfans used in the ventilation system and their advantages and disadvantages in different scenarios. The modeling techniques for simulation of ventilation by jetfans have been also studied recently (Król and Król 2018).

While experimental investigation of traffic effects on the ventilation system is quite difficult and rare (Kurtenbach et al. 2001), computational methods can be a very useful and achievable tool for the investigation of this condition (Bari and Naser 2010; Eftekharian et al. 2014). The stationary vehicles in the traffic causes a drop in the performance of the longitudinal ventilation system due to the associated pressure drag. The congested traffic may even lead to a 50% reduction in the air flow rate the jetfans can induced in the tunnel comparing with the no-traffic condition (Eftekharian et al. 2014; Sekularac and Jankovic 2018). This, consequently, may affect the function of the ventilation system in removing the heat and pollutant generated by the vehicles working at idle in the tunnel.

In this work, CFD simulations of longitudinal ventilation in a long tunnel under traffic jam condition are performed. The 3D geometry model which includes about 1100 stationary vehicles and jetfans of the ventilation system is generated and then meshed. The effects of traffic distribution, particularly the interaction of large vehicles with jetfans, is investigated. The performance of the ventilation system in removing the heat and pollutant generated by the vehicles is studied.

2 NUMERICAL METHODS

2.1 *Model description*

The tunnel of study has been constructed as one of a 16-tunnel system in the Tehran-North freeway project, Iran. This is the longest tunnel of this project with about 4892 m length. The tunnel is ventilated through two different mechanisms; one is a longitudinal ventilation by the function of straight jetfans and the other is a semi-transverse system which uses two axial fans at the both portals of the tunnel, in addition to the jetfans which help the development of airflow in the tunnel. A ceiling duct region was constructed inside the tunnel as a pathway for the airflow sucked by the axial fans. This ceiling duct is separated from top of the tunnel cross-section by a thin wall with the thickness of 11 cm. The semi-transverse system comes to the action only in an incident which involves fire and heavy smoke generation. As the focus of this work is the ventilation in a congested traffic condition without any fire, the longitudinal ventilation which is carried out by jetfans is simulated. Considering this, the effective cross-section of the tunnel below the ceiling duct is 65.5 m^2 with maximum height of 5.7 m and hydraulic diameter of 8 m. Figure 1 shows the cross section of the tunnel and the ceiling duct.

The longitudinal ventilation is performed through a set of straight jetfans installed in the left side of the tunnel (considering the direction of vehicle movement). The installation position of jetfans is illustrated in Figure 1. The spacing between two consecutive jetfans is about 80 m, and therefore, 60 jetfans are distributed along the tunnel – only 44 jetfans are installed in the real case scenario and the remaining is just modeled here for the purpose of investigation. The inner diameter of jetfans is 71 cm, while the casing thickness is 7.5 cm. The length of cylindrical model of jetfans is 2.41 m which includes a central fan region and two silencers at both sides of the fan. The approach for the modeling of jetfans is the same as our previous works (Eftekharian et al. 2014) by applying a constant velocity of 38 m/s inside the jetfan to provide the desired flowrate of 15.04 m^3/s.

The aim of this research is to investigate the effects of stationary vehicles in a congested traffic scenario on the performance of the longitudinal ventilation system. Therefore, the vehicles should be included in the computational domain. According to Piarc standard (PIARC 2012), the vehicles in a full traffic condition in 100 meters of a 2-line road tunnel includes 20 passenger cars, 2 buses and 2 trucks. The passenger cars modeled in this study are supposed to be all the same model, being similar to those introduced in our previous work (Eftekharian et al. 2014), with some aerodynamic modifications to have a more realistic drag coefficient (Eftekharian, Abouali, and Ahmadi 2015). In summary, the width, length and maximum height of modeled cars are 1.7, 4.4 and 1.41 m, respectively. Figure 2 shows a side view of the real car and also the sketch of the simplified version. Buses and trucks are modeled as single boxes attached to the ground with dimensions of 2.5 × 12.5 × 3.8 m and 2.5 × 16.3 × 3.7 m, respectively. These lead to the spacing of 1.65 m between two successive vehicles. Total number of modeled vehicles in the tunnel is 1176.

To include the effects of topography of the region on the ventilation system, two computational boxes at the portals of the tunnel with the approximate dimensions of 50 × 30 × 30 m were considered. By this, more realistic inlet and outlet boundary conditions can be applied at the portals of the tunnel. In addition, the concrete walls of the tunnel with the approximate thickness

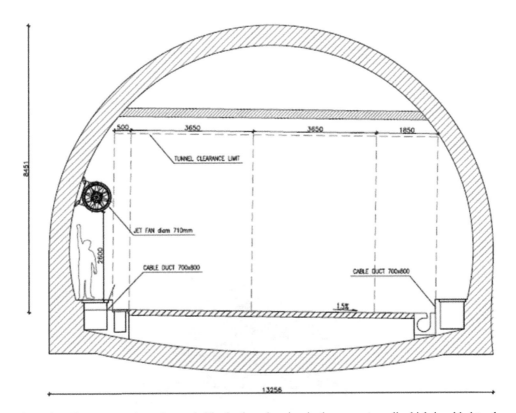

Figure 1. The cross-section of tunnel. The hachured region is the concrete wall which is added to the computational domain. The dimensions are in mm. The installation location of a jetfans is also illustrated.

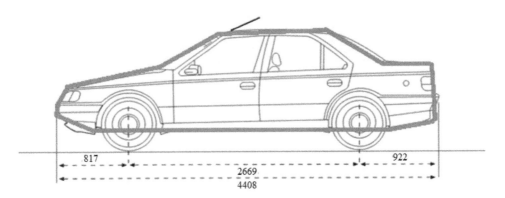

Figure 2. The simplified model of car shown with orange colour. A constant width of 1.7 m was considered for this car.

of 0.5 m were added to the computational domain (Figure 1). This solid region is of a great importance, particularly when the heat released by vehicles is simulated - it acts as a heat sink.

The 3D model of the domain was constructed in a modeling software and then imported to the meshing package. Due to the complexity of the domain and different required resolutions in different regions, a multi-zone technique was employed for grid generation. A finer grid was used near jetfans and vehicles to be able to capture the flow-field in these regions more precisely. Three different grids were generated for the domain and, finally, the one with about 21 million hexagonal cells was found to be the most appropriate mesh for the simulations. Figure 3 illustrates the grid generated for some regions of the domain.

2.2 *Governing equations and boundary conditions*

The governing equations which are solved in the numerical simulations include continuity, momentum and energy equations:

$$\frac{\partial(\rho u_i)}{\partial x_i} = 0 \tag{1}$$

$$\frac{\partial(\rho u_i u_j)}{\partial x_i} = -\frac{\partial P}{\partial x_i} + \frac{\partial}{\partial x_i}\left[(\mu + \mu_t)\left(\frac{\partial u_i}{\partial x_j} + \frac{\partial u_j}{\partial x_i}\right)\right] \tag{2}$$

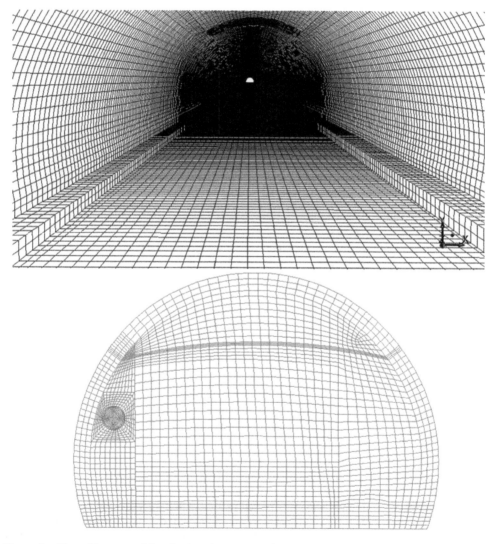

Figure 3. The grid generated for the domain. Top: a selected view of the tunnel walls; Bottom: tunnel cross section.

$$\frac{\partial}{\partial x_i} \cdot (u_i(\rho E + P)) = \frac{\partial}{\partial x_i} \cdot \left(k_{eff} \frac{\partial T}{\partial x_i}\right) + S_i \tag{3}$$

Where ρ, u_i, x_i, P, μ, μ_t, E, k_{eff} and T are, density, velocity vector components, location vector component, pressure, kinematic viscosity, turbulence viscosity, total energy, effective conductivity and temperature, respectively. The effective conductivity is the sum of fluid thermal conductivity and turbulent thermal conductivity. The term S_i is the source term in energy equation.

The average elevation of tunnel from sea-level is 2244 m which causes the average atmospheric pressure to be about 76 kPa. That is, the air density at temperature of 300 k is about 0.9 kg/m^3. In this work, where the heat released by cars is not considered in the simulations, the density is treated constant. Otherwise, the ideal gas assumption is employed for relating the temperature and the air density.

The standard k-ε turbulence model was used in the simulations. The roughness of 3mm was applied for the tunnel wall, while that of 1cm was used for the tunnel road. This height of roughness seems to be unrealistic, but it is used to incorporate the effects pressure losses associated with other small obstacles (firefighting boxes, lightning system, ...) in the tunnel. The P1 model was used for simulation of radiation heat transfer, where the energy equation is solved – it is added to equation (3) as a source term.

No slip boundary condition is applied on all solid walls of the tunnel. The zero-gauge pressure is also considered on the surfaces of the portal boxes at two sides of the tunnel. The air is sucked from the inlet box to the tunnel and exhausted to the outlet box. As mentioned, the jetfans are modelled by applying a fixed x-velocity zone of 38 m/s inside of them. The inlet air is supposed to be 300 k and also the outer surfaces of the tunnel solid wall have a fixed temperature of 300 k.

It is supposed that the stationary vehicles in the congested traffic condition which work at idle release some heat into the tunnel. This comes from the cooling system of vehicles (radiators) and also the air conditioning systems. In this condition, it is assumed that each car and a large vehicle release 14.4 and 60 kw heat into the domain, respectively. This leads to about 26 MW total heat due to the function of all stationary vehicles in the tunnel.

Another important issue in a congested traffic condition is the distribution of pollutants generated by the engines of the idle vehicles. Here, we only consider the production of CO which is the most hazardous gas exhausted by the engines. The distribution of CO is simulated by solving a scaler transport equation:

$$\rho u_i \cdot \frac{\partial \varphi}{\partial x_i} - \Gamma \frac{\partial^2 \varphi}{\partial X_i^2} = S_\varphi \tag{4}$$

Where φ and Γ are the concentration and diffusion coefficient of the scaler (CO). The diffusion coefficient is sum of laminar and turbulent diffusion coefficient (Eftekharian et al. 2014). The right term, S_φ, is the pollutant source term. The rate of CO generation for each car and large vehicle is supposed to be 15.5 and 14.6 gr/hour. This is added to the domain through a plane at the back of each vehicle.

The governing equations are solved in a coomertial flow solver which is a very powerful CFD package. The details of the solution procedure are not reprduced here for brevity (there are many references elsewhere addressed these details).

3   RESULTS AND DISCUSSION

3.1   *Cold tests*

As said, in the real case scenario, only 44 jetfans are installed in the tunnel. The distribution of the jetfans suggested by the designer is not uniform. Moving from the inlet to the outlet of tunnel, after a 100 m gap, 6 jetfans are installed. Then, there is a 400 m of no-jetfan zone which is followed by 16 jetfans. After another 800 m gap, there is another set of 16-jetfan.

Finally, there is a set of 6 consecutive jetfans after a 240 m gap. This arrangement for the jet-fans is for saving in the cabling system.

Here, for the evaluation of the ventilation system, simulations are performed without considering the heat and pollutants released by the vehicles in the tunnel. As there is no heat, and therefore, no temperature raises in the domain, these simulations are called "cold tests".

The momentum injected to the domain by 44 jetfans leads to an induced flow rate of 118 m³/s in the congested traffic condition. This is equivalent of an average air velocity of 1.76 m/s. However, the velocity distribution is not uniform in the tunnel cross section and is higher in the left side where the jetfans are installed. Due to relatively short height of jetfan location, it is likely that the high velocity stream exiting a jetfan has a direct interaction with a large obstacle (buses and trucks) which are at the left side of the tunnel. This causes a significant loss in the domain and negatively affects the performance of jetfans. To investigate this phenomenon, another computational domain was developed in which the large vehicles were located only in the right row of the tunnel. That is, no buses or trucks is in front of the jetfan streams. The number of vehicles, however, is the same as the previous model. Figure 4 compares the distribution of vehicles in these two models.

The results of the simulation in the new model showed that the induced flow rate and average air velocity are 142 m³/s and 2.12 m/s, respectively. That is, the arrangement of vehicles in the tunnel and eliminating the chance of direct interaction of high velocity stream of jetfans and back of large vehicles can change the flow rate in the tunnel up to 20%. However, left side of the tunnel experience a larger air velocity, while the jetfans cannot thoroughly induce a flow at the right. This is because of the dense distribution of large vehicles at the right side of the tunnel. Figure 5 illustrates the velocity distribution on a horizontal plane in the tunnel. This clearly shows two high and low velocity regions at left and right sides, respectively.

Figure 6 also compares the stream lines exiting from jetfans in two simulated vehicle distributions. The buses and trucks at the left of tunnel cause a deviation in the flow path lines which is clear in the figure.

3.2 *Simulation of heat and pollutant dispersion in a congested traffic*

As discussed, the stationary vehicles in a congested traffic condition release some heat and pollutant, particularly CO, in the tunnel. The temperature distribution and CO concentration along the tunnel is an important safety factor in such a condition. Here, we assume that the traffic lasts for a sufficiently long time that the steady-state simulation can be a good approximation of the final condition.

In this scenario which involves temperature change due to the heat released by vehicles, and consequently, density drop, the air flow rate and average velocity at the tunnel inlet, induced by the function of 44 jetfans, are 115 m³/s and 1.76 m/s, respectively, while those of tunnel outlet are 128 m³/s and 1.95 m/s, respectively. This shows that the heat generated by cars does

Figure 4.   Distribution of vehicles in the tunnel in two simulated models: Left) large vehicles are in bot left and right rows. Right) Large vehicles are only at the right row.

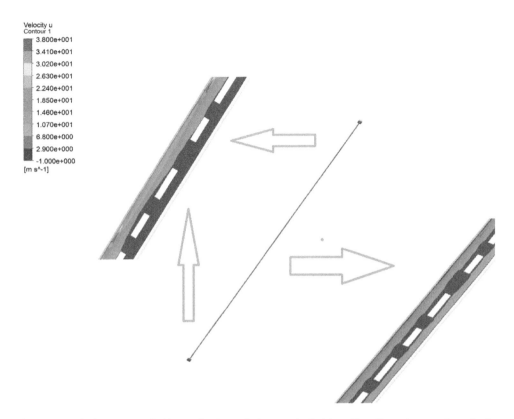

Figure 5. Velocity contour (m/s) on a horizontal plane at the height of 2 m from the tunnel road, when the large vehicles are at the right of tunnel. The left figure shows a zone in which the jetfans are installed. The right figure is in the no-jetfan region.

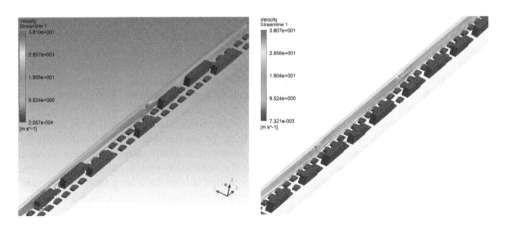

Figure 6. The flow path lines coloured by velocity in two traffic arrangements: Left) large vehicles are in bot left and right rows. Right) Large vehicles are only at the right row.

not significantly affect the performance of jetfans and the tunnel air flow rate remains nearly constant. The mass flow rate of the tunnel is 101 kg/s. It should be noted that in this simulation the traffic arrangement is the same as left part of Figure 4 in which the large vehicles are distributed in both rows.

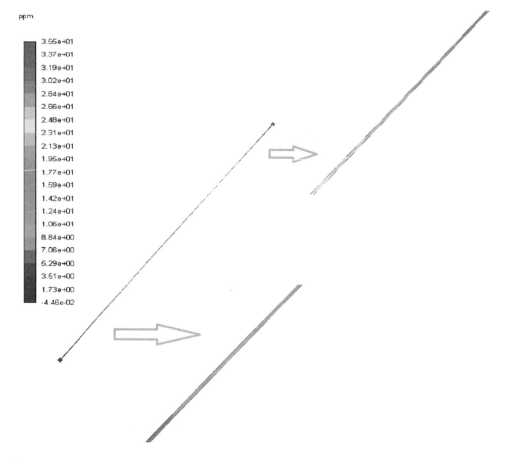

Figure 7. CO concentration (ppm) on a horizontal plane at 3m distance from the tunnel road.

The results of ventilation simulation showed that the average concentration of CO at the outlet of tunnel is 27 ppm. Figure 7 shows the concentration of carbon monoxide in ppm on a horizontal plane at 3 m distance from the tunnel road. As seen, along the tunnel, the concentration of the pollutant builds up and at the outlet reaches the local maximum of 35 ppm. This is safely below the maximum allowable concentration of CO which is 70 ppm, and therefore, the ventilation system function is acceptable in this regard.

The temperature profile in the tunnel, however, rings some alarms. The average outlet temperature is 332 k (59 °C), which is beyond the safe range for humans. Figure 8 shows the temperature contour on a horizontal plane in the tunnel. The temperature increases along the tunnel and reaches 338 k. These results show that the ventilation system fails in removing heat from the tunnel in a congested traffic condition, provided it lasts for a long time. This situation may be corrected by employment of the semi-transverse ventilation system in such a condition.

4 CONCLUSIONS

In this paper, the CFD simulations of longitudinal ventilation in a 5-km tunnel in a congested traffic condition were presented. The ventilation is performed through 44 straight jetfans which were nonuniformly distributed in the tunnel. A significant number of stationary vehicles were added to the computational domain to model the full traffic condition. Two different arrangements for traffic were modelled to see the effects of large vehicle distribution along the tunnel and their interactions with the jetfans.

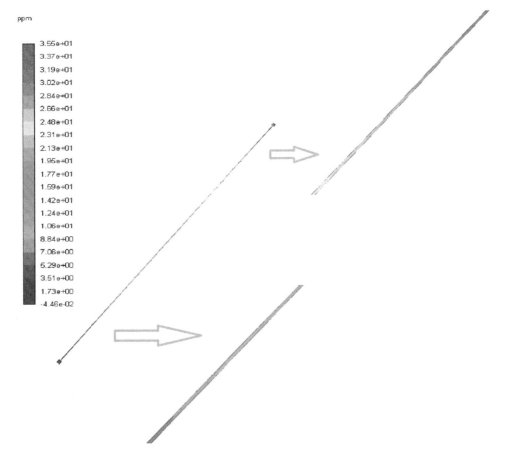

ppm

3.55e+01	
3.37e+01	
3.19e+01	
3.02e+01	
2.84e+01	
2.66e+01	
2.48e+01	
2.31e+01	
2.13e+01	
1.95e+01	
1.77e+01	
1.59e+01	
1.42e+01	
1.24e+01	
1.06e+01	
8.84e+00	
7.06e+00	
5.29e+00	
3.51e+00	
1.73e+00	
-4.46e-02	

Figure 8.   Temperature contour (k) on a horizontal plane at 3m distance from the tunnel road.

The results showed that large vehicles at the left side of tunnel, near jetfan location, can significantly affect the performance of ventilation system, such that the air flow induced in the tunnel increased by 20% when large vehicles were not allowed to move in the left side of tunnel.

The length of tunnel, and consequently, number of vehicles inside the tunnel in a full traffic condition, causes the heat released into the domain to be very high. The longitudinal ventilation system failed in removing the heat generated by cars in a long-time traffic. The average outlet temperature could reach about 60°C which is beyond the safe limit for human. In terms of pollutant removal, however, the ventilation function was acceptable and could provide a safe condition for the passengers inside the tunnel.

ACKNOWLEDGMENT

The work of the authors was financially supported by Tehran-North Freeway company and this support is gratefully acknowledged.

REFERENCES

Ang, Chin Ding (Edmund), Guillermo Rein, Joaquim Peiro, and Roger Harrison. 2016. "Simulating Longitudinal Ventilation Flows in Long Tunnels: Comparison of Full CFD and Multi-Scale Modelling Approaches in FDS6." *Tunnelling and Underground Space Technology* 52 (February). Pergamon:119–126.

Bari, S., and J. Naser. 2010. "Simulation of Airflow and Pollution Levels Caused by Severe Traffic Jam in a Road Tunnel." *Tunnelling and Underground Space Technology* 25 (1). Pergamon: 70–77.

Betta, V., F. Cascetta, M. Musto, and G. Rotondo. 2010. "Fluid Dynamic Performances of Traditional and Alternative Jet Fans in Tunnel Longitudinal Ventilation Systems." *Tunnelling and Underground Space Technology* 25 (4). Pergamon:415–422.

Du, Tao, Dong Yang, Shini Peng, Yingli Liu, and Yimin Xiao. 2016. "Performance Evaluation of Longitudinal and Transverse Ventilation for Thermal and Smoke Control in a Looped Urban Traffic Link Tunnel." *Applied Thermal Engineering* 96 (March). Pergamon:490–500.

Eftekharian, Esmaeel, Omid Abouali, and Goodarz Ahmadi. 2015. "An Improved Correlation for Pressure Drop in a Tunnel under Traffic Jam Using CFD." *Journal of Wind Engineering and Industrial Aerodynamics* 143 (August). Elsevier:34–41.

Eftekharian, Esmaeel, Alireza Dastan, Omid Abouali, Javad Meigolinedjad, and Goodarz Ahmadi. 2014. "A Numerical Investigation into the Performance of Two Types of Jet Fans in Ventilation of an Urban Tunnel under Traffic Jam Condition." *Tunnelling and Underground Space Technology* 44 (September). Pergamon:56–67.

Król, Aleksander, and Małgorzata Król. 2018. "Study on Numerical Modeling of Jet Fans." *Tunnelling and Underground Space Technology* 73 (March). Pergamon:222–235.

Kurtenbach, R., K.H. Becker, J.A.G. Gomes, J. Kleffmann, J.C. Lörzer, M. Spittler, P. Wiesen, R. Ackermann, A. Geyer, and U. Platt. 2001. "Investigations of Emissions and Heterogeneous Formation of HONO in a Road Traffic Tunnel." *Atmospheric Environment* 35 (20). Pergamon:3385–3394.

Musto, Marilena, and Giuseppe Rotondo. 2014. "Numerical Comparison of Performance between Traditional and Alternative Jet Fans in Tiled Tunnel in Emergency Ventilation." *Tunnelling and Underground Space Technology* 42 (May). Pergamon:52–58.

PIARC Technical Committee Road Tunnels. 2012. "Road Tunnels: Vehicle Emissions and Air Demand for Ventilation." *2012 PIARC Technical Committee on Road Tunnels. France*, 72.

Sekularac, Milan, and Novica Jankovic. 2018. "Experimental and Numerical Analysis of Flow Field and Ventilation Performance in a Traffic Tunnel Ventilated by Axial Fans." *Theoretical and Applied Mechanics*, no. 00:10–10.

Wang, Feng, Mingnian Wang, S. He, and Yuanye Deng. 2011. "Computational Study of Effects of Traffic Force on the Ventilation in Highway Curved Tunnels." *Tunnelling and Underground Space Technology* 26 (3). Pergamon:481–489.

Yan, Tong, Shi MingHeng, Gong YanFeng, and He JiaPeng. 2009. "Full-Scale Experimental Study on Smoke Flow in Natural Ventilation Road Tunnel Fires with Shafts." *Tunnelling and Underground Space Technology* 24 (6). Pergamon:627–633.

Tunnels and Underground Cities: Engineering and Innovation meet Archaeology,
Architecture and Art, Volume 9: Safety in underground
construction – Peila, Viggiani & Celestino (Eds)
© 2020 Taylor & Francis Group, London, ISBN 978-0-367-46874-3

Transverse ventilation system applied to unidirectional road tunnel: Practical case of S2 Tunnel in Warsaw

N. Faggioni
Imgeco Srl, Genoa, Italy

F. Bizzi
Astaldi SpA, Rome, Italy

M. Bringiotti
Geotunnel Srl

ABSTRACT: This paper concerns a practical case of transversal ventilation system applied to a unidirectional road tunnel and how the functional design has enabled the optimization of the tunnel geometry, reducing the estimated implementation costs and thereby promoting the award of the Contract with the subsequent realization of the works. The ventilation system described is related to the twin-tube tunnel (2.4 km long) located on the new roadway S2-Southern Bypass in Warsaw – Task A. It is then described the starting assumption about the performance of the ventilation system, the draft presented by the Contracting Authority during the tender phase and the optimization proposals, these last ones as results of the analysis of the required performances. Furthermore, It has been demonstrated the applicability and effectiveness of the Computational Fluid Dynamics (CFD) and the Fire Safety Engineering for the validation of the transversal ventilation system in a road tunnel with unidirectional traffic.

1 INTRODUCTION

Generally, ventilation system has an important impact in the main geometry and functionality design of a road tunnel. Interferences increase especially when it comes to transverse ventilation and where additional ventilation ducts or chambers installations are usually required.

The scope of this paper is to show how a careful optimization of ventilation system design can lead towards new technical solutions that reduce impact on civil works. In particular, in the case of the WARSAW SOUTHERN BYPASS S2 TUNNEL Project, the use and application of a new design configuration allowed the optimization of the tunnel geometries by restricting the excavation sections and consequently reducing the cost analysis; this allowed, during the tender phase, the awarding of the works to the contractor.

The innovative design solutions have been verified, in terms of efficiency and reliability, during the design building permit design using the Computational Fluid Dynamics and Fire Safety Engineering Techniques.

2 TENDER REQUIREMENTS

2.1 *Technical e geometrical data*

The functional elements of S2 road tunnel (as prescribed by Tender Functional-Utility Program) are fixed for every tube:

- traffic lanes: 3 × 3.50 m
- emergency lane: 1 × 3.75 m
- roadsides: 1 × 0.25 m
- sidewalk: 2 × 1.0 m
- nominal speed (traffic): 80 km/h
- gauge (height): 4.70 m
- transversal slope (road): 2.5% (min.).

All the functional elements were annexed to a unique cross section inside tender project as reported below.

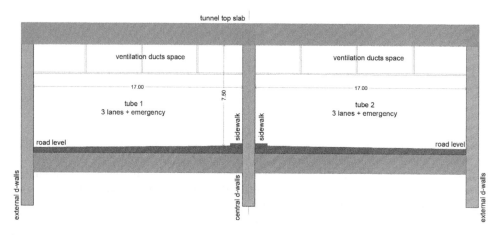

Figure 1.   Typical cross section with main ventilation elements.

2.2   *Ventilation system*

About ventilation system performances, mandatory requirements expressed by Regulation nr. 63 art. 735 of the Ministry of Transport and Polish Maritime economy, foresaw the use of transversal ventilation in the tunnel.

In fact, unlike what is required by the legislation of the other EU member states and the technical reference literature (see for example PIARC and RABT), the Polish law provides for the application of transverse ventilation systems in all road tunnels longer than 1000 m without distinction between functional and traffic types (one-way or two-way tunnels). Some basic elements of transverse ventilation were already contained in the tunnel concept design in which the use of a channel at the top of each tube was described to indicate the presence of one or more ventilation ducts.

According to the main reference manuals and guidelines for road safety design, such as the German RABT 2006 (Table 1), transverse ventilation should be applied mainly to two-way traffic tunnels longer than 1000–1500 m while it could be applied to unidirectional traffic tunnel with a length of over 3000 m, only when the results of the risk analysis (in terms of expected damage) do not give sufficient results just with the simple application of a longitudinal system.

Hence the main problem of colliding the regulatory and contractual requirements with tested and approved methods that do not provide the application of transverse ventilation in one-way traffic tunnels. In these tunnels, in fact, the main velocity component of the ventilation flow has longitudinal direction and is mainly produced by the presence of the circulating traffic and the related piston effect. A transverse system, to be defined as such, must be able to cancel the longitudinal component of the airflow velocity and create by itself a main transverse flow to the tunnel axis.

Table 1. Comparison between RABT and Polish law requirements.

Regulation - Guidelines	Tunnel length	Two-way tunnels	Unidirectional tunnels
RABT 2006	< 400 m	Natural ventilation	Natural ventilation
	400 – 600 m	Longitudinal ventilation Risk analysis method: - Longitudinal	
	600 – 1200 m	- Longitud. intermediate shaft - Semitransverse	Longitudinal ventilation
	1200 – 3000 m		
	> 3000 m	- Semitransverse ventilation - Fully transverse ventilation	- Longitud. intermediate shaft - Semitransverse ventilation
Polish Law nr. 63 poz. 735	< 1000 m	Longitudinal ventilation	
	> 1000 m	Transverse ventilation	

3 FIRST TENDER HYPOTHESIS

A first project idea was identified in the tender phase assuming a new geometry that included the two tunnel tubes separated by two ventilation ducts for air supply and exhaust. The solution seemed immediately winning as it allowed a strong reduction of the excavation sections and reduced the d-walls sizing.

In the first instance (Figure 2), it was decided to create a semi-transversal system with un upper exhaust duct and with a lower exodus path (escape tunnel) installed between two central

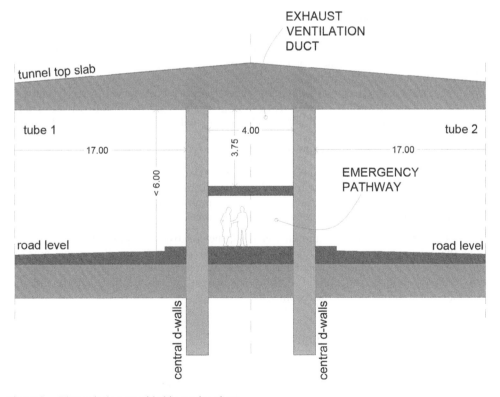

Figure 2. First solution provided in tender phase.

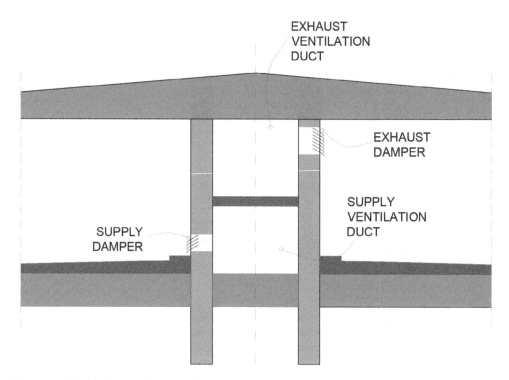

EXHAUST
VENTILATION
DUCT

EXHAUST
DAMPER

SUPPLY
VENTILATION
DUCT

SUPPLY
DAMPER

Figure 3.   Final design solution provided in tender phase.

structural diaphragms. The proposal was shared as the best possible solution in terms of safety and efficiency of operation while ensuring the presence of a transversal system in the tunnel.

The mandatory necessity to create a purely transverse system (not semi-transverse – only with exhaust duct) then guided the Competitor towards the definitive design solution (Figure 3) which provides for the expansion of the escape tunnel and the relative replacement with a real own air supply duct channel.

The creation of a single pair of ventilation ducts to service both tunnel tubes, installed on the middle position between two road carriageways, has allowed a significant reduction in the estimated costs of construction without introducing any defects in efficiency and safety, thus ensuring the awarding final contract.

The final tender solution provided for a total of:

- the reduction of excavation volumes of around 300 000 m^3
- raising the level of the road with a total reduction in the excavation depth of approx. 2.00 m
- the construction of only two ventilation ducts common to the both tunnel tubes, together with the elimination of the false ceilings inside the tunnel.

With regard to the removal of the false ceilings, the constructive optimizations introduced during the design phase determined in any case the need to keep the top slab at the same original level provided by concept design and to create an intermediate closing slab in some sections. thus the double function of limiting the height of the traffic compartment and containing the deformations of the diaphragms.

4   VENTILATION SYSTEM DESIGN

The new S2 Tunnel transverse ventilation system consists essentially of a pair of rectangular ducts (13–15 m^2 net section) for the supply and extraction of air from the tubes (whose cross-sections are now equal to 110 m^2). Ventilation airflows are powered by a couple of ventilation station installed above the track, on the tunnel slab, in the technological and dedicated areas near the portals.

The air is supplied and extracted from the tunnel by means of motorized steel fire dampers (600 °C – 2h resistance) installed inside with 50 m spacing. Exhaust dampers have a total unit area of 5 m² and are installed in the upper part of the central diaphragms, while the supply dampers with 1.5 m² of unit area are installed in the lower level of the diaphragms, just above the road platform (Figure 4).

The ventilation system has been sized to guarantee the correct dilution of the pollutants produced by the vehicular traffic present in the tunnel in all the operating scenarios, contemplated in the traffic forecasting study, and at the same time, guarantee the management of those critical scenarios that foresee, the presence of congested or blocked traffic. Not only that, the system must guarantee, as required by the ventilation systems, the control of heat and smoke produced by a hypothetical event of fire in the tunnel, ensuring the safety and exodus of the involved users. At the scope to ensure compliance with the performance requirements and to validate the proposed ventilation model, during the tender phase and building permit design, Computational Fluid Dynamics (CFD) and the Fire Safety Engineering approach was used by performing the related fire simulations in the tunnel. The simulations were conducted in the hypothesis that an outbreak of fire with total heat release of 100 MW is present along the tunnel route. Fire scenarios and CFD simulations were performed ensuring the analysis of the "worst" locations keeping into account, among the others, the road slope level. In particular, as shown in Figure 5, simulations related to both ascend and descend tunnel section were performed – Fire 1 and Fire 2 scenarios.

The use of 100 MW fire with total smoke production equal to 200 m³/s, is a very hard test for the verification of ventilation ducts, their sizing and relative spacing of the dampers as well as for the overall effectiveness of the ventilation station.

The innovative geometric and functional approach of the ventilation adopted for the tunnel suggested the completion of the study analysis extending the CFD approach, firstly required only for the analysis of the fire scenarios, also to the sanitary ventilation phases. So, new analysis were conducted in presence of critical operating scenarios with highly congested and/or jammed traffic in order to verify the effectiveness of pollutants dilution along the entire cross-section of the tunnel.

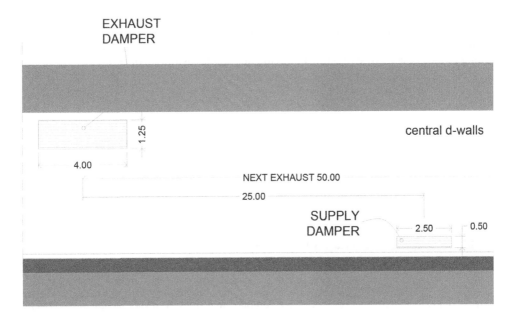

Figure 4.   Transversal tunnel section (view of the central d-walls).

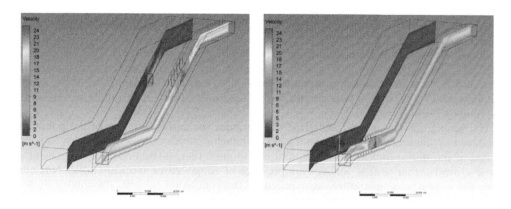

Figure 5. Fire 1 and Fire 2 velocity result (tunnel and exhaust duct).

5 CFD RESULTS

The results obtained by means of CFD analysis confirmed the correctness of the proposed solution both for tunnel sanitary and fire scenarios.

In particular, for the sanitary scenarios, the results have provided the expected pollutant dilutions confirming the effectiveness of the cleaning effect over the entire cross section of the tunnel. Specifically, as shown in Figures 6 and 7, the results the capacity of the system with overlying channels and vertical dampers to operate the full ventilation of the tunnel section from the central diaphragms up to the opposite side of the carriageway.

During fire events, for all the analyzed scenarios, the results showed the correctness of the proposed solution and allowed positive evaluations regarding the effectiveness of the ventilation. In particular, they highlighted the full capacity of the ventilation system to operate control of fire events with total heat release up to 100 MW and 200 m³/s smoke production, with particular regard to the ability to extract smoke from tunnel and maintain tenability criteria (temperature pollutants and visibility) along the walkways and the emergency paths. For this aim, in the next figure FIRE 1 simulation results related to time step 500 s from the fire start are fully shown.

In particular you can see how temperature and visibility contours highlight the free smoke area in the lower layer above the road level; velocity contours show how 5 active dampers are working in the fire zone by extracting and removing smoke by the carriageway.

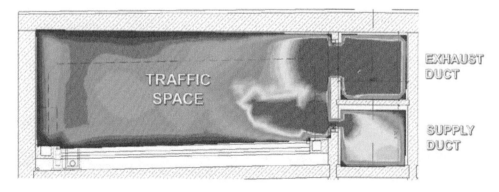

Figure 6. CFD Sanitary model result. Colored contours on a transversal slice section of air velocity - scale 0 (blue) – 5 m/s (red).

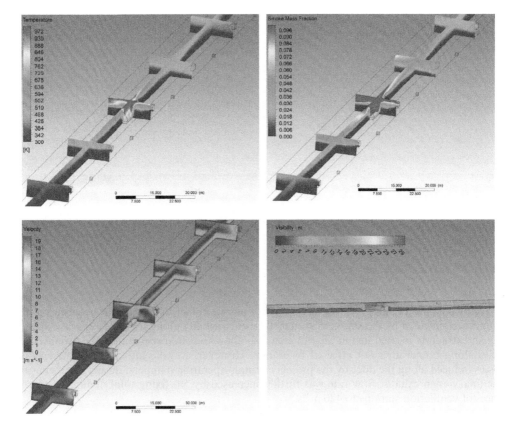

Figure 7. CFD fire model results. Colored contours of temperatures, smoke concentrations, velocity and visibility.

6 FINAL DESIGN

The fluid-dynamic analysis allowed also to determine the correct extraction capacity of the ventilation system and to make any necessary adjustments related to the presence of a vertical duct exhaust system. Furthermore comparative case studies are available in technical literature concerning positioning of ventilation dampers and louvers and they give some important advices about ventilation system behavior and their efficiency in presence of horizontal, vertical and inclined damper openings.

RABT 2006 and other technical guidelines suggest, for traditional transversal ventilation systems, to adopt an exhaust flow rate at least equal to one and a half times the smoke flow rate (reference fire scenario). This assumption should lead to choice a total extraction flow rate of 300 m³/s but considering lower efficiency of exhaust system due to the presence of vertical dampers (instead of horizontal), the extraction capacity of the system has been increased by more than 30% and set up to 400 m³/s.

Overall, tunnel transverse ventilation design includes:

– structural geometric configuration as proposed during the tender phase
– nominal exhaust flow rate equal to 400 m³/s (huge!)
– nominal air supply capacity equal to 100 m³/s
– transverse ventilation working along the entire length of the tunnel during sanitary ventilation scenarios
– transverse ventilation working just on the outbreak of fire tunnel area (sector operation) during fire ventilation scenarios.

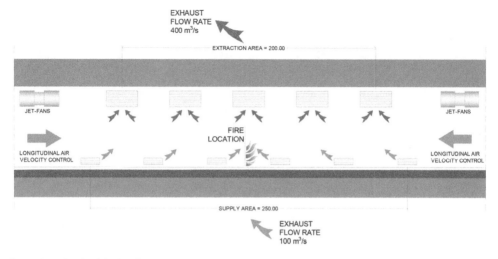

Figure 8. Nominal design flow rates.

Sector operation strategy during fire scenarios allows in fact to contain the extension of the tunnel area invaded by the smoke ensuring sufficient visibility along the exodus paths. Operating an extraction located just in the fire area also allows not to disperse the total exhaust capacity along tunnel area not affected by smoke. However, to take into account the leakage losses induced along the duct by the presence of non-operating dampers, during fire scenarios, the final design exhaust flow rate was further increased by 5% fixing total nominal performances of ventilation stations to 420 m³/s.

7 JET-FANS AND VELOCITY CONTROL

In order to maximize the effectiveness of the transverse ventilation system, a series of jet-fans have been provided in the tunnel; jet-fans task is to keep controller the longitudinal velocity of the residual ventilation air flow. Air velocity control is required for both sanitary and fire ventilation scenarios and it is necessary to balance the overpressure produced inside tunnel by piston effect and meteo-climatic conditions. During sanitary mode, longitudinal air velocity control prevents the pollutants propagation in the longitudinal direction by reducing the part not extracted from the ventilation system and the consequent emission to the environment (from the portals).

In the event of a fire, the longitudinal speed control ensures maximum efficiency on smoke extraction operation. Ventilation control must ensure, in the fire zone, the presence of a

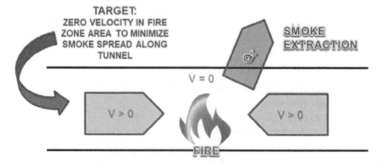

Figure 9. Flow rate balancing and Fire Ventilation strategy.

Figure 10.   Smoke stratification and lower free smoke layer.

(quite) zero longitudinal velocity obtained as a natural balancing of the involved ventilation flows. The sum of the ventilation flows upstream and downstream of the fire, controlled by the installed jet-fans, must be equal to the total exhaust flow rate fixed in the fire zone according to the following scheme.

Longitudinal velocity control during the first stages of a fire event also ensures that the natural smoke stratification induced by fire is maintained as long as possible in its upper layer. Smoke stratification (Figure 10) is a basic phenomenon that assumes a main role in fire protection and fire-fighting operations. The buoyancy thrusts induced by fire can generate a warm layer (upper layer) which is self-supporting in contact with the tunnel ceiling and allows the coexistence of a lower free smoke area (lower layer) that assumes a primary scope in the users exodus capability.

8   CONCLUSIONS

In the practical case described above, applicability and efficacy of a transverse ventilation system to a unidirectional road tunnel has been demonstrated and verified. The performance approach by means of Computational Fluid Dynamics (CFD) and Fire Safety Engineering allowed ventilation system design optimization confirming the design choices made during the first design phases.

Simulations showed that the new ventilation system proposed is adequate, in all the tested operating conditions, to ensure:

– sanitary air exchange during normal and peak traffic scenario
– smoke extraction and temperature control during fire scenarios.

REFERENCES

VV.AA. Road Safety in Tunnels. 1995. PIARC Committee on Road Tunnels, C5
VV.AA. Fire and Smoke Control of Tunnels. 1999. PIARC Committee on Road Tunnels, C5
RABT Guidelines for equipment and operation of road tunnels (Richtlinien für die Ausstattung und den Betrieb von Strassentunneln)
Journal of laws of the Republic of Poland, nr. 63. Regulation of the Minister of Infrastructures Poz. 63
Faggioni, N., Bizzi, F., Bringiotti, M. 2016. La Ventilazione Trasversale applicata alle gallerie artificiali a traffico monodirezionale: il caso della galleria S2 di Varsavia. Gallerie 119.
Devia, F., Faggioni, N., Franzese, E. 2015. Un esempio di approccio cfd nella verifica di sistemi di ventilazione complessi a servizio di gallerie stradali. Le Strade 1506 April 2015. La Fiaccola

Tunnels and Underground Cities: Engineering and Innovation meet Archaeology,
Architecture and Art, Volume 9: Safety in underground
construction – Peila, Viggiani & Celestino (Eds)
© 2020 Taylor & Francis Group, London, ISBN 978-0-367-46874-3

Fire ventilation systems in a metro station equipped platform screen doors

N. Faggioni
Imgeco Srl, Genoa, Italy

S. Franzoni
SIIP Srl, Naples, Italy

F. Devia
University of Genoa, Italy

ABSTRACT: Scope of this paper is to analyze a way to investigate an admissible solution for fire ventilation systems installed in a metro station equipped with Platform Screen Doors (PSD). Fire ventilation is related to fire scenarios involving generic events on the rolling stock present in the station. In order to evaluate capability and efficiency of desmoking systems, computational fluid dynamics analysis (CFD) have been developed, using Fire Dynamics Simulator software (FDS). Analysis have involved simulations in a model where different system configurations are applied. In all these models the railway platform area is delimited from the truck area by means of PSD system. System configurations assumed during the design provide for the possibility to operate ventilation from the train side using side and upper ventilation as well as platform level ventilation. Final ventilation strategies using under platform ventilation obtained excellent results in term of smoke and compliance of tenability criteria.

1 INTRODUCTION

The scope of this study is to show a way to investigate an admissible solution for ventilation system to be installed inside a metro station equipped with PSD in order to manage generic fire scenario involving events on the rolling stock. In the study fluid dynamics analysis (CFD) have been used in order to evaluate the efficiency of the proposed ventilation systems with particular reference to its behavior in fire regime.

  In particular the study consists in the development of a fire simulation in a typical model of Metro Station in which an Under Platform Extraction system (UPE) is adopted in order to manage the ventilation in case of fire. Preliminary analysis showed that the best results in terms of ventilation efficiency are guaranteed by a UPE system compared to other ventilation systems (such as Over Truck Extraction or Tunnel Ventilation System) and in the following we want to highlight the good results obtained with the simulations concerning to the present case.

2 SUBJECT

2.1 *Models and Fire Scenarios*

The fire scenarios, covered by the simulations, refer to a generic event of fire on the rolling stock present in the station. It wants, in fact, to determine the behavior of ventilation systems during an emergency scenario involving this convoy in the station platform.

The railway platform area is delimited from the rail area by means of a system of platform screen doors (hereinafter PSD). The system configurations assumed during the design provide for the possibility to operate a ventilation from the under platform level as shown in the following figure.

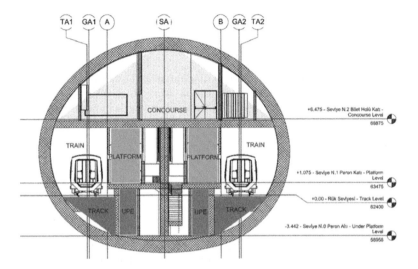

Figure 1. Typical cross section with main ventilation elements.

In particular, there is a train side ventilation system (obtained underneath the seat of the platform tunnel and station), defined UPE (Under Platform Extraction), that provide to extract air or smoke from train areas.

Ventilation strategies adopted in case of fire on train inside station area are schematically described in next figure where main fluxes of UPE system is highlined. Blue dot lines and red arrows show fully airflow path from extraction point to external environment. UPE system provide exhaust ventilation at both ends of the station, in one track side, using ventilation shaft and UPE Duct installed under platform. Extraction at both ends is actuated using two vertical smoke damper installed on train side.

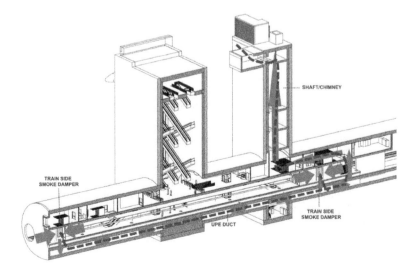

Figure 2. Typical cross section with main ventilation elements.

2.2 *CFD analysis method and objectives*

As mentioned in the previous paragraphs, CFD approach was used in this simulation to evaluate efficiency of smoke management system that are designed for (Ref. [1]):

– Allowing fire department personnel sufficient visibility to approach, locate, and extinguish a fire
– Limiting the rise of the smoke layer temperature and toxic gas concentration and limiting the reduction of visibility
– Maintaining of tenability for people involved.

Simulation results allow to understand, in a qualitative and quantitative way, the behaviour of the ventilation systems during an event of fire in accordance with the ventilation strategy chosen. Factors that should be considered in a tenability analysis include the following:

– Heat exposure
– Smoke toxicity
– Visibility.

3 DESIGN FIRE SIZE

3.1 *HRR*

Fire scenarios are characterized by a designed fire scenario. The design of fires is most easily and commonly expressed in terms of a Heat Release Rate (HRR) versus time curve for the progress of the fire. A HRR curve is a simple approximation of real fire behavior and is generally represented, in CFD models, with an unsteady fire with a heat release rate that varies with time.

The growth phase of the fire shall be described using t-squared fire growth model with $Q = \alpha t^2$ where α is the fire intensity coefficient; α value can be considered as 0.012 kW/s^2 as well as for Medium growth up fires.

The radiative portion of the heat release rate of the fire is in general determined from equation $Q/Q_r = \xi$, where Q is the total heat release rate of the fire, Q_r is its radiative portion (kW) and ξ is its radiative fraction (dimensionless). In practice a value of 0.35 should be used for the radiative fraction and it has been used in these simulations. Given the above, we considered in our CFD simulations, following parameters for design the train fire size:

– peak HRR = 12 MW
– peak specific heat flux = 240 kW/m^2 (train geometry)
– peak Time = 1000 s
– α growth rate = 0.012 kW/s^2
– ξ radiative fraction = 35%.

3.2 *Fuel and Exhaust*

The combustible matter inside the carriage mainly consisted of floor, walls, ceiling, seats and luggage Ref. [5]. The metro carriage is made of steel. Finally there is a thin steel plate with a insulation board behind the existing interior combustible materials.

It is possible to reduce the complexity of simulating the behavior of such a large number of materials making, in favour of safety, the hypothesis that any combustible matter has a polymeric like behavior. In particular the polyurethane is a plastic material used in the inner linings of motor vehicles and in the form of foam also for the upholstery of chairs and generic furnishings and contributes to a considerable production of smoke combined with an equally important production of carbon monoxide. Valid references are available in literature for the soot and CO yields, produced by fuel during fire reaction, estimated according to experimental results. Both soot and CO yields vary significantly during the tests but can be approximated, to the purpose of this analysis, to values represented below:

– CO Yield = 0.10 kg/kg$_{fuel}$
– Soot Yield = 0.07 kg/kg$_{fuel}$

4 CFD CALCULATION DOMAIN

4.1 *Geometrical and Physical model*

In the next figures are reported the sketches of some elements that have been modelled as essential part of the train fire simulations. The main components of the calculation domain are reported in the following list:

– tunnel an station geometry (platform level)
– train geometry
– second platform geometry
– elevator, escalators and stair in concourse areas (and their open connections with upper level)
– fire geometry and conditions
– open condition at tunnel boundaries
– open condition at PSD boundaries (second platform)
– smoke ventilation design parameters.

In the geometrical and physical model, following parameters are been considered and setted in the processor code.

– Height of ceiling inside tunnel and station zones
– Size and location of fire in plane underneath the train with surface = 50 m^2
– Open vents at tunnel boundaries
– Open vents at concourse interface (stairs, escalators, ...)
– Open vents at PSD interface in the second train platform (to consider failure)

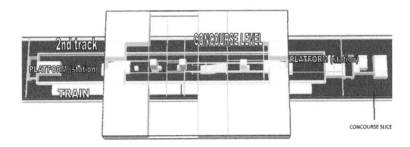

Figure 3. FDS 3D Model Calculation Domain. Highlighted the main horizontal slice view used for result plotting at platform level.

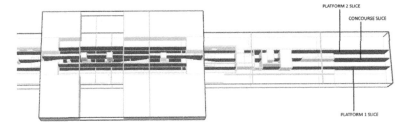

Figure 4. FDS 3D Model Calculation Domain. Highlighted the main vertical slice view used for result plotting at platform and concourse level.

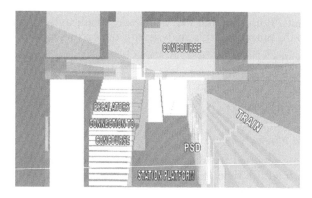

Figure 5. FDS 3D Model Calculation Domain. Inside the station, a view from platform level.

- Inert condition for model wall and ceiling
- Initial temperature of 20° C
- Initial tunnel air velocity = 1.5 m/s (train running direction).

The inclusion of platform edge screens is a design option that is effective for comfort control in stations as well as for smoke control in tunnels. For that reason, PSD system at both platform has been included in model geometry to evaluate ventilation system efficiency. At simulation time = 0 s (initial state), train is stopped at station platform; its doors and station PSD are opened.

Summarizing in table below, main geometrical characteristics of domain has been shown.

Table 1. CFD Domain Characteristics.

Domain Characteristics*	Tunnel	Station	Other platform	Concourse area
Geometry	5.2 x 5.8 x 160 m	9.8 x 5.8 x 130 m	5.2 x 5.8 x 160 m	40 x 5.4 x 20 m
Volumes	4 800 m^3	7 400 m^3	4 800 m^3	4 300 m^3
Medium cell dimension	20 cm	20 cm	20 cm	20 cm
Medium cell volume	0.008 m^3	0.008 m^3	0.008 m^3	0.008 m^3

* Total number of cells approx. 5 millions.

4.2 Ventilation Systems and Equipment for smoke control

The emergency ventilation system is designed (according to Ref. [2]) to accomplish to the following tasks:

- Provide a tenable environment along the path of egress from a fire incident in enclosed stations and enclosed railway
- Be capable of reaching full operational mode within 120 seconds.

The main parameters which has a strong impact on the design of the ventilation system are the following ones:

- The heat release rate produced by the combustible load of a vehicle and any combustible materials that could contribute to the fire load at the incident site
- The fire growth rate
- Station and railway geometries
- A system of fans, shafts, and devices for directing airflow in stations and railway.

For this purpose, an under-platform (UPE) ventilation system has been considered, in the model, for the extraction of heat from the traction and braking devices and for de-smoking.

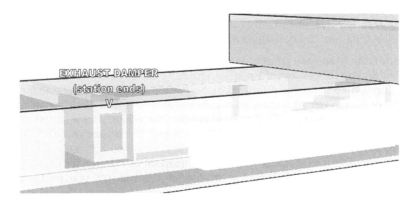

Figure 6. Vertical exhaust damper at station ends.

Table 2. Ventilation system characteristics.

Ventilation system	Tunnel
Nr. of exhaust dampers	2
Dampers Geometry	3 000 x 2 000 mm
Distance between dampers	105 m (approx.)
Total vent area	11.2 m^2
Nominal flow rate (total)	360 000 m^3/h

UPE system is located below the platform level. UPE System is designed considering a concentrated extraction system with vertical exhaust dampers installed at both ends of the metro stations as shown in the next figure.

A fixed delay of the activation for the ventilation system has been considered in every scenario. The selected delay time is to 60 s from outbreak of fire. Delay Time is the minimum time required for the detection of fire, coherent with the HRR (t^2 growth). Time response of the ventilation systems is modelled with a 30 s ramp which is the time needed for the full development flow regime; thus nominal flow rate value is achieved 90 s from the beginning of the simulation.

5 TENABILITY CRITERIA

The effects of a fire are crucial to the assessment and management of evacuation procedures. The flames and even more the fire effluents can seriously affect people health during evacuation and they can cause a reduction of escape capability; it can upraise up to a total incapacity of movements that eventually takes to fatalities.

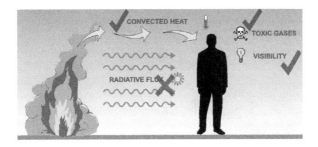

Figure 7. Adopted design tenability criteria.

According with Ref. [2], main factors that affect the tenability of a fire environment, for periods of short duration, can be defined as follows:

– Air temperatures: Combined effect of intensity and duration of exposition to a thermal stress are combined using the concept of Fractional Effective Dose (FED). Maximum allowed temperature during evacuation was estimated by NFPA 130 using FED criteria that consists in a calculation of the time to incapacitation under conditions of exposure. Limit value for incapacitation are fixed (in NFPA) equal to 80°C per first few minute (3.8 minutes) and 60°C for first 10 minutes during fire. This value (60°C) is adopted in our scenario to evaluate human capacity to perform the evacuation during the entire duration of the simulation (1200 s)
– Air carbon monoxide (CO) content: as well as for temperature maximum values for CO exposition during fire is fixed to of 2000 ppm in first few minutes and 1000 ppm for the first 10 minutes. For these reasons 1000 ppm value is fixed for maximum reference value during evacuation
– Smoke obscuration levels: this parameter must be continuously maintained below the point at which doors and walls are discernible at 10 m.

Other main tenability criteria regard Heat exposure due to radiation but this effect is ignored in these set of simulations because of two distinct reasons. Direct radiation from fire is more relevant in the very proximity of fire in case negligible shields, screens (i.e. PSD) or shadowing effects, which are relevant in this case. Radiation from smoke layers becomes relevant for long period of exposure, longer than design evacuation time.

6 FINAL RESULTS

To give full understanding of the results some graphical representations of the main variables examined within the computational domain have been reported, over time, in the following figures. The representations are provided in the form of section planes (SLICE plane) longitudinal and transversal with respect to the domain, in which the reproduction of the results (scalar values or vectors) is expressed in the form of color gradation. Graph trend of most important variables along platform are also represented.

In particular, the analysis of temperature and visibility on the median planes station allows, qualitatively, demonstrate fully compliance of fire ventilation strategy with design tenability criteria. The results are useful to determine in graphical form and among others, the thermal stress produced by the fire along the walkways of the platform and the visibility along the said walkways.

6.1 *Qualitative Results – Slices colored contours*

In order to better clarify the full compliance of the tenability criteria during fire scenario guaranteed by the adopted ventilation strategy, in the next figures final results relative to 600 s time simulation are shown.

Figure 8. Spread of smoke at platform level (perspective view).

Figure 9. Visibility contours along platform areas in the horizontal slice plane at 2.00 m height - scaled from 0 m of visibility (blue) to 10 m and above (red).

Figure 10. Temperature contours along platform and concourse areas in the vertical slice plane - scaled from 20 °C (blue) to 60 °C and above (red).

Figure 11. Temperature contours along platform areas in the horizontal slice plane at 2.00 m height scaled from 20 °C (blue) to 60 °C and above (red).

6.2 Quantitative Results – Graphical trends

The graphs depicted in the next report the longitudinal profiles of the main quantities and variables along tunnel, at the station platform and in the escape walkways. Values of interest are the ones highlighted in the following image.

In particular in the next graphs you can analyze the values, which have been registered, during a fire scenario simulation. The most important two values are the maximum temperatures and the toxic gases concentrations along station platform and its walkways. Graphlines confirm the qualitative results described in the previous pictures and shows that maximum values assumed in order to maintain tenability criteria inside station area were not exceeded during first 600 s of simulation. Thus results are expressed as 3d plot with trend of temperatures and pollutants along the platform over time (0–1200s). Peak temperatures along the platform exceed limits values starting from 800 s since outbreak of fire. In the last graphs

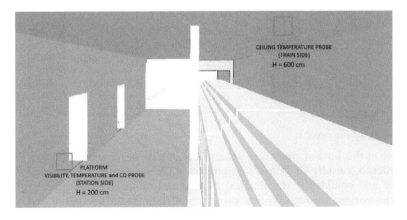

Figure 12. Thermocouple and probe position inside domain.

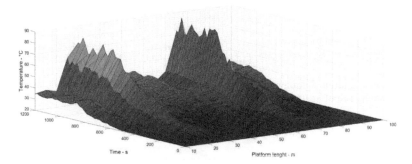

Figure 13.   Temperature trend above station platform (h = 2.00 m).

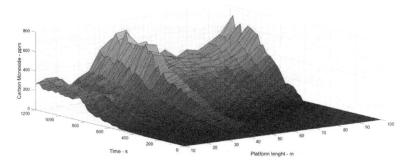

Figure 14.   Monoxide carbon concentrations trend above station platform (h = 2.00 m).

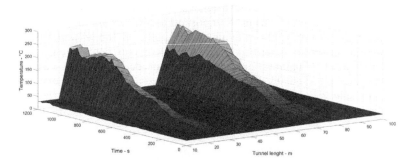

Figure 15.   Temperature trend along tunnel.

temperatures in the tunnel area are shown. The UPE effect underline 2 temperature peaks close to the extraction area.

6.3   *Confidence of the results*

The objective of this kind of study is to achieve a high level of reliability and confidence in the results in order to consider CFD analysis as a part of the design of the ventilation system.

Credibility and confidence are obtained by demonstrating acceptable levels of error and uncertainty as assessed through verification and validation. Determining this precisely is often difficult, but in general CFD simulations are however affected by intrinsic error, produced by as example:

- Errors exist because continuum flow equations and physical models represented in a discrete domain of space (grid) and time but can be reduced by user analysis and model settings
- Level of error is function of interactions between the solution and the grid and boundary conditions that can change during real phenomena.

Any user of the numerical model must be aware of the assumptions and approximations being employed. In fact, the growth of the fire is very sensitive to the thermal properties of the surrounding materials for both real and simulated fires. Moreover, even if all the material properties are known, the physical phenomena may not be simulated due to limitations in the model algorithms; for all these reasons, an uncertainty and intrinsic error can be introduced during model settings and processing and this can be transferred in the obtained results. Given the above, it can be said that using modelling techniques and procedures described inside FDS software Verification and Validation Guides, that uncertainty can be reduced and considered reasonable given the scope of analysis.

7 CONCLUSIONS

As a general conclusion, this paper describes the activities aimed at setting up a reliable Model and Fire scenario for the simulation of an event of fire and to the subsequent tenability of environmental condition for people involved. A special attention has been payed to model the details of the model boundary condition, of fire size and its Heat Release curve and of the tenability reference conditions.

The insertion of the selected ventilation system in this model has demonstrated, throughout the final simulation, that it can be effective in the management of smoke produced by the fire.

In conclusion, the qualitative and quantitative results demonstrate fully capability of the investigated system to fulfill tenability criteria, in the target area, for the entire duration of the analysis.

REFERENCES

Standard for Smoke Control Systems. 2018. National Fire Protection Association nr. 92
Standard for Fixed Guideway Transit and Passenger Rail Systems. 2017. National Fire Protection Association nr. 130
Ingason, H. Design fire curves in tunnels. Fire Safety Journal 44.
VV.AA. 2017. SFPE Handbook of Fire Protection Engineering. Morgan Hurley Springer SFPE
Li, Y. Z. 2015. CFD modelling of fire development in metro carriages under different ventilation conditions. SP Technical Research Institute of Sweden.
Lonnermark, A. Lindstrom, J. Li, Y. Z. Claesson, A. Kumm, M. Ingason, H. 2012. Full-scale fire tests with a commuter train in a tunnel. SP Technical Research Institute of Sweden.
Robbins, A.P. Wade, C.A. 2008. Soot Yeld Values for Modelling purpose - Residential occupancies. Branz Study Report 185
McGrattan, K. Hostikka, S. McDermott, R. Floyd, J. Weinschenk, C. Overholt, K. 2013 Fire Dynamics Simulator Technical Reference Guide. Volume 2: Verification. NIST Special Publication 1018–2 Sixth Edition.
McGrattan, K. Hostikka, S. McDermott, R. Floyd, J. Weinschenk, C. Overholt, K. 2013 Fire Dynamics Simulator Technical Reference Guide. Volume 2: Validation. NIST Special Publication 1018–2 Sixth Edition.

Tunnels and Underground Cities: Engineering and Innovation meet Archaeology, Architecture and Art, Volume 9: Safety in underground construction – Peila, Viggiani & Celestino (Eds)
© 2020 Taylor & Francis Group, London, ISBN 978-0-367-46874-3

Fire safety system design in double bore railway tunnels: Bologna city undercrossing for high speed line and recent experiences

A. Falaschi, S. Miceli & A. Pigorini
Technical Department, Italferr SpA,, Rome, Italy

ABSTRACT: The double bore tunnel for under passing Bologna has connecting tunnels (bypass) at most every 250 m. In order to consider one tunnel a safe place compared to the other, safety system has been studied to prevent in fire scenarios the smoke passing from one tunnel to the other. The system has the following subsystems: bidirectional, auto-closing, fire and pressure resistant doors, system for door position detection, pressurization system (fans), real time supervision/control. The main design issue described in the paper are the design of bidirectional fire and pressure resistant doors, the pressurization system in order to guarantee an adequate overpressure or air flow in the bypass depending on the position of the doors (closed or open) by setting the operating rotation speed of fans. The system has been developed and built for the first time in the Bologna twin tunnels (2012) and nowadays is the basis for new application.

1 INTRODUCTION

The fire strategies, which are considered for the Italian railway network for tunnels, don't provide for the use of longitudinal fan systems to full fil the PUSH – PULL ventilation strategy. In other words, instead of keeping the smoke moving to its critical velocity (which is defined as the velocity that prevent the Back-Layering phenomenon) its stratification is preferred.

For long tunnels, according to the Italian Low (DM Sicurezza nelle Gallerie ferroviarie – 28/5/2005) and the international law (STI SRT 2008 and 2014) to avoid the construction of lateral access tunnels it can be designed a double bore configuration with connecting tunnels at least every 500m.

Therefore the Italian fire strategy for long tunnels is to consider a smoke disconnection system in the bypass connection tunnels.

A smoke disconnection system is also foreseen when passing from a single bore tunnel to a double bore tunnel.

In this work the experiences coming from the undercrossing of the city of Bologna and from the equipment provided for the bypass will be described.

2 FIRST APPLICATION: UNDERCROSSING OF THE CITY OF BOLOGNA

For the construction of the undercrossing of the city of Bologna and of the new High Speed station, which was completed in 2012, the project of a long tunnel (more than 10 km) has been developed. The tunnel was bored with different construction lots, in two different configurations: double bore single trach tunnel and single bore double track tunnel (Figure 1).

Also different excavation techniques have been used: mechanized excavation using TBM-EPB (Tunnel Boring Machine - Earth Pressure Balance) for double bore tunnel, conventional excavation for double track tunnel and cut and cover tunnel for double track tunnel with shallow overburden.

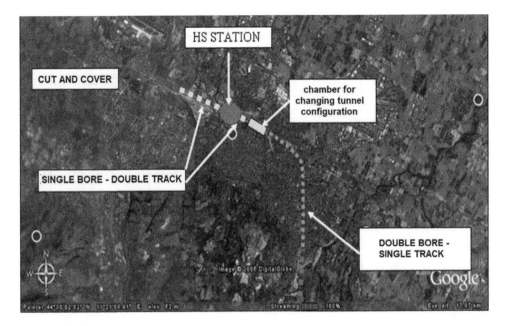

Figure 1. Underpass of Bologna, different tunnel configuration.

Tunnel safety systems have been built with a specific technological and equipment contract after the end of the civil works.

The part of the tunnel built in the southern section consisted of two single-track bore tunnels 6.0 km long, in which connecting tunnels have been provided every 250 m. In this way, in the event of fire, passengers egress will be towards the other bore, deemed to be a "safe place".

The two tunnels are quite close to each other in fact the distance between the two bores is about 15m from axis to axis (Figure 2). This choice has been made to place the two tunnels just beneath the existing railway line - placed in embankment – so to reduce the settlement effects comparing to the one occurring beneath the city buildings (Pigorini et al. 2007).

The safety study defined the mechanical equipment required for the bypass for preventing smoke passing from the on fire tunnel to the "safe bore" (not on fire) during egress. Namely, to guarantee the above mentioned safety criteria, it was decided to apply a configuration featuring with the following subsystems:

- passive protection system (fire doors size 1.2 m x 2.1 m);
- fan system capable of providing an overpressure value of at least 30 Pa with closed doors and an at least 2 m/s air flow from doors during egress;
- opening doors detection system;
- real-time supervision system capable of managing and commanding the systems, as well as of detecting their functional status, even remotely.

In the following paragraphs the subsystems performance and the design choices will be described.

3 DESIGN

3.1 Passive protection system

The function of the fire doors envisaged by the design is to protect the bypass and they must be able to be opened in both directions (Figure 2).

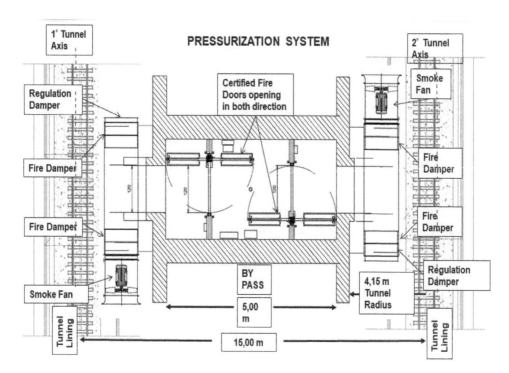

Figure 2. Scheme of bypass.

The previous experience gained with the Firenze-Bologna high speed railway tunnels, showed that it was not possible to use normal fire doors because the passage of trains generates excess pressure that hits the door causing mechanical damage and reduced functionality. The excess pressure induced by the high speed train travelling in tunnel has not been investigated until safety doors for passenger egress have been introduced in tunnel. Starting from that moment the problem showed all its seriousness.

In addition to these conditions, there was the need to guarantee the egress in both directions, namely from the on track bore to the other track bore or vice versa, according to fire scenarios.

3.2 Bypass pressurization system

In fire scenarios, bypass pressurization is necessary to prevent smoke from invading the way-out towards the safe tunnel. The project envisaged a system consisting of one axial fan placed at each side of the bypass, ducts, dampers and central switchboard set up to control the system as well as to set fan speed in function of door status. As per design, the system had to guarantee an overpressure value of 30 Pa with closed doors and at least 2 m/s air flow exiting from the doors during egress. The system, which can be activated via the supervision system, is made even more flexible by means of the installation of regulation dampers.

During construction, the main issues encountered were:

- the construction of the systems in very confined spaces;
- the manufacturing and certification of fire-rated double direction opening doors;
- the setting and testing of a complex and strongly interrelated system.

3.2.1 The construction of the systems in very confined spaces

The bypass have been built during the excavation of the TBM tunnels due to a specific request of the fire brigade so to allow the passage from one tunnel to the other during the construction, about 500 m form tunnel face. The main issues faced during bypass construction have been:

- ensuring the excavation stability namely in loose water bearing alluvial deposit (sand and silty sand) and minimizing settlements;
- ensuring the stability of the tunnel pre-casted lining, as the bypass opening can cause a reduction of structural resistance in the tunnel lining.

The first issue has been solved by using grouting and waterproofing injections. These injections were carried out on the bypass contour, with injection drillings driven from one tunnel towards the other one.

The second issue was solved by placing a special lining segment (Figure 3) composed of:

- a metallic frame around the opening;
- temporary beams used as structural reinforcement during the TBM drive;
- bolted sheets steel to simplify the removal.

However, due to the 1.5m length of the pre-casted lining, the width of the bypass access was limited to 1.2 m.

Besides the difficulties posed by the limited dimension of the by-pass excavation (Figure 4), other system-related issues concerning the definition of the mechanical equipment for the bypass have been faced. Because of the confined spaces, the installation of the pressurization system (fan and switchboards) has been placed externally to the bypass in the main tunnels (Figure 5). It was therefore necessary to protect fans and ducts from fire by using fire-resistant panels and relating support structures.

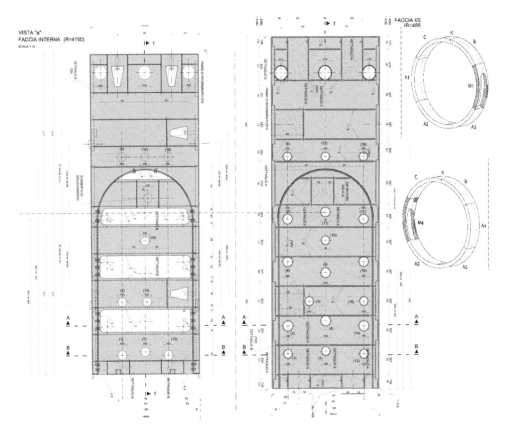

Figure 3. Special metallic segment used to speed up the demolition phase for by-pass opening.

Figure 4.   Bypass during excavation and in the final configuration.

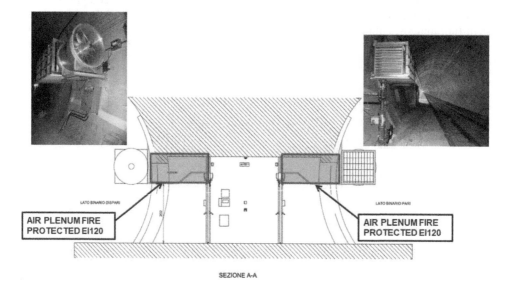

AIR PLENUM FIRE
PROTECTED EI120

AIR PLENUM FIRE
PROTECTED EI120

SEZIONE A-A

Figure 5.   Example of bypass equipment: configuration of doors and layout of mechanical units.

3.2.2 *The manufacturing and certification process of fire-rated double direction opening doors*

To guarantee the bypass resistance to fire, the performance required to the doors was the following:

- fire rated doors EI120 as per regulation EN 13501;
- suitable for being installed in an aggressive environment such as a tunnel;
- capable of withstanding the stresses induced by the pressure of train passages;
- capable of being opened in both directions of egress without the need of excessive opening force.

During the design phase, such doors were not available and no manufacturer was able to guarantee and to certify the above performance characteristics all together. It has been necessary to define the test guideline to test the doors in certified laboratory, to have the certification to fire, overpressure and fatigue resistances.

To obtain the certification, a test protocol was defined.

Namely in addition to withstand the pressure peak, fatigue parameters have been established consisting in repeating 500,000 times (value based on train passage frequency) the application of a ±1.5kPa pressure (defined according to the maximum train speedy of 160 km/h). To reach this performance level, the initial door was totally redesigned, with the implementation of technical solution capable of increasing the fatigue resistance of the door. The new door configuration obtained the certification for fire, mechanical and fatigue resistance. It achieved the EI120 fire resistance rating certification.

3.2.3 *The setting and testing of a complex and strongly interrelated system.*

Testing was conducted in various phases, featuring the aspects of interfacing between door opening/closing and air flow regulation. Namely the setting of the system had to guarantee:

- 30 Pa with closed doors;
- an air flow velocity through the open door of 2m/s;

It was therefore necessary to set up fans rotation speeds according to the various configurations (closed doors, one open door, two open doors). During the system setting, two other parameters have been taken into consideration:

- it shall be possible to open the doors without excessive effort (220N maximum);
- the doors shall not close too quickly to avoid hitting people during the exodus and to allow the correct closing of the door itself (avoiding the opening in the opposite direction given the lack of the frame).

4   RECENT APPLICATIONS

The experience coming from the Bologna undercrossing tunnel has become the standard for the new projects.

Compared to the Bologna bypass, given the new regulations, the new standard for by-pass shorter than 10m provides that each side of the bypass has to have 2 doors (0.9 m x 2 m) with a filter zone in the middle The doors will be bidirectional, EI120 certified and will be resistant

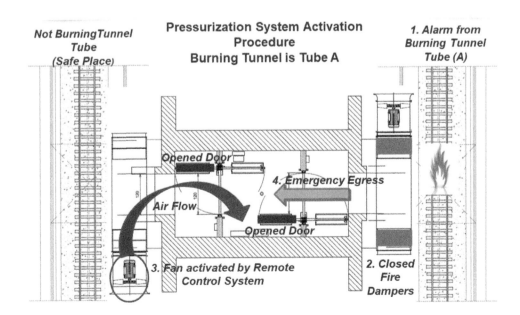

Figure 6.   Pressurization system activation procedure. Fire on right tunnel.

to overpressure fatigue. The overpressure values are defined by STI INFRA 2014 in ± 5.5 kPa. To date, many manufacturers can produce doors with these characteristics.

The new standards consider also the case of by-pass length more than 10 m: in this situation two filter areas separated by the central intermediate area, are foreseen (Figures 7–8).

System regulation in function of different operational scenarios (open doors, closed doors, several open doors) it is provided by fan speed regulation inverters and by the dampers for venting air overpressure in the operational scenarios:

- with all open doors, the air flow speed must be at least 2 m/s through each doors (overpressure damper closed);
- with all filter area closed doors, the system's function shall guarantee internal overpressure of at least 50 Pa (overpressure damper open);
- in the intermediate operational scenario, featuring the opening of only several of the 4 doors of the filter area, the overpressure damper shall switch to the closed position and the fan to an intermediate speed setting.

The detection of the various operational scenarios will be possible via the use of pressure sensors and of micro-switches on the doors to indicate the number of open doors. The data coming from these sensors will be sent to the management and control switchboard.

This approach has become the standard that has been recently applied to all the projects under development such as the long tunnels under Apennines mountain chain foreseen in the Naples - Bari new high speed railway line. The Figure 9, shows a schematic of the egress through the bypass in case of fire in one of the 2 tunnels of the double bore configuration.

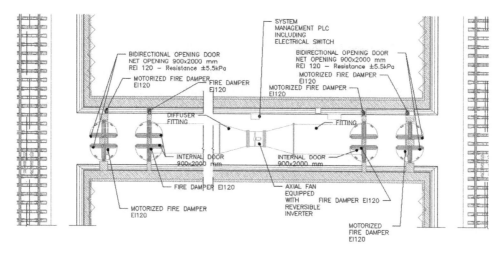

Figure 7.   Layout of new standard bypass for length > 10m.

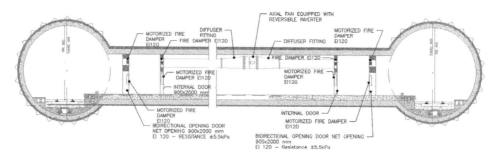

Figure 8.   Section of new standard bypass for length > 10m.

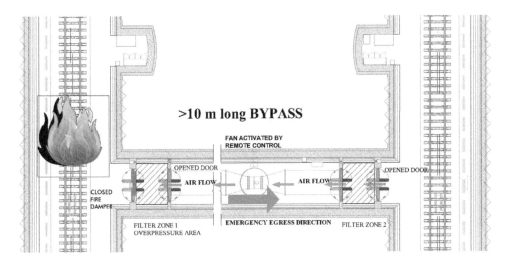

Figure 9.   Functional scheme of a bypass of length > 10 m.

5   BIM DESIGN APPROACH

The new **BIM** design methodology (Building Information Modeling) applied to infrastructures and namely to the equipment design in tunnel by-pass, permitted to optimize interface between civil structures and safety equipment, therefore optimizing the excavation volume, the by-pass shape and reducing interferences between equipment and civil structures (Figure 10).

The tridimensional by-pass model (Figure 11) can also be used to simulate a fire event in one tunnel and to verify the effectiveness of pressurization system. Furthermore, during design phase, it's possible to verify fan technical features for each functional condition related to the number of opening doors.

In addition the **BIM** methodology major advantages can be observed during building, management and maintenance phases during operation.

Once the major interferences have been found and solved during the design phase, it's possible to reduce delay caused by design errors and variations. In addition, this methodology

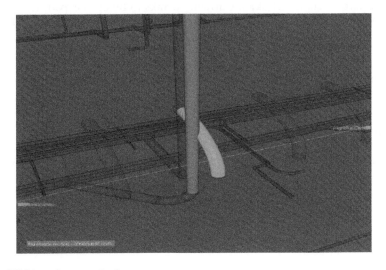

Figure 10.   BIM interferences check.

Figure 11. By-Pass BIM Model.

simplifies during the construction phase the monitoring of the work progress and the and the As Built BIM model.

The control and management during the operation of the systems can be improved by using as-built model realized with BIM methodology. The BIM model can be used as graphic interface to manage alarms and signals coming from sensors and safety components inside the system. Moreover, it's possible for each system to manage alarm inputs due to scheduled operations or out of order machines.

Using advanced components to control and to check the equipment it's possible to run BIM model to make the predictive maintenance and prevent break down of the safety equipment.

6 CONCLUSIONS

Starting from the first approach of designing and commissioning of the safety systems for the bypass in Bologna under-passing, progress has been made especially for doors resistance to fire and fatigue, considering that there were no testing protocol and the resistance was limited to an overpressure resistance ±1,5kPa. Currently thanks to Bologna experience and thanks to feedback from that installation, the door overpressure resistance can fulfill STI standards being able to resist at least at an overpressure of ±5,5 kPa.

For connecting tunnels longer than 10m a double filter zone is foreseen by the new standard.

Full integration with BIM model allow a predictive maintenance management with full advantages in the availability of the all system.

REFERENCE

Pigorini, A. Iannotta, F. Martelli, F. Pedone, E.M. & Sciotti, A. 2007. Il nodo ferrovario AV/AC di Bologna. Scelte progettuali: le gallerie con scavo meccanizzato (Lotto 5). *Gallerie e grandi opere in sotterraneo n° 81 – marzo 2007*. Milano: Società Italiana Gallerie.

*Tunnels and Underground Cities: Engineering and Innovation meet Archaeology,
Architecture and Art, Volume 9: Safety in underground
construction – Peila, Viggiani & Celestino (Eds)
© 2020 Taylor & Francis Group, London, ISBN 978-0-367-46874-3*

New generation of safety solutions for underground construction

D. Faralli, G. Fioravanti, F. Bonifacio & C. Salvador
Advanced Microwave Engineering Srl., Florence, Italy

ABSTRACT: EGOpro Safe Tunnel is a complete, integrated and reliable solution, that allows various modules to be managed such as access control and tracking operators, managing communication in tunnels, closed circuit video camera systems, environmental sensing and managing emergencies and alarms. The solution is based on identifying and controlling the presence of personnel and detecting and storing the number and position of operators, in real time, inside the tunnel. Everything is done automatically for more efficient and more timely management of emergency procedures and evacuation procedures. One distinctive element of the EGOpro access points, in addition to the ability to activate them at great distances, is the multi-reading of operators. The system, running on web-based software, provides the control room with an overview of the situation and displays, in real time, the position and the identity of each operator, as well as monitoring the status of any alarms.

1 INTRODUCTION

Major underground infrastructures are strategically important in the world of national and international logistics. By streamlining the road network systems and their environmental impact, they offer a new perspective and a new horizon for both the transport of goods and passengers.

Over the recent years, like other countries, Italy has been promoting the introduction of major infrastructures in order to improve and optimize the traffic flows. In this context, the construction of tunnels is fundamental: for example the Brenner's underpass will be the world's longest underground railway link connecting Italy to Austria with a total of ca. 230 km of tunnels.

Ensuring safety conditions during the tunnel construction phases is a challenging test for this type of major infrastructure work. Safety in an underground working environment is a complex issue, the personnel involved is exposed to particularly high risks combined into different types: the risks occurring at underground construction site in general and the risks of working underground, i.e. all those risks deriving from the excavation systems adopted (explosives, TBM, traditional excavations, etc.).

Therefore, careful planning of a construction site's safety strategy is clearly a very important phase in the execution process. This is even more crucial as far as major infrastructures are concerned, as they require careful coordination among the operating companies. Underground construction site are constantly evolving during the construction of tunnels. A constant and real time risk-analysis and assessment is essential, especially in relation to any possible scenarios concerning accidents or occupational diseases.

As for building site safety, the objective is to have a constant, real time overview of what is happening on the construction site. This means to have the possibility of detecting the presence and the position of operators inside a tunnel giving them the option of being able to communicate outside the tunnel and vice versa at any event, to allow faster intervention in case of emergency and, in the worst cases, to evacuate the site as quickly as possible.

2 TECHNOLOGY

During the past years, wireless and RFID (Radio Frequency Identification) technologies have been expanding and have seen a growing interest. Their application is evolving in every field, especially as far as safety is concerned. Several Italian and foreign companies currently communicate and exchange data using this type of infrastructure as a basis. The system's flexibility and invisibility make it a successful technology and undoubtedly competitive in the market.

The LNXessence technology is a property of Advanced Microwave Engineering for the automatic identification of objects and people and it is based on the capability to store, access and exchange data. LNXessence is an active transponder RFID technology that, in addition to simple automation operations, allows to exchange and manage data and information. LNXessence is the operating principle that stands at the base of the EGOpro Safe tunnel solution.

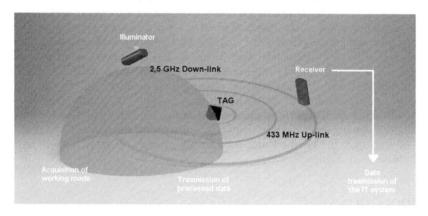

Figure 1.   LNXessence technology.

LNXessence is an identification system composed of three devices:

– ILLUMINATOR: a microwave interrogation unit;
– TRANSPONDER (Tag): a double frequency transponder;
– RECEIVER: a radio frequency reception unit.

Our technology's operating principle is intuitive: Tag lies in a state of quiescence until it enters the activation area, i.e. the area covered by the illuminator, and is "awakened". Once activated, it interprets the signal transmitted by the illuminator, it performs the necessary operations and transmits its own code and results to the receiver. The system allows to exchange data and information among these three devices.

3 THE CONCEPT OF ACTIVE SAFETY AND RISK OBJECTIVATION

The new Italian regulatory guidelines suggest and increasingly reward the introduction of innovative approaches that imply the implementation of new technologies concerning risk reduction in the working areas. To date, the safety approach in workplaces is based on the static assessment of latent hazards in order to produce a risk assessment document (Documento di Valutazione di Rischio, DVR). On these grounds, preventive or protective measures (Operational Safety Plan - POS) to minimize the probability of injury are introduced. This process is known as "passive" approach. However, in order to grant full effectiveness, an advanced system of prevention and protection against accident risks must grant an evaluation of the risk function with a dynamic and objective approach in real time. Then, based on this assessment, it must trigger any appropriate intervention to solve the condition of danger (minimization of the risk function through an "active" approach).

$$R = f(D, P) \tag{1}$$

Determining the risk function f means defining a model for the exposure of the personnel to a certain danger, putting in relation the extent of the expected damage with the probability that such damage will actually occur.

This happens under any operating condition:

R is the risk, i.e. it represents the consequences of the damage;

P is the probability that the event and the consequences will occur;

D is the value that the element exposed to danger gets in terms of human lives involved, or in damage caused.

In every environment, it is essential to analyze the chemical, physical and biological risks related with the work tasks, the machines, the equipment, and the substances used or otherwise present in the work environment. Implementing any preventing measure is a priority, as well as training the personnel in order to eliminate or minimize the risks and to reduce the probability of a hazardous event.

The remaining risk deriving from this risk reduction is called Residual Risk, which remains very high in an environment such as an underground construction site.

For these reasons, next to reducing the risk probability, actions are also taken against any unavoidable residual risks by operating on the emergency measures' setup phase. In case of danger in a tunnel, a fewer seconds in the intervention can be vital. This is why it is so fundamental to produce a real time overview of the personnel, the position of the means of transport, the exact emergency site location as well as being able to communicate with the operators underground.

The active RFID LNXessence technology allows to contextualize an event in space and time by associating it with an identification. Compared to other technologies (Fibro Laser, Cameras, UHF, etc.), it allows the automatic identification of "goods/people at risk" even under difficult conditions such as reduced visibility or the presence of obstacles between the sensor and the operator. Furthermore, the LNXessence technology allows a simultaneous identification of multiple on-site operators/workers.

This technology enables a real time assessment of the dynamic risk level in a construction site, with an operating scenario that is constantly evolving.

The innovation carried by this identification technology in the market of safety in workplaces opens the doors to the revolutionary concept of "active safety", producing a real time and constant personnel control and the identification of the possible surrounding dangers and their prevention allowing a dynamic, active and continuous risk reduction.

4 THE EGOPRO SAFE TUNNEL SOLUTION

The EGOpro Safe Tunnel Solution allows, through a single web-based software platform, to monitor six security modules in real time:

- Access control: management of entry permits to the underground construction site and in the tunnel through a Multi-technology identification (vehicle plate recognition system, badges, video surveillance, active TAG)
- Tracking: real time monitoring of the presence and position of personnel in the areas of the tunnel.
- Voice communication: active communication throughout the tunnel through GSM, radio and telephone lines in order to enhance voice communication in the tunnel.
- Data infrastructures: the setup of a reliable, redundant communication network capable of transferring data on the various systems (tracking, voice communications, CCTV, etc.) in all areas of the tunnel. Generally, this infrastructure is created through an optical fiber.

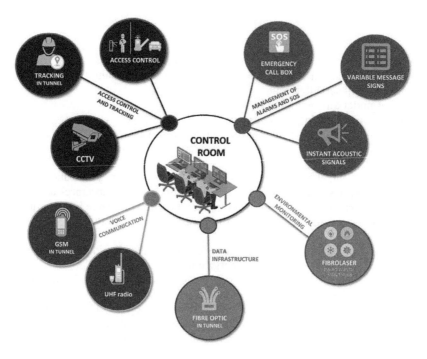

Figure 2.   Integrated module system.

– Management of alarms and SOS: management of emergency call boxes (including emergency calls and alarms and their activation), of variable message information panels and of acoustic warning systems, in particular for the management of drill-and-blast excavations.
– Environmental Monitoring: detection of environmental parameters in the field and management of the resulting alarms.

The three elements that make this solution truly innovative and relevant are:

1. The integration of all safety modules, managed from a single web-based platform and accessible to all authorized users (user/password) from any location connected to the network at any time. This allows, in case of emergency and management control, to obtain a real time 'photograph' of the operating site. It also allows sending alarms to the person in charge in case of danger in the tunnel (via e-mail, via message, etc.) and to carry out the first operations from a remote location.
2. The tracking of operators in the tunnel through the application of the proprietary LNXessence technology. Every operator entering the tunnel is equipped with an active TAG emitting a continuous signal every few seconds. This device can be configured (usually set on ca. 7 seconds). Thus, the personnel units moving inside the tunnel constantly transmit their presence and therefore their position to the system. Sensors/receivers are installed at the entrance and inside the tunnel, normally every 500 meters; since the sensor's distance range can reach a maximum of 900 meters, the distance between sensors can be certainly suited to each environment. The sensor creates an area where the data sent by the active TAG is detected and managed, thus identifying its presence. All the data received by the system is processed by the concentrator module, which selects only the relevant data, i.e. the TAG/operator's zone change, and sends it to the management software without overloading the infrastructure. This allows a constant detection of the operator's position and of the vehicles available within the various areas of the tunnel. Moreover, by associating a registry to an active TAG, it is also possible to receive any data concerning a single operator, such as name, company, safety courses attended, tasks/responsibilities.

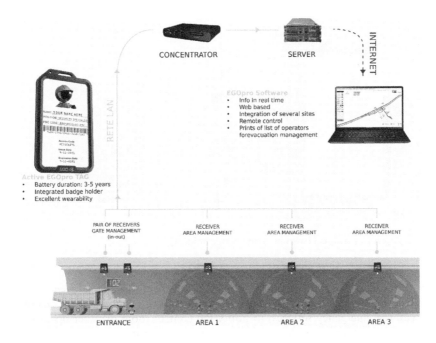

Figure 3. EGOpro Safe Tunnel Architecture.

3. The MULTI-READING feature of the active TAG/operators, which is the ability to detect several people at the same time, also inside a moving vehicle, without having the personnel doing anything voluntary for their identification. All this data allows, in case of emergency on a specific site in a tunnel, to detect the operators in that specific area. An activation of cameras in the tunnel is also possible, in order to view the events unfolding in a specific area, thus triggering acoustic and visual alarms in order to report the event. In addition to this, it enhances phone and radio communication with the personnel involved in tunnels in order to make them aware of their operational needs. Therefore, the EGOpro Safe Tunnel solution allows starting the evacuation procedures having a real time emergency scenario, reducing the time of intervention and consequently lowering the damage to its minimum.

5 CASE HISTORY

Among the various examples of case history regarding the installation of the EGOpro Safe Tunnel solution, it is worth to mention the most relevant case history in terms of major infrastructures: in fact, the EGOpro Safe Tunnel system integrated with its concerning safety modules was used on the construction site of the Brenner Base Tunnel - where, starting from the analysis of requests from the PSC, an infrastructure was built and created with the integration of all the safety modules described above.

A number of sensors have been installed inside the Brenner Tunnel which create the areas to be monitored and each operator who enters the site must wear an EGOpro active Tag. In this way, the operator's or the visitor's movements inside the tunnel can be tracked in real time in order to optimise and speed up any intervention or evacuation in the case of an emergency. One distinctive element of the EGOpro access points, in addition to being able to active them at great distances, is the ability to MULTI-READ operators. This is the ability to detect many people inside a moving vehicle at the entrance to the tunnel, without the operators having to do anything to be recognised.

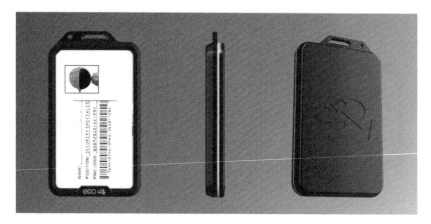

Figure 4. EGOpro Active TAG (with integrate badge holder).

Thanks to the EGOpro Safe Tunnel solutions, in the event of an emergency in the tunnel, operators can communicate with the outside (using radios, telephones and emergency call boxes) to call the emergency services, to notify the control room or other operators, and to activate local or general alarms. At the same time that the alarm or notification is received by the personnel monitoring the site (in the control room), a real time overview of the situation in the tunnel is displayed (showing every operator's position, the position of the active alarms as well as the type of alarm), thereby allowing the control room personnel to take the most appropriate safety procedures.

6 CONCLUSIONS

In times like these, when workplace accidents, even fatal ones, are becoming more commonplace, it is essential to invest in safety by adopting the best technologies the market has to offer, like the EGOpro Safety solutions. The EGOpro Safety solutions, based on proprietary technology and using the very latest technological tools, allow superior levels of worker safety to be achieved.

The integrated system "EGOprosafe TUNNEL", allows to give, through a web based software, three types of information essential for the safety of the worker in order to optimise and speed up any intervention or evacuation in the case of an emergency:

– information of programmed alarms (Variable Message Signs, Acoustic Alerts)
– information about unexpected alarms in the event of an emergency (Emergency call boxes/ Alarm columns, environmental monitoring systems such as fibrolaser)
– information of the position of the operator within the tunnel in real time.

REFERENCES

R.M. Choudhry & D. Fang, S. Mohamed, 2007. The nature of safety culture: A survey of the state-of-the-art. *Safety Science* 45: 993–1012
Decreto Legislativo 81 del 9 aprile 2008 e sue modifiche e integrazioni
A.R. Hale, F.W. Guldenmund, P.L.C.H. van Loenhout & J.I.H. Oh, 2010. Evaluating safety management and culture interventions to improve safety: Effective intervention strategies. *Safety Science* 48: 1026–1035
R. Sabatino, 2015. La progettazione della sicurezza nel cantiere. INAIL Dipartimento Innovazioni Tecnologiche e Sicurezza degli Impianti, Prodotti ed Insediamenti Antropici

Tunnels and Underground Cities: Engineering and Innovation meet Archaeology,
Architecture and Art, Volume 9: Safety in underground
construction – Peila, Viggiani & Celestino (Eds)
© 2020 Taylor & Francis Group, London, ISBN 978-0-367-46874-3

Safety in the Italian railway tunnels

P. Firmi, F. Iacobini & A. Pranno
Rete Ferroviaria Italiana – Infrastructure Department, Rome, Italy

ABSTRACT: Italy is the European country with the highest number of railway tunnels whose cover about 10% of the entire infrastructure. Rete Ferroviaria Italiana (RFI) is the company in charge of almost all the Italian railway network management and it ensures the safety conditions on over 1.500 km of tunnels, which have been built until 150 years ago and are still in service.

In this respect, RFI applies the current regulations of the sector, such as Italian Ministerial Decree - Safety in railway tunnels - 28/10/2005 and EU Regulation Safety in Railway Tunnels Technical Specify of Interoperability (SRT TSI).

These national and European regulations are relatively recent, since they have been developed in the last 20 years.

The paper will illustrate the criteria through which the Italian infrastructure manager defines of the safety requirements of tunnels in service and new projects.

1   INTRODUCTION

The particular orographic conformation has made Italy the European country with the largest number of railway tunnels. RFI, as the manager of the national railway infrastructure, therefore has to guarantee the safety of the most conspicuous underground asset among all European nations.

The network managed by RFI has an extension of approximately 16800 km. The tunnels over 100 meters are about 1518 km long, of which over 50% double track.

Due to the evolution of the construction techniques in the last years, the existing tunnels are highly heterogeneous from the point of view of the geometry of the section and the materials

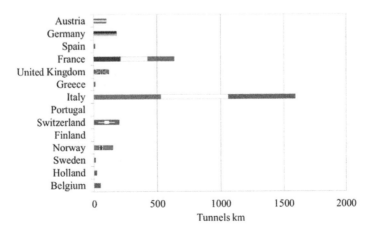

Figure 1.   Kilometres of railway tunnels, comparison at european level.

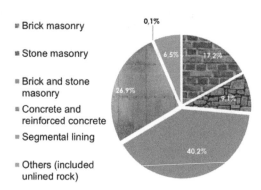

■ Brick masonry

■ Stone masonry

■ Brick and stone masonry

■ Concrete and reinforced concrete

■ Segmental lining

■ Others (included unlined rock)

Figure 2. Construction type of RFI tunnels.

used for the lining (brick masonry, stone masonry, brick and stone walls, concrete and reinforced concrete), as shown in the graph in Figure 2.

Most of the concrete tunnels were built in the post-1960, while most of the masonry/stone tunnels precede the Second World War and are mainly developed over a period from the Unification of Italy (1861) to the early '30s. Tunnels made by masonry generally have a variable and irregular section with piers of stone blocks, often not cemented, and a brick vault. 38% have no invert, a large part (about 70%) is simple track. As for concrete tunnels, it is observed that 13% have no invert and only 10% are single-track.

The high number of tunnels and the occurrence of some accidents in the underground of the transport infrastructures have meant that the issue of security becomes central in the functional design of tunnels.

2 REGULATORY FRAMEWORK

The first reference in the field of safety in tunnels is the "Guidelines for the improvement of safety in railway tunnels" edited in 1997. It was prepared by a group made up of members of the National Corps of Firefighters and FS, illustrated the design criteria for improving safety in the design, planning and management phase in the tunnels of length between 5 and 20 km. The document distinguished the following three case studies:

– Existing tunnels, in which the application of safety requirements had to be evaluated case by case in relation to existing constraints;
– Tunnels in progress, in which it was not yet completed, could be applying certain safety requirements;
– Tunnels of future realization, where the guidelines were applicable in their complexity.

Except for the guidelines, the first regulatory text regulating the approach to safety in tunnels by those who are involved in various ways (users, transport companies, network operators and agencies responsible for the rescue) is Ministerial Decree 28/10/2005 "Safety in railway tunnels".

The Ministerial Decree of 28/10/2005 describes the preventive, mitigating and facilitative safety measures/predispositions of the exodus and/or the rescue to be adopted in the various stages of construction, operation and maintenance of tunnels by defining minimum requirements and additional requirements. The field of application is that of railway tunnels with a length of more than 1000 m located on the railway infrastructure and on non-insulated regional networks. It also applies to tunnels with a length between 500 and 1000 m, limited to some requirements considered necessary such as lighting, emergency signage and walkways.

Subsequently, in Europe, the "Safety in railway tunnels" Technical Specification (SRT TSI) was published, approved by the European Commission Decision of 20 December 2007 (GUCE of 07/03/2008), force in the EU Member States from 1 July 2008 and subsequently updated with Regulation (EU) no. 1303/2014 of the Commission dated 18/11/2014, in force since 1 January 2015. The standard indicates the essential safety requirements for the harmonization on a European scale of tunnel safety conditions in order to achieve interoperability on European networks. The field of application is that of railway tunnels longer than 100 m.

Therefore the entry into force of the SRT TSI in 2008 gave rise to a dual regulatory regime in Italy. On one hand the Technical Specification of Interoperability, to be applied to the new, restructured or renewed subsystems of the European railway network, on the other side the Ministerial Decree 28/10/2005, to be applied to the new subsystems and to those already existing at the entry into force of the Decree. The coexistence of both standards has also created considerable overlaps and misalignments concerning the different methodological approach to safety and the safety requirements to be achieved.

Regarding the requirements to be applied, the main differences between the TSI and the Italian Decree for the "Infrastructure" subsystem are related to: exits and side emergency accesses, fire-fighting system inside the tunnels, t fire resistance of the structures, emergency telephone line (live/voice) and public address system, triage area, helipads and availability of rescue equipment.

With regard to the different methodological approach to safety, it should be noted that the SRT TSI is characterized by a prescriptive approach, prevalently non-retroactive. On the contrary, the Ministerial Decree is characterized by a double prescriptive/performance approach, strongly retroactive. In particular, the decree provides for an integral application of all the safety requirements prescribed for both new and existing infrastructures. The SRT TSI, on the other side, provides for an integral application of the safety measures prescribed only in the case of new tunnels. The European law for existing tunnels envisages a strategy of gradual implementation of the level of safety to be implemented, immediately, through the introduction of operational procedures (emergency plans and exercises) and subsequently through interventions on the infrastructures to be carried out during the renewal and/or renovation projects.

Finally, the reference regulatory framework was further amended by the provisions issued by the Law of 24/03/2012 n. 27 on "Urgent measures in the field of competition, liberalization and infrastructure".

In particular, the art. 53.2 reports: "They cannot be applied to the design and construction of new national railway infrastructures, as well as to adjustments to existing ones, more stringent technical and functional parameters and standards than those provided for in the European Union agreements and standards."

Therefore, for the railway existing tunnels, the Law of 24/03/2012 n. 27 has effectively blocked the implementation of the interventions for the adaptation to the safety requirements of the Ministerial Decree, which is subordinated to the harmonization currently underway of the same Ministerial Decree with the SRT TSI, by the Ministry of Infrastructure and Transport.

3 SAFETY PROVISIONS

The incidental risk scenarios to be considered for the definition of safety provisions are described in the SRT TSI, which makes the distinction in:

- 'hot' incidents (fire, explosion followed by fire, emission of toxic smoke or gases.), where the confined environment of the tunnel can cause an increase in the risk for passengers due to the sharp increase in temperature and products of combustion that make survival difficult;
- 'cold' incidents (collision, derailment), where the tunnel does not necessarily increase the risk to passengers;
- prolonged stop (an unplanned for longer than 10 minutes) which, although is not by itself a threat to passengers and staff, can generate panic phenomena and lead to a spontaneous and uncontrolled evacuation.

For these scenarios, the SRT TSI provides safety provisions for the "Infrastructure" subsystem as well as for the "Rolling Stock" and "Operation and traffic management" subsystems.

3.1 Infrastructural measures

The main infrastructural measures prescribed by the SRT TSI for the design/construction of railway tunnels have been accepted by RFI in the *RFI Civil Works Design Manual - Section IV - Tunnels* (2017).

3.1.1 Mitigation of the consequences of accidents: fire resistance of tunnel structures, reaction to fire of building materials, fire fighting points

The SRT TSI prescribes fire resistance requirements for load-bearing structures of tunnels longer than 100 m, so that the self-rescue of passengers and the intervention in safety of Fire department's teams are guaranteed.

In general, the fire resistance checks of the tunnel liner must be carried out assuming a temperature of 450 ° C for a duration of at least 120 minutes. In case of tunnels, whose collapse could involve superficial or adjacent structures, the load-bearing structures must be verified with fire curves representative of the type of traffic allowed to circulate, for a period of time sufficient to allow the evacuation of the risk areas of the tunnel and adjacent structures. Generally, these checks are carried out considering the nominal curve RWS (MD 28/10/2005) and a fire duration of 120 minutes.

With reference to reaction to fire, tunnel building material shall fulfil the requirements of classification A2 of Commission Decision 2000/147/EC, non-structural panels and other equipment shall fulfil the requirements of classification B of Commission Decision 2000/147/EC. Materials that would not contribute significantly to a fire load are allowed to not comply with the above.

The update of the 2014 SRT TSI introduces the definition of fire-fighting points, i.e. areas for the stop of the burnt train in order to allow the escape of passengers, the intervention of rescue services, the earthing of the contact line by fixed devices and the fire treatment of the train itself.

These areas must be placed at the entrance of tunnels longer than 1000 m and anyway at a distance variable between 5 and 20 km depending on the category of rolling stock allowed to circulate. The escape of passengers must be ensured through a walkway 2 m wide and long as the train of maximum capacity. In addition on walkways there must be a fire fighting system for water supply, in order to guarantee a total flow of 800 l/min for 2 hours. In particular, RFI standards require a fire fighting system constituted by: a network of UNI 45 hydrants, a primary supply line made of plastic material and motor pump connections compliant with the Fire Brigade standards. The system must be powered by a 100 m³ water reserve able to guarantee a minimum pressure of 2 bar for the hydrant placed under the most unfavorable conditions.

3.1.2 Escape facilitation: walkways, emergency lighting, escape signage, safe area, access to safe area

In order to facilitate the escape of passengers in case of emergency in tunnel, the SRT TSI defines the requirements of walkways and emergency lighting, for tunnels longer than 500 m, and escape signage, for tunnels longer than 100 m.

In single-track tunnels, walkways must be made on at least one side and in multiple-track tunnels on both sides, ensuring accessibility starting from each track. On the RFI network, the walkways must have a minimum width of 120 cm, if placed on one side of single-track tunnels or a minimum width of 80 cm, if they are on both sides of multiple-track tunnels. The height of the walkway, which varies according to the type of armament, must consent an easy use by passengers descending from the train (+55 cm for ballast, +25 cm for slab). A continuous handrail must be installed at a height between 0.8 and 1.1 meters above the pavement, which serves as a guide to the safe area.

For emergency lighting systems in the escape routes it is necessary to guarantee a minimum illumination value equal to 1 lux at the level of the walking surface.

The escape signage in the tunnel have both the function of facilitating the exodus of passengers, for example indicating the location of the emergency exits, and the function of facilitating the work of the rescue teams, for example indicating the position of the emergency equipments or power points. In any case, the signage must not be made using lighting fixtures, which could constitute potential obstacles to train conductors, but must be photoluminescent, i.e. made on aluminum supports covered with a film that allows the visibility of the sign, both under normal lighting conditions and in dark conditions. For the arrangement of the signs, in order to facilitate their reading, it is necessary to avoid the contemporaneity of similar signs and the crowding, complying with the maximum distance of 50 m.

For all tunnels longer than 1000 m, a safe area must be provided to allow the escape of all passengers of the maximum capacity train allowed to circulate, guaranteeing their conditions of survival for a time necessary to complete evacuation from the safe area to the final safe place. In single-tube tunnels the safe area consists in the area of the lateral access placed beyond the 'transition chamber' and the square placed on the surface. In case of double track tunnels, the access to the safe area from the opposite track must be guaranteed by a pedestrian subway or, in cases of difficult, by a fire-proof rubber pedestrian level crossing. On the other hand, the safe area in twin-tube tunnels equipped with transversal connections (cross passage), consists in the tube adjacent to the tube site of accident. Cross passage must be placed with inter-distance of no more than 500 m and, according to RFI standards, must be maintained at an overpressure of 30 Pa by smoke control systems.

For single-tube tunnels, in order to guarantee accessibility to the safe areas, it is necessary to have lateral or vertical access (windows) to the outside every 1000 m. For RFI standard tunnels longer than 4 km, a driveway access must be made with a distance of approximately 4 km. The windows for pedestrians and vehicles must be accessible through REI 120 doors and at the end of the tunnel side they must have a larger area with smoke control systems ('transition chamber') where it must be ensured an overpressure of 30 Pa.

3.1.3 *Rescue facilitation: segmentation and earthing of overhead line, emergency communication*

For tunnels longer than 5 km the SRT TSI requires that the electric traction power system in tunnels must be divided into sections not exceeding 5 km. This requirement applies only if the signalling system allows more than one train in the tunnel to be present on each track simultaneously. In addition, remote control, switching of each «sectioner», and communication and lighting devices must be provided to allow safe manual operation and maintenance of the disconnection system.

For tunnels longer than 1 km, earthing devices must be provided at the access points and, if the earthing procedures allow the grounding of a single section, close to the sectioning points. RFI has issued a Technical Specification that provides for the remote control of the earthing in SIL 4 (Safety Integrity Level). This will avoid the visual check of the earthing devices, reducing the response times.

Regarding to emergency communication in tunnels longer than 1 km, it must be possible to communicate via radio between the train and the control center of the infrastructure manager with the GSM-R system. In order to guarantee the communications of the emergency teams with their command structures, RFI has delivered over 600 new equipment to the Fire Brigades who works in tunnels longer than 1000 m.

3.2 *Operating Procedures*

3.2.1 *Tunnel emergency plan and exercises*

For all tunnels longer than 1000 m, the SRT TSI requires the issue of a tunnel emergency plan that defines tasks and responsibilities of the different bodies involved in the management of the emergency in tunnel, such as local authorities, railway companies or rescue services. The document must provide indications on:

– location of the tunnel and accessibility conditions;
– safety provisions available for self-rescue, evacuation, fire-fighting activities and rescue;

- accident scenarios;
- methods of intervention of each subject involved in the emergency;
- operating procedures to be implemented in the phases immediately following an incidental event;
- telephone contacts of personnel involved in the emergency.

The plan must be consistent with the internal emergency plans prepared by each subject and in particular with the internal emergency plans prepared by the Infrastructure Manager on the basis of the current RFI guidelines.

In order to verify the efficacy of the plan, in terms of completeness of the assumed emergencies and adequacy of the prepared resources, it is necessary that exercises will be planned for all personnel involved in the management of the emergency. Different levels of exercise must be foreseen, with frequencies and methods depending on the length of the tunnel and the safety equipment. For tunnels shorter than 5,000 m, exercises with a frequency of less than two years must be carried out. These exercises consist of: on-site inspections, simulations of access to the tunnel and practical tests of the functioning of the emergency plants. To make all the subjects familiar with the infrastructure, the involvement, even non contemporary, of all the categories of personnel potentially involved in the rescue operations must be ensured, with particular attention to the personnel of RFI intervention teams and rescue teams of external bodies. For tunnels longer than 5,000 m, in addition to two-year frequency exercises, complete exercises with a four-year frequency must be performed, including evacuation and rescue procedures, with the participation not only of RFI personnel, but also of all organizations involved in emergency management.

4 EXISTING TUNNELS

As described above, the entry into force of the TSI has produced a situation of coexistence of two standards that are not always aligned and coherent with each other. Furthermore the Law Decree 24/01/2012 n. 1 expresses clearly that standards which are more restrictive than those defined at European level cannot be applied.

The change in the reference regulatory framework has raised the level of attention on the issue of safety tunnels, resulting in an increase in the level of knowledge of the existing tunnels.

Specifically, it was decided to run on risk analyses for the existing tunnels longer than 1000 m in operation on the date of entry into force of Ministerial Decree 28/10/2005. The risk analyses were conducted considering the characteristics of the infrastructure, rolling stock and traffic (speed, entity and composition). The results of the risk analysis made it possible to classify the galleries into four groups, based on the comparison between the cumulative risk of the tunnel and the tolerability limits for risk acceptance defined by the MD (Annex III). Specifically, the following groups are defined:

- GROUP 1: consisting of tunnels whose cumulative risk figure falls within the upper tolerability limit;
- GROUP 2: consisting of tunnels whose cumulative risk chart falls within the attention area (ALARP), between the upper and lower tolerability limits;
- GROUP 3: consisting of tunnels whose cumulative risk figure falls within the lower tolerability limit;
- GROUP 4: consisting of tunnels whose cumulative risk figure falls below the lower tolerability limit.

Figure 3 shows the fatality/frequency of incident plan in which the cumulative risk graph is plotted, with indication of the limits of tolerability and the identification of the areas of acceptability, attention and unacceptability.

The risk analyses represent an important decision-making tool as they allow to define the priority of the interventions for the improvement of the safety level of tunnels, to be realized

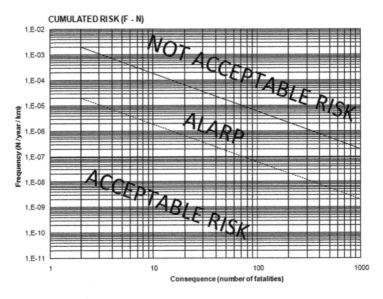

Figure 3. F-N plane with ALARP diagram.

through infrastructural predispositions (realization of access roads, handrails, signage, water reserve) and/or technology (earthing and emergency lighting).

5 CONCLUSIONS

Among the fundamental objectives of the mission of RFI there is the need to guarantee the safety in the railway network, with particular attention to the protection of human lives. RFI therefore constantly invests in research of solutions aimed to raise safety standards and participates in European technical tables to define more efficient regulations.

In this paper was illustrated the RFI standard for new and existing tunnels.

REFERENCES

2017. RFI Civil Works Design Manual - Section IV - Tunnels
2014. Technical Specification for Interoperability SRT (Commision Regulation (EU) No 1303/2014 of 18/ 11/2014)
2005. Ministerial Decree 28/10/2005 - Safety in railway tunnels

Tunnels and Underground Cities: Engineering and Innovation meet Archaeology,
Architecture and Art, Volume 9: Safety in underground
construction – Peila, Viggiani & Celestino (Eds)
© 2020 Taylor & Francis Group, London, ISBN 978-0-367-46874-3

The methodology of quantitative risk analysis in the application of the Legislative Decree n.264/06

A. Focaracci & G. Greco
Prometeoengineering.it Srl, Rome, Italy

ABSTRACT: The Legislative Decree n.264/06 'Implementation of Directive 2004/54/EC on safety in tunnels of the trans-European road network', defines a set of minimum safety requirements for tunnels and requires that a quantitative risk analysis is carried out according to the IRAM methodology (Italian Risk Analysis Method) defined by the same decree, when the transportation of dangerous goods through a tunnel is allowed, for tunnels with special characteristics or if compensative measures need to be adopted where certain structural requirements can only be satisfied through technical solutions which either cannot be achieved or can be achieved only at disproportionate cost. The legislator has in fact intended to over-come the DG-QRAM (Dangerous Goods – Quantitative Risk Analysis Method) method that, as declared by the PIARC, is applicable only when it is necessary to assess the level of risk in a tunnel determined by the transit of dangerous goods. The IRAM methodology, probabilistic in the formulation and quantitative in the determinations, has already been applied in the Safety documentation of more than 300 tunnels.

1 INTRODUCTION

The development of risk analysis in tunnel safety design has supplied an innovative answer to the methods required for a substantial improvement of tunnels safety, a need that was strongly felt by public opinion after the serious accidents over the last few years (Mont Blanc, Gottardo).

The laws, both at European and national levels, have been developed to allow to quantify risks in tunnels; therefore, the fundamental principles to start a large intervention program of safety implementation for road, railway and metropolitan tunnels were created. The safety measures will contribute to place Italy – that by itself has the 44% of European tunnels – at the avant-garde in tunnelling, world-wide, once more (Figure 1).

The necessity to use quantifying assessment criteria and to adopt a systematic approach method, stems from the fact that hazard perception - in other words the psychic perception of danger - is subjective because it is related to the degree of familiarity that a subject has with the system employed (equipment, plant, structure, etc.).

Therefore, it is not possible to resort to a simplistic listing of provisions to implement for the infrastructure safety, unless the most likely hazards and admitted safety standards have been defined beforehand.

Such method offers the advantage to avoid that tragic events, involving the emotional sphere, could induce disproportionate safety interventions, with added financial costs of a scale hampering a country's investing potential and subtracting resources that could be devoted to top priority works. On the contrary, a correct and coherent approach to the genuine requirements of a certain infrastructure allows to optimize investments, and this makes the achievement of large-scale safety objectives more realistic.

A further advantage achieved when adopting targeted interventions to a specific tunnel system is that this avoids implementing earlier standard solutions, evenly applied to all cases, while concurrently promoting the research for innovative technological solutions.

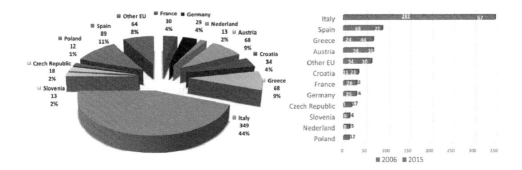

Figure 1. Road Tunnels in UE.

The 2004/54/CE Directive identifies the target safety objectives, a system of safety parameters to enforce, it fixes the sets of minimum safety requirements to satisfy and defines the risk analysis as the instrument to employ for the assessment of a tunnel's safety standards.

The methodological principles, as indicated by the Directive to the effect of achieving the safety objectives, have been taken as the technical basis of the new Italian legislation in road tunnel safety (Legislative Decree n°264 dated 05.10.2006, transposition of the 54/2004/CE Directive), and in railway tunnels safety (Interdepartmental D. dated 28/10/2005: "Safety in Railway Tunnels").

2 ACQUISITION OF GEOMETRICAL, STRUCTURAL AND PLANT CHARACTERISTICS DATA RELATED TO THE WORKS AND TO ACCIDENT PROBABILITY; SETTING OF DATA-BANK ACCORDING TO CODED CRITERIA

2.1 The natural tunnels

As all design project, safety design cannot overlook the compilation of a database as follows:

– The works' geometric characteristics, with specific reference to length, cross section shape (number, lanes width and direction, height or overall dimension, footpath, etc.); the road layout geometric characteristics and – for existing works – typology and year of construction. For projects at design stage these parameters represent the initial hypothesis, and are subject to modifications deriving from the safety check of the works.
– The environmental characteristics of the surrounding context concern the weather/climate conditions, mainly at the tunnel entry-exit, the tunnel accessibility and the possibility of localizing rescue teams.
– The traffic characteristics in terms of volume and type of traffic, traffic regimen (high-speed curves) and expected standard of service.

3 ANALYSIS OF STRUCTURAL VULNERABILITY

Upon completion of the database follows the elaboration initial phase consisting in assessing the tunnel system vulnerability, identifying the potential hazards related to the tunnel system and the possible hazardous scenarios.

The Vulnerability Analysis allows the identification of possible safety parameters, inconsistencies as well as underperformance related to the minimum, legally-set requirements; it enables to identify the risk analysis procedure to apply in the subsequent phase and to achieve an outline of the type of risk of the tunnel system, to subsequently proceed with the individuation of the most appropriate design solution.

4 INDIVIDUATION AND DESIGN OF THE SAFETY REQUIREMENTS IN STRUCTURAL AND SYSTEMS ENGINEERING TERMS

From the vulnerability analysis, the safety designer can comprehend which safety instruments must be selected among the preventive, protective or mitigating measures (escape-facilitating), the geometric and structural measures and the systems requirements current norms define as a minimum - either compulsory of optional - in certain conditions.

In particular, the Transports and Infrastructures Ministerial Decree, in cooperation with the Internal Affairs Ministry, dated 28th October, 2005 (G.U. n. 83 dated 8th, April, 2006) defines the safety requirements (minimum and integrative requirements) to adopt for the safety of more than 2000 railway tunnels existing in our Country and for those under construction and at design stage.

Such norms refer respectively to the sub-groups: Infrastructure, road and rail network materials and Operation Procedures. Within this context, the rules directives recommend to minimize the infrastructure interventions – traditionally expensive and with high running costs – to the benefit of plant systems and new technologies such as, for example, fire extinguishing systems able to contrast the onset of fire.

Likewise, the n. 264 Legislative Decree dated 5th, October, 2006, in function of the 54/2004/CE Directive, identifies the minimum requirements for new and existing tunnels, listing them as structural and plant requirements. The safety design and the risk analysis must individuate alternative solutions that guarantee a safety standard equal or higher in case such requirements are impracticable or only possible at disproportionate cost.

Therefore, safety design for both road and railway tunnels, entails the identification of structural solutions, plants equipment, managerial provisions - including innovative ones enabling the achievement of safety targets – and the subsequent check of the solutions selected using a quantifying risk analysis.

5 RISK ANALYSIS TO CHECK THE ACHIEVEMENT OF SAFETY TARGETS

The safety design process proceeds with the study of hazardous events starting from the causes possibly generating events at the origin of a process that transforms a potential hazard into a real risk, to the individuation and categorization – in terms of probabilities of occurrence and damage - of the end-of-emergency scenario.

The representation of possible accident causes and the identification of occurrence probabilities of the original critical events are illustrated by the causal tree technique (FMA – Failure Modelling Analysis). The causal tree also allows to represent the action as supplied by the preventative measures for originating events that can develop into incidental scenarios.

The group of accidents scenarios related to the tunnel system is defined using the events tree technique, whereby each branch represents a possible incidental scenario. The actions aiming at conditioning the development of an accident scenario are supplied by the protection and mitigation safety measures.

The quantitative risk analysis used in the safety design process can be clearly illustrated by the so-called "butterfly" diagram.

The diagram structure shows two different sections individuating the field of application of the FMA (Failure Modelling Analysis) technique and of the ETA (Event Tree Analysis) technique as separate from the Original Critical Event.

The butterfly diagram right hand sector concerns the group of the trajectories of the system's probable accidents, the causes potentially inducing the system development on different accidents trajectories, the conditioning action that the protection and mitigation measures apply on the achievement of the end-of-emergency situations.

The critical accident scenarios for the tunnel systems relate to events characterized by a low-occurrence probability and high consequences satisfying the utmost statistics, i.e. with low probability of occurring and significant consequences.

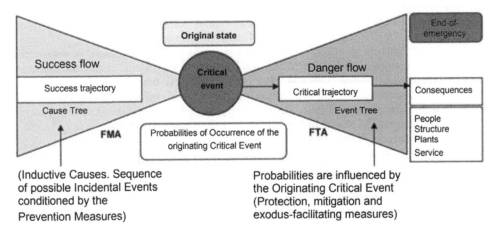

Figure 2. Butterfly diagram.

The iterate application of the risk assessment model - as defined by the Legislative Decree n. 264/06 - to the range of Italian roads tunnel systems, allows the formulation of a simplified risk analysis procedure (risk calculator). Such procedure defines the risk level of a safety project using one single, focused operation between distribution functions representing the occurrence ratio of hazardous events and the expected consequences on the sensitive elements of the tunnel system, supported by the comparison of results supplied by models with the trend lines derived from the statistical analysis of real data.

For what concerns the analysis of consequences deriving from an accident, the literature identifies two different methods. One is the risk analysis using a criticality matrix, compiled via the introduction of occurrence probability of hazardous events, assessed on the basis of a statistical analysis of tunnel accidents and adopting the experts' assessment to fix the expected consequences on the tunnel system's sensitive elements. This method individuates the classes of risk expressed as a uniform probability distribution.

The other method is represented by the risk analysis using the event tree: this supplies a deterministic analysis of the consequences and it is carried out assuming the existence of a dimensioning event and assigning an ideal reliability and efficiency rating to safety barriers. The outcome is the identification of risk levels expressed as distribution of continuous probabilities (complementary collective curves).

The Legislative Decree adopts this latter method (event tree) whereas the one defined by the criticality matrix, a somehow empirical and approximate formula in determining the risk level related to a safety project, is not considered by the Decree.

6 RISK ANALYSIS - THE IRAM METHOD (ITALIAN RISK ANALYSIS METHOD)

The risk analysis method, whose application methods are hereunder detailed, concerns the accidental events considered as critical in the specific and constricted road and railway tunnel environment, in other words: fire, collisions with fire, derailing, flammable material and toxic and hazardous substances discharge.

For what concerns events connected to road accidents as such, related to the infrastructure's geometric characteristics, not induced by the tunnel specific environment and not involving additional risks to those related to road traffic, in order to ensure prevention, they are to be considered and tackled within the road traffic rules and road design.

The victims of this latter type of accidents are to be computed as road casualties.

The Risk Analysis carried out according to the classic Bayes' method, using the uncertainty analysis, is the correct analytical method identified acknowledged as appropriate to define the risk level specific to the existing road and railway Italian tunnels, as according to the recommendations by the 2004/54/CE Directive and Ministerial Decree 28/10/05.

Thanks to the systemic approach employed, it is possible to define the users' survival rate in possible escape scenarios considered as critical and consequent to accidents, within the tunnel specific environment, and to quantify the risk related to the individual tunnel over a set time-frame.

As previously illustrated, the safety design procedure for road, railway and metropolitan tunnels entails a previous analysis of the infrastructure vulnerability that establishes a univocal relation between homogeneous groups of minimum safety requirements as set by the norms, and the limit safety parameters as defined by the statistical analysis in historical data of accidents.

The vulnerability analysis allows to identify anomalies of safety parameters and shortfalls of minimum parameters as set by the norms, and to identify the structures requiring to undergo risk analysis.

For what concerns road tunnels, the risk analysis must be carried out for every tunnel that does not comply with minimum parameters and thus requires the adoption of alternative safety measures in order to demonstrate that they can guarantee an equivalent or higher safety level; in other words, every tunnel having unusual characteristics as compared to the parameters set by the law.

The safety minimum requirements are mainly set to specifically provide for:

- tunnel system's protection and mitigation for overall hazards deriving form critical events, such as reduction of safety systems' intervention time, reduction of fire hotbed temperature, control of fumes dispersion;
- facilitation of self-rescue escape operations such as emergency exits, improved visibility and communication means;
- facilitation of emergency rescue operations such as road accesses, improved communication means and water supply.

Some of the above-mentioned requirements also hold a preventive role during standard working conditions.

The risk analysis must demonstrate that the overall preventative, protection, mitigation measures for the tunnel's overall hazardous situation deriving from critical events and the exodus and rescue facilitation can ensure that the structure risk level remains below the satisfactory risk level and cannot be further improved other than with unrealistic works or at a disproportionate cost (cost-safety analysis).

The major points of the risk analysis are summarized in the conceptual diagram below.

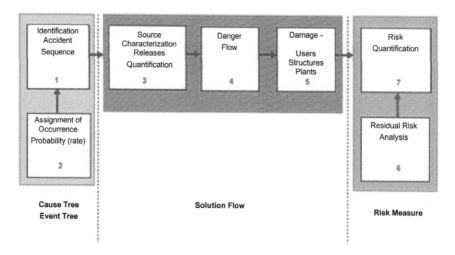

Figure 3. Risk analysis flow diagram.

6.1 *Risk analysis methodology*

The risk calculation methodology must comply with Legislative Decree 264/06, as reported in ANNEX 3 (provided for by article 13, paragraph 3) of the decree itself and as described in the Glossary to Annex 1 (envisaged by 'art.1, paragraph 2).

That it is, at least the following elements must be included and specified in the analysis:

– since Annex 1 to the glossary defines the Risk Analysis "[...] Risk assessment is a process that involves identifying the sources of danger and determining the exposure of the population to danger and includes estimating the uncertainties. connected. [...] "and the risk as" Analytical link between the probability of occurrence of an event and the extent of the consequences deriving from it, including the uncertainties related to the estimation of the definition quantities" it is necessary to highlight the methods for defining the estimation of uncertainties, the sources statistics for the definition of the probability of occurrence of the critical event and the distribution functions used in the determination of the population exposed to the danger;

– in compliance with par 1 of Annex 3 which states that "the level of detail of the Quantitative Risk Analysis methodology must allow the determination of the salvability of users for scenarios deriving from critical events", it is necessary to specify the number of scenarios analysed and the simulations carried out for the determination of the consequences induced by the initiator events;

– since in Annex 1 the scenario is defined as "A sequence of events that describes, starting from a given initiator event, the modalities conditioned by the security measures adopted, which induce certain consequences", it is required to describe the modalities in which the security measures present have been considered in the risk calculation procedure;

– since paragraph 3 of Annex 3 requires that the "accident scenarios and their evolution in tunnel in terms of danger" are "represented by models including the tree of the causes, the critical event initiator and the tree of the events. [...] The event tree is characterized in terms of the probability of occurrence of critical initiator events and of conditioned probabilities of evolution along the individual branches, as an expression of the reliability and efficiency of the installed or planned safety measures ", to give evidence of the tree of events used, methods for determining the reliability and efficiency of security measures and how they are an expression of the conditioned probabilities of evolution within the tree of the events itself;

– in compliance with par. 3 of Annex 3 which states that "The salvability of users in a specific tunnel is determined by quantifying the flow of danger within the structure. The modelling of the danger flow is implemented with different levels of detail depending on the needs and using the best known and available techniques. The results of the modelling of the flow of danger constitute the input data for the simulation of the exodus process of the users from the structure ", illustrating the thermodynamic models adopted for modelling the flow of danger and how the results constitute the inputs of the exodus models;

– since par. 3 of Annex 3 requires that "the number of users involved in the exodus process" is "determined through the formulation and solution of suitable models of queue formation in the analysed gallery", highlighting the models of queue and exodus formation adopted;

– as specified in par. 3 of Attachment 3, to illustrate how in the method adopted "The risk connected to a tunnel" is "defined as the expected value of the damage or as distribution of the probability of exceeding predetermined damage thresholds (cumulative distributions reported on the so-called F - N plan)".

6.2 *Fault Tree and Event Tree Analysis*

The accidents scenarios are illustrated by models inclusive of the cause tree, the critical event, the event tree.

The critical event is defined in terms of occurrence probability and potential hazard based on statistical evidence for tunnel systems as a whole, possibly integrated by data available for the individual tunnel under consideration, with reference to occurrence rate detected and the tunnel design specifications.

The event tree is defined in terms of occurrence probability of critical events and of probability of development along specific branches, as conditioned by the safety systems action quantified in terms of their reliability and efficiency.

The event tree branches finish off in end-of-emergency scenarios, defined by the number of permutations mutually excluding the conditioning actions implemented by the mitigating procedures provided for.

The figure 4 shows and example of application of the event tree technique to define fire-prevention safety in a tunnel assumed to be equipped with safety systems as follows:

- Monitoring-detection,
- Communications,
- Ventilation,
- Lighting,
- Firefighting,
- ...

6.3 *The coding of risk sources and scenarios*

In order to ensure a universal basis of input data for the application of the risk evaluation assessment method to the experts involved in the safety plan program, it seems necessary to code the sources and hazard scenarios.

The fire scenarios in tunnels are defined on the basis of the chemical-physical parameter of the fire sources thermal power. The parameter used for thermal power generated by a fully developed fire is estimated starting from the energy of the type of fuel feeding the fire and using a semi-empirical formula derived from experimental data attained during tests carried out in real-life situations, with natural ventilation, within the Eureka Project and the Memorial Tunnel experimental program.

The thermal energy released rate during the event is defined by modelling the development phase, the still phase and the extinguishing phase, using analytical functions appropriate to reproduce the thermal energy release path as shown in the graphs of the test's details and reported by public domain literature.

The effect of tunnel ventilation on the fire-generated thermal power development is taken into account introducing an appropriate ventilation factor as defined by the application of the

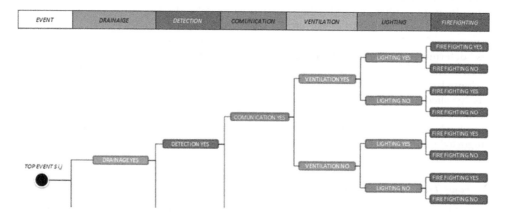

Figure 4.   Event Tree.

Bayes' analysis to the experimental data detected during real-life tests or in laboratory tests and reported by public domain literature.

The combustion phenomenon is defined by a simplified model, introducing the relevant parameters as in the subsequent characterization of the micro-climate occurring in a tunnel during a fire (oxygen consumption, carbon dioxide and carbon monoxide production, smoke curtain), that are appropriate parameters in function of the fire-generated power.

The description of the combustion phenomenon using advanced model is admitted as far as they produce an outcome that can be comparable or improved as compared to the simplified approach results.

The statistical definition of fire scenarios in terms of occurrence probability and intensity of fire, as defined by the historical tunnel-accident series analysis attained from public domain literature, is carried out using the event tree technique.

The data variation about thermal power released from fire hot-beds of different nature, inclusive of those with a highly flammable potential, is taken into account introducing specific distribution functions. The proposed coding of hazard sources and scenarios can be adjusted, providing it is documented with detailed references to scientific publications related to approved innovations concerning occurrence modalities and events of fire development in tunnels. A similar coding is adopted for hazardous goods sources and related hazardous scenarios.

6.4 Risk Flow Modelling

The risk flow is outlined as a sequence of situations defined by the evolution of chemical and physical phenomena consequent to the occurrence of an hazardous event, whose development is conditioned by the safety barriers restraints (emergency trajectories of the tunnel system).

The risk flow is defined applying the thermo-fluid-dynamic method, quantifiable at differently detailed levels, and represented using the event tree technique.

Among the models employed to quantify the risk flow (concentrated parameter models, zoning models, network models, field models) the most widely adopted are the field models.

The value of a risk flow simulation is defined by the models reliability and by the accuracy of solutions.

The outcome of risk flow simulation, carried out employing the models adopted with statistical techniques – in order to include the consequences of epistemic uncertainties related to the safety barriers' performance and the description of hazardous phenomena – defines the paths of space and time variables characterizing the hazardous phenomena (temperature fields, toxic and hazardous substances concentration fields).

The risk flow defines the hostile environment that construction materials, the structure's cladding and the emergency management systems are exposed to, within which users implement the escape course of action and rescue teams put into operation the intervention procedures. The fields of variables characterizing the risk flow are used to define, using appropriate statistical models, the consequences on the tunnel system sensitive elements (damage vector).

6.5 Characterization of tunnel traffic and escape simulated scenarios

In case of accident, the time required to close the tunnel must be compared with the time required by the vehicles queue's travelling time. Figure 6 shows a possible diagram of queue formation process.

The users' escape from a tunnel is a course of actions carried out by groups of individuals with specific behaviours, moving along a rough terrain and in a hostile environment.

The elements used to define a simplified simulation method for escape scenarios are:

– Geometric parameters appropriate to define emergency exits and obstacles on the exodus path, the users' spatial distribution in the event initial stage;

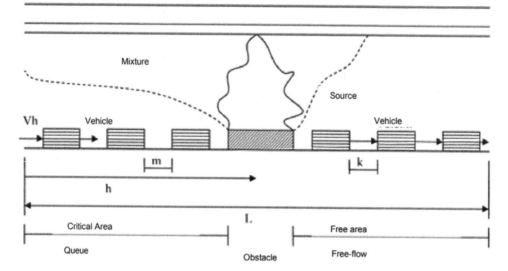

Figure 5.   Queque formation in tunnel.

– Distribution functions appropriate to define the time phases within which the exodus procedure unfolds (perception time, reaction/response time, evacuation time):
– Chemical-physical parameters appropriate to define the hostile environment within which the escape procedure takes place (the field of variables representing the risk flow).

The exodus process simulation can be designed developing models of formal complexity, increasing in function of the size of the escape path taken as representative, of the approach used in the formulation, of the adopted solution techniques. Among the models that can be employed (mono-dimensional, bi-dimensional, Lagrange or Euler models, deterministic or statistic techniques) the one apparently offering the best outcome is the Lagrange method, that allows to define the movements of each person involved in the event, and that can include the interaction among individuals, using appropriate interaction prospectives. The Montecarlo method generates a statistically significant sample of persons involved in an escape.

The curves in the illustration 8 graph represent the distance of users from the nearest emergency exit in function of the time: in the simulation, the victims are represented by the curves tending to become asymptotically horizontal, evidence of the end-of-escape due to environmental conditions incompatible with survival.

6.6   The risk evaluation

According to the consequences, the risk extent can be classed in:

– Victim's number: individual risk, social risk;
– Financial damage: direct costs, indirect costs.

The variable taken as representative when defining risk magnitude within a tunnel is the number of victims consequent to a critical accident event.

Social risk is normally calculated assessing the event frequency over a year ("f") and the number of victims ("N") related to each individual event identified and possible consequences.

Each "f-N" combination can be represented by a dot on a graph, thus generating histograms knows as "f-N-curves".

The area between the Reasonable Risk curve and the Acceptable Risk curve defines the area of application of the "ALARP" (As Low As Reasonably Practicable) principle; such principle, taken as the guiding criteria for the costs-safety analysis, establishes that the

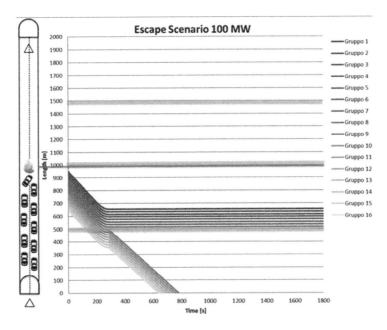

Figure 6.   Escape scenario simulation.

lessening of risk level in a specific tunnel must be compatible with the structural project's intrinsic, technical and financial restrictions. The optimal design solution originates from the combination between the preventive and protective safety measures deemed appropriate to ensure an acceptable risk level for the tunnel under consideration.

The risk level intrinsic to a generic tunnel is defined by the cumulated complementary curve (C.C.C.) of the "f-N" plan, that allows to represent the risk as a complete distribution of potential loss, highlighting the uncertainty effect related to malfunction or to the inadequacy of the safety systems adopted. The area subtended by a cumulated complementary curve originates a global risk indicator appropriate to define the conditions of equivalence among various design solutions for a tunnel system, as it defines beforehand the appropriate comparison criteria that take into account the uncertainties related to the system.

In case of road tunnel, the curves F-N (fig. 7) represent - on a logarithmic scale - the function:

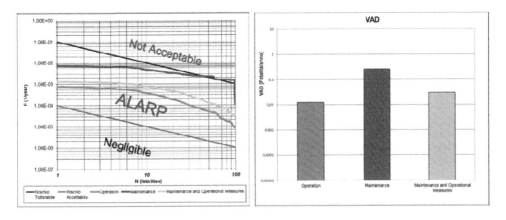

Figure 7.   The Road Tunnel F-N Diagram.

$$1 - F_N(x) = P(N > x) = \int_x^\infty f_N(x)dx$$

Where $F_N(x)$ is the probability distribution function of victims's number per year, $f_N(x)$ is the probability density function of victims's number per year.

7 CONCLUSIONS

The safety design procedure replaces the concepts of accident scenario and of dimensioning event (deterministic analysis of consequences) with the concept of probabilistic group of an escape scenario and of damage expected distribution (probabilistic approach) correlated by a risk flow simulation, and of the escape course of action within a given structure.

Such procedure adopts the clear-cut, analytical IRAM (Italian Risk Analysis Method) risk analysis method, acknowledged as appropriate to define the risk level intrinsic to Italian road and railway tunnels. Moreover, the analysis method provided for represents a reference for the new norms in tunnel safety matters.

Thanks to the systematic approach introduced, it is possible to define the users' rescue rate in possible escape scenarios consequent to accident events considered as critical, and to quantify the risk related to the individual tunnel over a set time-frame.

Such method allows to tackle the safety issue in engineering terms, via a logical sequence of analyses and assessments of numerical and quantitative nature; this avoids that, due to the emotional spur generated by a serious accident, the interventions are excessively allocated, implying a financial burden able to seriously hamper a country's investing capability; therefore it is possible to design the infrastructures' management safety similarly to the design of the works' structural design.

REFERENCES

Focaracci A. 2004. Nuovi orientamenti in tema di normative di sicurezza per gallerie stradali e ferroviarie, *Gallerie e Grandi Opere sotterraneo n 73*; *Bologne August 2004*

Cafaro E., Focaracci A., Guarascio, Stantero L. 2006, Quantitative Risk Analysis for Tunnels, *Tunnel Management International Conference*; *Turin May 2006*.

Focaracci A. 2006. Relazione del Presidente del Comitato C.3.3 sulla gestione delle gallerie stradali, *XXV Road National Symposium AIPCR*; *Naples 4-7 October 2006*

Focaracci A. 2007. Progettazione e realizzazione della sicurezza nelle gallerie stradali e ferroviarie, *Strade & Autostrade n.1/2007*; *Milan*

Focaracci A. 2007. Progettare la sicurezza - Italian Risk Analysis Method, *Le strade n.4/2007*; *Milan*

Angelozzi E.; Bandini C.; Doferri Vitelli M.; Focaracci A.; Grassi F. 2007. Il progetto del potenziamento appenninico, *Le strade n.6/2007*; *Milan*

A. Focaracci; Tozzi G. 2007. L'applicazione del D.Lgs. n.264/2006 alle gallerie di Autostrade per l'Italia (ASPI), *Le strade n.11/2007*; *Milan*

Focaracci A.; Cafaro E. 2009. Lo studio del livello di rischio di una galleria, *Strade & Autostrade n.1/2009*; *Milan*

Fire and Smoke Control in Road Tunnels, *PIARC Committee on Road Tunnels* 1999

Road Tunnels: Emissions, Ventilation, Environment, *PIARC Committee on Road Tunnels*; *1999*

ISO 13387–1999 Fire Safety Engineering Parts 1-8, *PIARC Committee on Road Tunnels*; *1999*

NFPA 502 Standard for Road Tunnels, Bridges and other limited access highways, *PIARC Committee on Road Tunnels*; *2004*

IEC International Standard 60300 –3 – 9, Risk Analysis of technological systems, *PIARC Committee on Road Tunnels*; *Geneve 1995*

NFPA 551 Evaluation of Fire Risk Assessments, *PIARC Committee on Road Tunnels*; *2004*

Tunnels and Underground Cities: Engineering and Innovation meet Archaeology, Architecture and Art, Volume 9: Safety in underground construction – Peila, Viggiani & Celestino (Eds)
© 2020 Taylor & Francis Group, London, ISBN 978-0-367-46874-3

"Terzo Valico dei Giovi, Italy": Safety and equipment innovative measures

A. Focaracci
Prometeoengineerign.it Srl, Rome Italy

N. Meistro
Salini Impregilo S.p.A., Milan, Italy

ABSTRACT: The 'Terzo Valico dei Giovi' is the new high-speed rail link between the port of Genoa and the Po Valley, and constitutes the outlet to the Mediterranean Sea of the Reno-Alps Corridor, one of the main corridors of the TEN-T strategic network. The Third Pass Line will develop on a route of about 53 km, of which about 47 km in twin-tube-tunnels, with transversal connections that allow each tunnel to be a safe place for the other. The railway system is expected to have a significant daily journey up to 76 passengers train and 139 freight trains. The article highlights how the compliance with the safety requirements defined by the most recent European regulations has required the upgrading of the safety layout of the entire project and a complex design of the tunnels systems whose performance, defined by three-dimensional thermo-fluid simulations and probabilistic risk analysis, will make the new line one of the most advanced European railway systems from the point of view of safety.

1   INTRODUCTION

The III Valico Tunnel Complex is composed by the tunnels of the Genoa-Milan High-Speed Railway Line, the Voltri Linking Section and the interconnections between these lines.

The rail line Voltri Linking Section is composed by the consecutive tunnels Doria, Montegazzo, Caverna Borzoli and Fossa dei Lupi for an overall length of 7,6 km; considering the project for the improvement of the existing line the overall length is 8,4 km.

From the Voltri Linking Section, through the interconnections with the III Valico line, from the entrance of the Doria Tunnel to the north portal of the Valico tunnel a rail route with more than 29km in tunnel develops through the following tunnels: Doria, Monte Gazzo, Interconnection Tunnel, and Valico.

The mandatory technical regulation governing interoperability across the European rail system and safety standard was introduced via the EC Directive 2008/57/EC which is supported by the Technical Specification for Interoperability (TSIs), Commission Regulation (EU) No 1302/2014.

The safety design, based on in-depth analyses carried out by means of fire and exodus models and probabilistic risk analysis, has led to the definition of the functional layout and performance specifications of safety systems, and the development of a final safety design that is among the mostadvanced in Europe from an operational safety point of view.

The method adopted for the safety design is the IRAM-RT methodology, and has been developed through the following steps:

– special risk analysis in order to evaluate the interactions between the different routes and the overall tunnel complex, above all considering the special safety requirements for tunnel complex longer than 20 km;

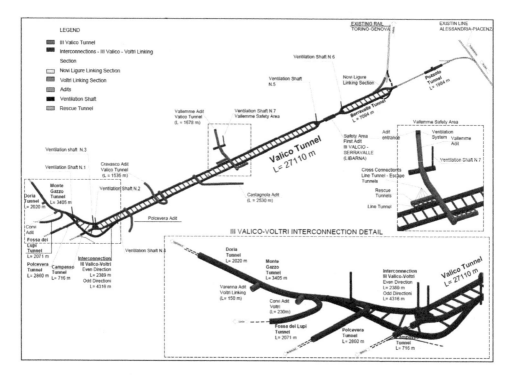

Figure 1. III Valico di Giovi and Voltri link.

- analysis of infrastructure vulnerability, starting from the acquisition of geometric, structural and equipment installation characteristics of the work, traffic and accident data;
- identification, structural and plant design of safety requirements that may prove to be necessary after the vulnerability analysis (Ministerial Decree of 10/28/05);
- risk analysis for the verification of safety targets achievements (Ministerial Decree 28/10/05);
- operating procedures and in particular the preparation of emergency management plans (Ministerial Decree 10/28/05).

In compliance with TSIs requirements, the construction of a safety area for passengers and freight trains is planned at a midpoint within the Valico tunnel. This safety area will be accessible by an adit through which will allow the controlled exodus of travelers and the intervention of rescue teams.

The definition of new safety standards determined the creation of an external safety area between Valico and Serravalle Tunnels. The above mentioned area is accessible by emergency vehicles through a special road. The safety area contains zones equipped with a Triage area, a technological building, a rescue helicopter port, and road-railway connection for use of bimodal rescue vehicles.

2 UTILITIES SYSTEM EQUIPMENT

The key works for which the new regulations have required a specific design are:

- the construction of a safety area inside the Valico tunnel provided with suitable exit routes, smoke extraction and automatic shutdown system;
- the construction an open-air safety area between the Valico tunnel and the Serravalle tunnels equipped with an automatic shutdown system;

- the construction of new cross-passages;
- the upgrading of the system services of the tunnel access;
- the construction of new ventilation shafts and adaptation of fire safety equipment.

2.1 Val Lemme Safety area

In compliance with STI (Technical Specifications for Interoperability) requirements, the construction of a safety area for passengers and freight trains is planned at a mid-point within the Valico gallery (about 27 km). This safety area will be accessible by an adit through which a conflagrant train can be driven and which will allow the controlled exodus of travelers and the intervention of rescue teams. The Val Lemme safety area consists of two evacuation tunnels, which extend 750 m from the axis of the adit, parallel to the tunnel axis, located 35 m between the even and odd track, respectively. The evacuation tunnels are accessible by the platform, through branches, placed at a center-to-center distance of 50 m and are connected by a walkway, placed over the two tubes, at the Val Lemme tunnel access connection. Access from outside the safety area takes place through the Val Lemme adit, 1592 m long; at the tunnel entrance and at the km 0+700.00 two ventilation chambers are placed.

The tunnel safety area will be equipped with the following systems to efficiently and effectively contrast the tunnel emergencies:

- ventilation system/smoke control;
- fire water system;
- automatic extinguishing foam system;
- hazardous liquid collection system.

The ventilation system (Figure 2) is designed according to the engineering approach to fire safety with reference to the international standards as NFPA 92B and NFPA 130 and analysing similar systems as those designed for the Turin–Lyon railway.

Using a probabilistic approach to safety design, the flow rate of the ventilation system was assumed to range from 200 m^3/s to 400 m^3/s. The design of the system was carried out with reference to the maximum capacity and, considering the uncertainties related to possible dysfunctions, the complex geometry and the behavior of airflow circuits in presence of hot smoke, a variability of about 100 m^3/s was estimated.

The system features a distributed fume extraction design whereby extraction is carried out by a set of extraction points located along the safety area, placed at the tunnel access and inside of six distributed cross-passages. The scheme above described, allows optimizing the extraction of fumes in relation to the most likely locations where the fire may occur within the safety area and to geometric characteristics of the same area. Fumes, once drawn and channeled, are conveyed into a false ceiling inside the Val Lemme tunnel access to be expelled through the shaft foreseen in the project (Figure 3).

The ventilation control unit is located in a tunnel made ad-hoc before the shaft, designed to accommodate four two-stage axial fans able to extract up to 120 m^3/s each.

The cross-passage system connecting the tubes of the train tunnel with the safety area leading to the exodus shaft is equipped with pressurization system, which will create an excess of pressure in the safety area so as to prevent the entry of fumes from the compromised tube. This is accomplished by means of a pair of fans (one backup) capable of preventing fumes from entering the safety area. Fresh air is drawn into the safety area from outside through a vent in the false ceiling along the adit, passing through a control unit located outside the tunnel access. In the case of malfunctioning fans, this control unit is also able to provide a minimum of excess pressure to the safety area. Moreover, the design of the external control unit includes a Saccardo ventilation installation, capable of pressurizing the entire tunnel access. In conclusion, the design includes an extraction system in the rescue vehicles parking area that picks up the fumes directly from vehicle exhausts.

The fire water system, in accordance with the provisions of the Italian Ministerial Decree 28.10.2005 and reference TSIs, will be composed of two pressurization control units, a storage

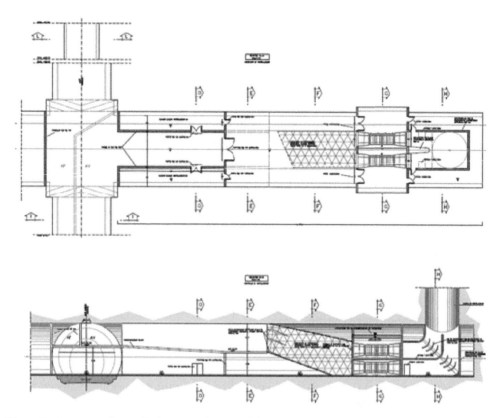

Figure 2. Layout to the ventilation control unit at Val Lemme.

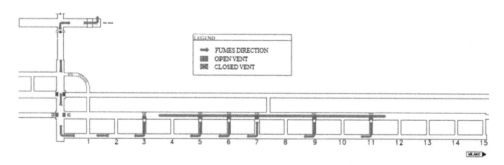

Figure 3. Layout odd platform for the fumes extraction.

tank and a network of fire hydrants equidistant to 125 m. For fires extinguishing of flammable liquids and combustible fuels the design proposes fire protection by means of monitors to AFFF foam additives (Aqueous Film Forming Foam), cooling agent and the formation of a protective film on any liquid fuel (B class). The protection system provides a high foam flow of up to 3000 l/min directly to the fire location, inhibiting combustion on the surfaces and subsequently cooling them. The use of foam allows better coverage of the wet surfaces. In case of spillage and fire of hazardous liquids the AFFF additive will cause the rapid formation of an impermeable liquid film on the surface of the spilled liquid. The designed system, thanks to the high flows and possibility to concentrate their action at the fire location, allow for significant mitigation of the force of the fire, such higher as earlier the system is activated. The presence of an automatic extinguishing system at the Val Lemme area reduces the uncertainties of risk management, by permitting the reduction of the fire power to less than 100 MW, resulting in a significant improvement of tunnel system performance in terms of emergency management.

Along the entire length of the safety area there will be a collection system of potentially hazardous liquids. The spilled liquids and waters discharged from the automatic shutdown will be channeled into a tank located in the lower point of the safety area where the flammable liquids will be separated.

2.2 *Arquata Libarna Safety Area*

The definition of new safety standards has necessitated the creation of an external safety area (Figure 5), with a length of 1166 m, located in the vicinity of the Arquata Libarna PC. This area is accessible by emergency vehicles through a special road. The safety area contains zones equipped with a Triage area, a technological building, a rescue helicopter pitch, and razed pathway for positioning the bimodal tack mechanism.

The area's external security systems are:

– fire water system;
– automatic foam extinguishing system;
– hazardous liquid collection system.

The fire water system, in accordance with the provisions of the Ministerial Decree 28.10.2005 and reference TSIs, has been designed based on what is already reported in the 2005 project considering the increase of pump capacity from 600 to 800 l/min. The automatic extinguishing system is the Monitors type similar to the one designed for the Val Lemme area but with monitors spacing equal to 50 m which can be active in groups of 3-6. The system facilitates the extinguishing operations and is able to handle high flow rates of foam. It is also effective against dangerous liquid fires with an AFFF foam additive able to extinguish class B fires. In addition, there will be a potentially hazardous liquid collection system. The figure below (Figure 4) illustrates the layout of the Arquata Libarna outdoor safety area with the automatic extinguishing system.

2.3 *Cross passages*

The escape system of the main line consists of a series of cross-passage link connections between the two single-track railway tunnels (odd and even) every 500 m approximately, in both the III Valico and Serravalle tunnels. Cross-passages are used for people escaping from one railway tunnel to a parallel railway tunnel; each cross-passage is compartmentalized into both galleries. The conceptual basis for the safety analysis is the consideration of the intact (unaffected) tunnel as a safe place. The ventilation system designed (pressurization of the

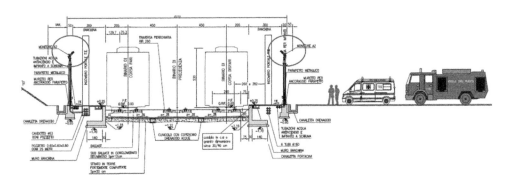

Figure 4.   Layout of the Arquata Libarna outdoor safety area.

cross-passage links road) allows keeping the exodus ways free from the fumes produced in the compromised tunnel, with the following basic criteria:

– ensuring effective pressure in the link road with respect to the compromised tunnel both when the access doors (to the compromised tunnel and the unaffected tunnel) are open, or closed;
– guarantee, even in minimum load conditions, a suitable flow rate of air replacement to the considerable possible presence of persons inside the cross-passage;
– determine the air velocity in the exodus zones with values compatible with the emergency situation of the passengers, hit by substantial flow of air;
– reduced start-up times of the fans (less than 30 sec) in order to reach, in the shortest possible time (about 35 sec), the standard overpressure expected for the volumes involved.

The system will still keep the cross-passages free from any fumes present in the line gallery where the accident has occurred.

The exodus system for the Interconnection lines consists of two pedestrian cross-passage crossing that are connected one to the other. Since at the starting point the crossings are rather long and of reduced section, a filter chamber beside the rail tunnel has been created. This chamber is pressurized in a way similar to the "transition chamber" in the adit, with the same considerations in terms of safety analysis (intact/compromised tunnel) and the conditions assumed in calculations.

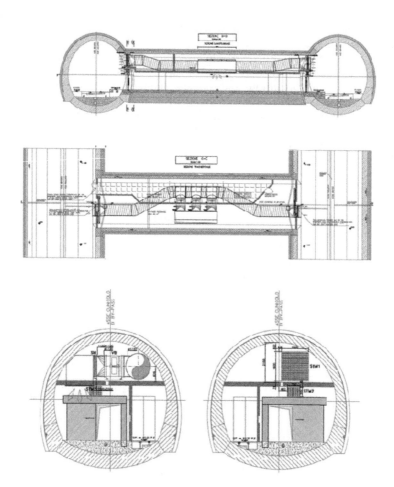

Figure 5.   Cross-passage link connections.

2.4 Access Tunnels (adits)

The exodus system with the adit allows for a widened area at the end of each side tunnel, which forms a space that is intended to allow for the reverse gear of rescue vehicles and to accommodate the beginning of the passenger flow from the tunnel towards the outside. Each of these areas, referred to as "transition chambers", is equipped with a series of doors (on the side railway tunnel and the on the side adit) and a ventilation system capable of keeping the same chamber in slight overpressure with respect to the tunnel. Furthermore, there is a second filter area, between the two rails tunnels (crossing the tracks) also equipped with a ventilation system, able to keep the same area in slight overpressure with respect to the tunnel. In case of tunnel fire, the ventilation system prevents fumes from entering the exodus adits, allowing the passengers to evacuate in safety. Finally, there is an extraction system designed for the adits in the emergency vehicle parking area, similar to that of the Val Lemme safety area, which directly picks up the fumes from the vehicle exhausts.

2.5 Ventilation Shafts

In line with the ventilation strategies adopted for Italian railway tunnels and with the provisions of Annex II of the Ministerial Decree of 28/10/2005, the points of transition from a twin-tube tunnel to a single tube tunnel are designed with measures preventing the circulation of fumes from the compromised tube to the unaffected tube, by means of ventilations shafts. Intake grids positioned on top of the tunnel will suck fumes into a circular segment section plenum. As a results of the design specifications based on the analysis of train fire scenarios and risks (where a thermal power of 10 MW was used for potential passenger train fires and 50 MW for freight train fires), the adjustment intervention of the III Valico final design includes the addition of new ventilation shafts and the modification of the extraction flow rate for those already planned. The ventilation shafts have been designed based on the thermos-fluid dynamic simulations results to allow flow rates extraction of approximately 200 m³/s.

3 RISK ANALYSIS

In accordance with Annex III of the Ministerial Decree of 28/10/2005, a risk analysis of the Valico tunnel system, considering together the Campasso tunnel, the Voltri Interconnection and the Voltri tunnel complex as one single system, was performed. The analyses conducted using the IRAM RT method have shown how the safety measures in the tunnel design allow for a level of risk which falls within the attention zone, primarily due to the high volume of freight trains predicted, as exemplified in the next figure.

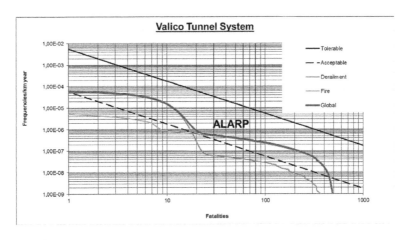

Figure 6. Global FN Curves.

The adoption of a training program aimed to limit the contemporaneity between freight trains and passenger trains would determine further risk reduction. In order to verify the functionality of the designed works in terms of both smoke management and exodus, numerous three-dimensional simulations for representative scenarios were carried out. The analysis made it possible to calibrate the statistical models adopted for the risk calculation, review the emergency management timelines, supporting the choices made with regard to safety systems such as power shutdown and ventilation. The execution of the simulations made it possible to reduce the uncertainties associated with the transportation system efficiency, in the occurrence of chaotic nature hazardous events such as fires. As an example, passenger exodus simulations and thermo-fluid dynamic simulations of smoke extraction carried out for the Val Lemme safety area are reported in the following sections.

3.1 Val Lemme Safety Areas – Thermo fluid dynamic smoke extraction simulations

The analysis of accident scenarios, using the three-dimensional Fire Dynamics Simulator calculation code (Figure 7), was performed by means of the simulation of the smokes spread generated by a stationary train at the Val Lemme safety area. The simulations target is the functionality and performance of the smoke extraction system verification. The proposed smoke extraction system involves the construction of n. 6 extraction points to be assessed based on the maximum distance between the ventilation outlets. The identified scenarios are summarized below:

– passenger train with a potential thermal power of 20 MW and suction capacity of about 200 m^3/s;
– freight train with a thermal power of 50 MW and a suction capacity of about 400 m^3/s.

The results of the simulations, expressed in terms of the distribution of the fumes in the tunnel, profiles of temperature, concentration of carbon monoxide and visibility for different time intervals, showed that:

– in both the analysed scenarios the ventilation system ensures environmental conditions at the dock compatible with the timelines, taken as reference for the exodus of the passengers and the train crew;
– the optimum configuration for the extraction points has been detected in a mixed combination, which provides for a distribution of vents every 100 m with the three control units every 50 m. The vents gathered centrally, where it is easier to happen the train on fire, nevertheless ensure the protection of 450 meters of tunnel that includes an entire passenger train.

In the case of freight train, even if fire occurs at the train end, the simulations have shown that the vents at 100 m distance are sufficient to ensure that the ventilation system guarantees the protection of drivers. In the case of freight train fire and in the phases of Firefighter intervention, the opening of the extraction vent localized on the graft of the adit has been designed, in order to allow the extraction to the maximum system capacity. The simulations conducted show that the security plan as a whole, thanks to the presence of the combined distributed extract ventilation system, guarantees a minimum level of security largely compatible with the current regulations at national and European level. The following figures summarize the three-dimensional analyses performed for the verification procedure of the tunnel access extraction system.

The simulation results have shown that the fume extraction zone is characterized by flows leading to high load losses in justification of the high fan performance necessary for the system operation phase.

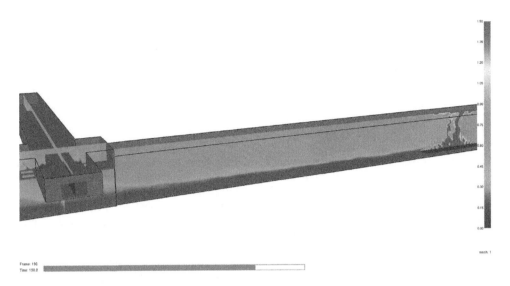

Figure 7.   Temperature Simulation.

3.2   *Val Lemme Safety Areas – Exodus Simulations*

The analyses of accident scenarios was performed using the three-dimensional Fire-Dynamics Simulator calculation code by the simulation of the smoke propagation generated by a passenger train stopped at the Val Lemme safety area, coupled with the simulation of the exodus process, conducted through the EVAC code (Figure 8).

The targets of the simulations consist of:

- simultaneous verification of the functionality and performance of the safety systems;
- verification of the escape times of the passengers in accident conditions.

The scenario analyses, with the primary purpose of verifying the emergency management, is characterized by a greater verisimilitude. The assumption adopted for the definition of reference fire, to check safety conditions of the exposed population, was characterized by a generated maximum thermal power close to 20 MW with a gradual increase in a time equal to 10 min.

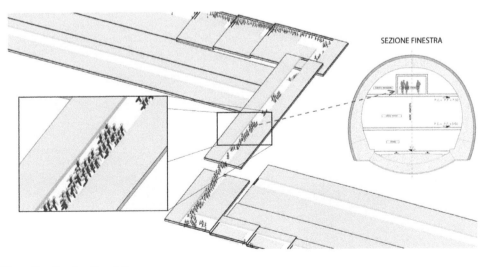

Figure 8.   Exodus Simulation.

The simulations carried out on the exodus process of 500 persons aboard a passenger train stopped at the Val Lemme safety area, show that, the combined fire simulation with the exodus simulation localized at the dock, in presence of train accident, all the passengers leave the train and enter in the connecting branches with a total evacuation time of about 3 minutes.

4 CONCLUSIONS

The safety of travelers is a major issue, which in recent years has become increasingly important also because of a significant change in the regulatory framework (Ministerial Decree of 28/10/2005 and Technical Specifications for Interoperability).

Based on these recent changes, the union of the primary and most competent engineering companies in Italy has allowed the creation of a consortium of design, which by increasingly advanced and cutting-edge technologies, has contributed to the current project of the III Valico railway design. The approach adopted in the design led to the definition of a performance benchmark for facilities that allow for emergency management in the most likely scenarios supporting the risk analysis results provided by the Ministerial Decree of 28/10/2005.

The obtained results at the design level discussed in this article, backed up by detailed analyses conducted by means of fire and exodus models and the analysis of probabilistic risk, have led to the definition of the functional layout, of the performance specifications of security installations and to the development of a final project that is among the most advanced in Europe from the point of view of safety in the operation phase.

REFERENCES

Focaracci, A. & Lunardi, P. & Silva, C. 2001. The Bologna to Florence Hight Speed Railways line: 92 km through the Appenines, *Progress in Tunneling after 2000 – Bologna 2001*
Focaracci, A. 2007. Designing Safety – Italian Risk Analysis Method. *Le Strade n°4-2007*
Focaracci, A. 2007. Commission Tunnels Safety to route and railways. *Conference Safety in Tunnels – Genova, 27–28 March 2007*
Focaracci, A. 2010. Italian Risk Analysis method IRAM. *11th International Conference Underground Construction Prague 2010 – Transport and City Tunnels*. 837–845.
Focaracci, A. 2010. Safety in Tunnels: innovation and tradition. *11th International Conference Underground Construction Prague 2010 – Transport and City Tunnels*.846–851.

Tunnels and Underground Cities: Engineering and Innovation meet Archaeology, Architecture and Art, Volume 9: Safety in underground construction – Peila, Viggiani & Celestino (Eds)
© 2020 Taylor & Francis Group, London, ISBN 978-0-367-46874-3

Investigation of the remaining life of an immersed tube tunnel in The Netherlands

K.G. Gavin & W. Broere
Technical University Delft, Delft, The Netherlands

M.S. Kovačević
University of Zagreb, Zagreb, Croatia

K. de Haas
COB, Delft, The Netherlands

ABSTRACT: In this paper we present issues related to the performance of immersed tube tunnels in the Netherlands. A range of issues are experienced for the ageing structures including long-term settlement at tunnel joints and consequent leakage. Because of the age of these structures some important aspects are often unknown thus creating uncertainty regarding remediation measures. Having discussed general issues the paper presents a case study of the Kil Tunnel which has experienced relatively large settlements over the past 40 years. In this case a lack of geotechnical engineering information for the soils below the tunnel was identified. A geophysical survey was undertaken and this provided ley insights into the ground conditions at the tunnel site.

1 INTRODUCTION

Worldwide, renovation of tunnels is becoming a huge challenge. Due to large costs and the need for accessible infrastructure, choices need to be made as to which tunnels will be renovated first or how to divide the renovation in affordable and practical parts, which renovations can be postponed and what should be the scope. Whilst a range of asset management strategies have been developed for road and rail infrastructure, the application of these approaches for tunnels is limited because of lack of data:

(i) From a structural perspective there is uncertainty in assessing the residual life span of the structure due to a lack of information on aging behaviour for joints, transitions and foundations.
(ii) The relationship between changes in the physical environment of the tunnel (soil, groundwater, changing river depths and widths, construction other structures) and the expected residual life span is uncertain.
(iii) Traffic loading is evolving and in conjunction with innovations in vehicle types this provides a further challenge in assessing future loading conditions in tunnels.

In this paper we consider the impacts of ageing on tunnels in soft soils common in the Netherlands. We focus on immersed tube tunnels as this was a very popular form of construction in the Netherlands from the 1960's onwards and a number of tunnels are due for major renovation in the coming years.

2 TUNNELING IN THE NETHERLANDS

2.1 Background

The majority of early road and rail tunnels constructed in the Netherlands were located away from urban areas. As a result, immersed tube and cut-and-cover methods were used exclusively from 1941 up to 1999. The need to minimize ground movements in the soft deltaic ground conditions with high water table levels during the development of metros lines in the major cities resulted in bored tunnelling technologies being deployed in approximately 25% of the transport tunnels constructed in the Netherlands since 1999. Given the predominance and relative age of the immersed tube construction form in the Netherlands the performance of these tunnels is the focus of the present paper.

2.2 Immersed Tube Tunnels

The first immersed tube tunnel constructed in the Netherlands was the Maas road tunnel in Rotterdam which opened in 1941. The system was a popular solution for Dutch ground conditions and topography with the result that to date almost thirty tunnels have been constructed using this technique, See Table 1. There was a particular boom in tunnel construction in the 1960's and 1970's. Given the design life of these structures many are due to undergo major retrofitting in the coming years. A number of these tunnels have exhibited signs of deterioration including corrosion, uplift of tension piles beneath approach embankments and leakage (Leeuw 2008) and van Montfort (2018). Leakage caused by differential is the largest problem facing tunnel owners and is the focus of the case study in this paper.

Table 1. Details of Immersed Tube Tunnels in the Netherlands.

Number	Name	Year Opened
1	Maas	1941
2	Coen Tunnel	1966
3	Benelux Tunnel	1967
4	Rotterdam Metro Tunnel	1968
5	IJ Tunnel	1969
6	Heinenoord Tunnel	1969
7	Vlake Tunnel	1975
8	Drecht Tunnel	1977
9	Prinses Magriet	1978
10	Kil Tunnel	1978
11	Hemspoor	1980
12	Botlek Tunnel	1980
13	Spijkenisse Metro	1984
14	Coolhaven	1984
15	Zeeburger	1990
16	Willemspoor	1990
17	Noord	1992
18	Grouw	1993
19	Schipol Railway Tunnel	1994
20	Wijker tunnel	1996
21	Willemspoor Tunnel	1996
22	Piet Heintunnel	1997
23	2nd Benelux tunnel	2002
24	Burgemeester Thomastunnel	2004
25	HSL-Zuid Oude Maas	2006
26	HSL-Zuid Dordtsche Kil	2006
27	2nd Coen Tunnel	2013

3 DESCRIPTION OF THE KILTUNNEL

3.1 Overview

The Kil tunnel with a length of 405m carries a two-lane highway and bicycle lane (each direction) under the Dordtsche Kil near Rotterdam. In cross-section the tunnel is a double tube concrete structure with a width of 31m and height of 8.75m. The immersed section is primarily composed of three elements, each approximately 113.5m long, with an end (land tunnel) of 35m at each end, See Figure 1. Each element was formed by joining five individual segments in a dry dock. The elements were then transported to site and connected in-situ at immersion joints. The system is sealed at the closure joint.

3.2 Performance to date

The tunnel in common with many others of this type constructed in the Netherlands has ongoing issues with settlement and leakage. Displacement measurements made throughout the operation of the tunnel are summarized in Figure 2. Linear variable displacement transducers were installed on the tunnel at the junction between segments. Initial movements were small, less than 5 mm and were concentrated in element 2. During maintenance work in 2001 significant quantities of sand was found in the pumping chamber and cracks were noticed in the wall between segments 2C and 2D, See Figure 3. Settlement measurements showed that significant settlements occurred between 1977 and 2001, forming a quite distinct pattern, whereby settlements were concentrated in elements 1 and 2, with element 3 exhibiting relatively small settlements. Unlike other tunnels, the settlements at the Kil tunnel were not concentrated at the immersion joints, rather on segment joints.

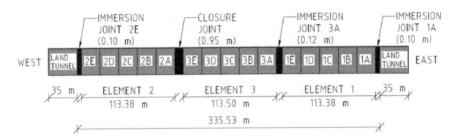

Figure 1. Long section through Kil tunnel showing the element locations, segment numbering and location of joints (from Rijkswaterstaat 1974).

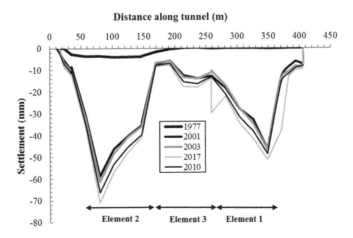

Figure 2. Vertical settlement of Kil Tunnel measured between 1977 and 2017.

Figure 3. Cracking at segment joint.

Due to the age and total design life of the tunnel a major remediation programme is imminent. Given the significant and ongoing settlement it is important for the tunnel owner to understand the mechanism driving this behaviour and, having established this, to determine if remedial action should be undertaken as part of the tunnel refurbishment. The Geo-Engineering Section at TU Delft and Rijkswaterstaat are lead partners in the EU Horizon 2020 SAFE-10-T investigating ongoing settlements at the nearby Heinenoord tunnel. The Dutch knowledge institute for underground space, Centrum Ondergronds Bouwen (COB) and the owners of the Kil Tunnel requested that this tunnel be included as an additional case study location in the project. Whilst a reasonably comprehensive ground investigation (in terms of number of locations) was conducted for the original construction, a review of the available data that comprised Cone Penetration (CPT) testing revealed that for the immersed tube section of the tunnel (Elements 1 to 3), very limited data was available for the soil immediately below the base of the tunnel. As a result, a geophysical investigation was planned in conjunction with SAFE-10-T partner the University of Zagreb to provide insight into the mechanisms controlling the settlement of Kil Tunnel.

4 GEOPHYSICAL INVESTIGATION

4.1 Test Method

The geophysical investigation technique chosen was multichannel analysis of surface waves, MASW. The method introduced by Park et al. (1999) and Xia et al. (1999) uses surface waves for the estimation of shear wave velocity (V_s) profiles. The method is an extension of the Spectral Analysis of Surface Waves (SASW) method (Nazarian and Stokoe, 1984), the most significant difference between the SASW and the MASW techniques, involves the use of multiple receivers with the MASW method (usually more than 12) which enables seismic data to be acquired relatively quickly when compared to the SASW method (Donohue et al 2011). A further advantage of the MASW approach is the ability of the technique to identify and separate fundamental and higher mode surface waves. According to elastic theory the small strain shear modulus, G_0, may be calculated from V_s, using the following equation:

$$G_{max} = \rho.V_s^2 \tag{1}$$

<u>Where:</u> G_0 = shear modulus (Pa), V_s = shear wave velocity (m/s) and ρ = density (kg/m^3).

MASW has been used extensively to estimate G_0 profiles when compared to other in-situ techniques such as cross-hole investigation (Donohue et al. 2003), as a quality assurance technique for ground improvement projects (Donohue and Long 2008) and to map changes in soil properties due to climate (Bergamo et al. 2016). The survey was undertaken in the cycle lane of the tunnel with both bike and road traffic remaining open during the survey, See Figure 4a. Due to the presence of large drainage culverts running perpendicular to the tunnel at immersion joint 1a and 2e respectively the survey could only be conducted on the soil beneath the three elements 1 to 3, See Figure 1 over a length of 335m. Four overlapping 100m long MASW profiles were acquired with overlap to ensure complete coverage across the tunnel section.

Whilst some embedment of the geophone in soils is necessary to ensure good contact, on road and concrete surfaces this is not possible and contact was ensured using a steel plates as shown in Figure 4a. A 6 kg sledgehammer was used to generate the surface waves which were in turn detected by 40 No. 10 Hz geophones placed at 2.5m centres, See Figure 4a. The source was located at the mid-point of the geophones, See Figure 4b. Data was recorded with a Smartsystem data logger at a sampling rate of 0.5 msec.

The Software SeisImager was used to obtain a dispersion curve from the phase velocity-frequency spectra, which was generated using a wave-field transformation method (Park et al., 1998). An example of a typical one-dimensional V_s profile measured at the site is shown in Figure 5.

Figure 4.   (a) Site test set-up (b) Test underway.

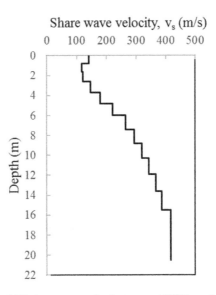

Figure 5.   Example of a typical 1D shear wave velocity trace at Kil Tunnel.

5 DISCUSSION

The MASW survey described in the previous section resulted in a series a 1D V_s profiles spaced at 2.5 m centres along the immersed tube tunnel. Using linear interpolation, a 2D Vs profile was created See Figure 6a. The dark blue colour indicates low velocity (soft soils) and the red colour is high velocity (stiff soils). The darker shade the blue colour is, the softer the soil. The main results are:

- There are very soft near surface soil zones (dark blue) evident in the profile, beneath element 2, Segments 2b-2E (See Figure 6b) and under element 1, Segment 1b. The location of the soft deposits correlates well with the observed settlement profile of the tunnel, See Figure 2 suggesting these low-stiffness zones are responsible for the unusual profile observed.
- Using Equation 1 to convert to shear wave velocity to soil stiffness suggest that the near surface stiffness of the soil beneath element 1 ($V_s \approx 270$ m/sec) is nine times higher than in these soft zones ($V_s \approx 90$ m/sec) which is generally in keeping with the magnitude of settlement evident in the different elements.

A preliminary interpretation would suggest that these soft zones are potentially old deeper channels of the river that were infilled with soft sediments or perhaps areas of poorly compacted sand that was backfilled into the excavated trench prior to tunnel placement. Given that MASW is a non-intrusive method of investigating the soil it has been recommended to perform addition CPT testing to provide verification of the interpreted ground model.

Considering the settlement pattern with time in Figure 7 and the settlement pattern in Figure 2, the accumulated settlement against time shows ongoing settlements that are highest in the region where the soft zones are evident and much lower where stiff soils exist. Settlements at all locations are continuing with time. Whilst the soft ground is clearly affecting the magnitude of settlement, the mechanism is not known. Two potential causes are considered, cyclic loading and creep. Cyclic loading of the Kil tunnel could arise from a number of sources, differential

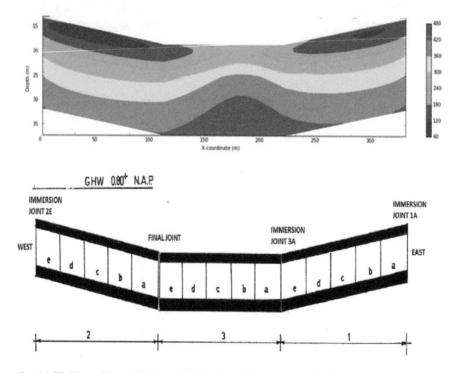

Figure 6.   (a) 2D Vs profile at Kil Tunnel (b) location of segments and joints (Rijkswaterstaat 1974).

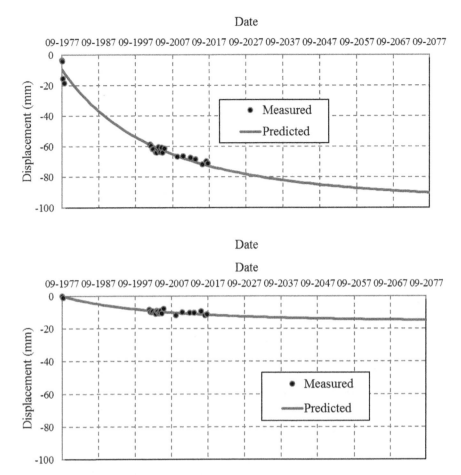

Figure 7. (a) Measured and predicted settlement for segment 2C and (b) segment 2E.

water pressures due to tidal effects, tragic loading, temperature etc. Creep or strain under constant load is a phenomenon affecting most materials, in this case it could arise due to creep of the foundation soil or concrete creep, particularly in the slopes segments 1 and 2 in the Kil Tunnel. For both cyclic loading and creep effects on soils the overall behaviour can be described by simple power-law expressions (Gavin et al. 2009 and Li et al. 2015) shown in Figure 7 assuming either plastic hardening during cyclic loading or time dependent creep reducing settlement with time. For all locations in Kil Tunnel these models suggest maximum total settlements in the range 10mm to 90mm up to the 100-year design life of the structure.

6 CONCLUSION

The paper presents a typical problem with ongoing settlement and leakage of ageing immersed tube tunnels in the Netherlands. MASW testing which is a quick, cost-effective and low-impact (no traffic disruption) technique provided vital information on the soil conditions that affect the displacement pattern observed at the Kil Tunnel. In particular it provided detailed information on the extent of soft soil zones underlying the tunnel and gives fundamental parameters (soil stiffness) for geotechnical engineering calculations. Whilst this is encouraging further work is required to (i) verify the ground model using direct in-situ testing and (ii) understand the mechanism driving the settlement observed at this and similar tunnels in the Netherlands.

ACKNOWLEDGEMENT

The work is part of the H2020 Project SAFE-10-T and the tunnel programme of the COB, Netherlands Centre for Underground Construction and Underground Space. The permission of the owners of the Kil Tunnel to publish the data is gratefully acknowledged.

REFERENCES

Bergamo, P., Dashwood, B., Uhlemann, S., Swift, R., Chambers, J., Gunn, D., & Donohue, S. (2016). Time-lapse monitoring of climate effects on earthworks using surface waves. Geophysics, 81(2), EN1–EN15. DOI:10.1190/geo2015-0275.

Donohue, S. and Long, M., (2008) Ground improvement assessment of glacial till using shear wave velocity. Proceedings of the 3rd International Conference on Site Characterization ICS'3: Geotechnical and Geophysical Site Characterization, Taipei, 825–830

Donohue, S., Gavin, K, Long, M. and O'Connor, P. (2003) Gmax from multichannel analysis of surface waves for Dublin boulder clay. In: Vanicek et al eds. Proc. 13th. ECSMGE Prague, pp.515–520

Donohue, S. Gavin, K. and Tooliyan, A. (2011) Use of geophysical techniques to examine slope failures, Journal of Near Surface Geophysics. Vol 9, No.1, February, pp 33–44, DOI: 10.3997/18730604.2010040

Gavin, K. Adekunte, A. and O'Kelly, B. (2009) A field investigation of vertical footing response on sand. Proceeding of ICE, Geotechnical Engineering, Vol.162, Issue GE5, pp 257–267, DOI: 10.I680/geng.2009.I62.5.257, October.

Leeuw, L. (2008). Lekkage in tunnels, Rijkswaterstaat Bouwdienst.

Li, W, Igoe, D. and Gavin, K, (2015) Field tests to investigate the cyclic loading response of monopiles in sand, Proceedings of the ICE Journal of Geotechnical Engineering (2015), Volume 168, Issue 5, October, pp 407–421.

Nazarian S. and Stokoe K.H. 1984. In situ shear wave velocities from spectral analysis of surface waves. Proceedings of the 8th World Conference on Earthquake Engineering, San Francisco, California, USA, Expanded Abstracts, 31–38.

Park C.B., Miller D.M. and Xia J. 1999. Multichannel Analysis of sur-face waves. Geophysics 64, 800–808.

Rijkswaterstaat (1974). Tekeningen Kiltunnel. Utrecht, Directie Sluizen en Stuwen.

Van Montfort, R. (2018) Insufficiency of immersion joints in existing immersed tunnels, MSc. Thesis TU Delft

Xia J., Miller R.D. and Park C.B. 1999. Estimation of near surface shear wave velocity by inversion of Raleigh waves. Geophysics 64, 691–700

Tunnels and Underground Cities: Engineering and Innovation meet Archaeology,
Architecture and Art, Volume 9: Safety in underground
construction – Peila, Viggiani & Celestino (Eds)
© 2020 Taylor & Francis Group, London, ISBN 978-0-367-46874-3

Fire safety engineering and risk-based tunnel safety design approach

S. Giua
IM Maggia Engineering SA, Locarno, Switzerland

L. Stantero
Studio Tecnico Luca Stantero, Torino, Italy

ABSTRACT: Fire Safety in tunnels became design primary issue to define the works' cost. Costs saving in energy, maintenance and safety management can be obtained adopting a risk-based approach. The paper presents this approach based on Fire Safety Engineering and probabilistic risk analysis. The CFD (Computational Fluid Dynamics) and egress simulations give the full integrated approach: ventilation and smoke extraction are evaluated and optimized, and egress pathways are verified by means of evacuation simulations. Risk-based design is an innovative approach of safety system dimensioning by adopting probabilistic methods to take in account different situations. The sizing of safety systems is no more based on the fire magnitude but on the global estimated risk. Alternative Risk Reduction Measures like automatic fire suppression, advanced ventilation management, traffic control measures, can be evaluated for a more effective approach. The risk-based approach can be applied dynamically, during operation, to define operational measures and maintenance priorities.

1 INTRODUCTION

The EU Directive 2004/54/EC introduced for the first time the risk-based design tunnel safety approach.

The new approach started from a fully prescriptive criteria by defining Minimum Safety requirements and introduced Risk Analysis as a design tool to allow derogation, evaluate integrative measures and adopt innovative safety systems.

This mixed approach between prescription and performance based pushed all the EU Countries to develop quantitative risk analysis methodologies that during the last 10 years have been improved and adapted.

The development and improvement process of methodologies and calculation methods has brought to a phase in which many aspects of the design have been considered in an even more complete way.

The main aspects of the design that emerged during the analysis are: the performances of safety systems and operational measures, the analysis of the evacuation scenarios, the risk quantification, the comparison with acceptability criteria and the cost-benefit analysis.

In this framework the authors have developed, adopted and verified risk analysis for hundreds of tunnels reaching a level of completeness of the design process that has been synthesizes in four points that allows to improve the safety design either for simple and for complex cases resulting a powerful design tool. The design tool allows the reduction of costs and/or the improvement of the real safety lever of the tunnel.

The four phases described in the paper are: systems performance analysis, system-based risk analysis, scenario-based risk analysis, quantitative risk evaluation.

2 RISK ANALYSIS LEVELS

The risk analysis for road tunnels is carried out on four different levels that complete each other to provide a complete picture of the risk associated with road tunnels. The four levels are:

1. Systems performance analysis by means of advanced methods (CFD),
2. Systemic approach- system based risk analysis,
3. Scenario based risk analysis by means of evacuation analysis,
4. Quantitative risk evaluation and cost-safety analysis.

The four levels have the function of analyzing the project from the risk point of view under its global and detailed aspects using the best available techniques and providing a result that allows the manager to have a comprehensive assessment of the safety level of the tunnel finalized to validate the design or define some design choices.

The combination of the four levels provides a complete and quantitative assessment of the safety level of the tunnel having considered:

1. the quantitative performance of safety systems,
2. the quantitative and probabilistic analysis of the global response of the tunnel system to the critical events connected to it based on its main characteristics,
3. the quantitative analysis of emergency scenarios with high level of detail to verify the possibility of saving lives,
4. the quantitative comparison with the acceptable risk limits.

2.1 *LEVEL 1: Systems performance analysis*

The Italian Transposition of the EU Directive on Road Tunnel Safety (D. Lgs 264/06) states:

> *"The minimum safety requirements of Annex II are predominantly designed to play a specific role in protecting, mitigating or inhibiting the potential increase in danger of the initiator event (e.g. heat output of the firebox, smoke propagation rate, etc.), as well as facilitating self-rescue actions for the exodus (e.g. emergency exits, visibility, opacity reduction, effective communication, etc.) and rescue in emergency conditions. Some of the above requirements also play a general preventive role in operating conditions."*

and referring to the performance of security systems:

> *"Safety subsystems determine the system's response to emergency conditions and consequently define the hazardous conditions for the population exposed to possible critical events."*

In order to evaluate the effectiveness of safety measures, the designer must analyze in detail the performance of the safety systems by defining the response of the systems for appropriate dimensioning scenarios to be able to meet the acceptable risk levels.

Specifically, the risk analysis must include the acquisition of the checks made by the designer regarding the systems in order to be able to analyze the scenarios with the methods described below.

For tunnel systems it is necessary to evaluate the following parameters:

- timing of incident events,
- closing times of the tunnel,
- air speed in all the branches of the tunnel and reaching times (including the ventilation strategy),
- smoke extraction flow rates (where a smoke extraction system is adopted),
- arrival times of the Fire Brigade,
- switch-off times and different fire ratings for automatic systems,
- reliability of all electrical and special installations,
- effectiveness of traffic management and communication devices to users,

- level of lighting of the evacuation paths and visibility of the signage,
- mitigation and suppression HRR and certified times for Fixed Fire Fighting systems.

About the installations that directly control the evolution of the emergency (i.e. ventilation and fire suppression) it's necessary to refer to the calculations and checks carried out in the context of the design. CFD modeling can be used conveniently in the following applications:

- optimization of aeraulic circuits (ventilation units, smoke extraction systems, air conditioning);
- verification of the fire rating of system components;
- support for the implementation of ventilation management systems;
- verification of the performances of ventilation and smoke control systems
- of detection systems performances
- verification and design support for Fixed Fire Fighting systems.

In some cases, the CFD model could allow the reduction of the uncertainties on the design solution, in others it allows to evaluate the performances of innovative systems to be proposed as integrative or alternative safety measures.

The value of the activities is strictly connected to the added value determined by the greater information coming from the model. The model can also be requested to better demonstrate a performance to the road manager or to the authorities. During the tender phase, the model leads to an enhancement of the technical offer and to a better explanation of the adopted solution.

2.2 LEVEL 2: Systemic approach- system based risk analysis

The Italian Transposition of the EU Directive on Road Tunnel Safety (D. Lgs 264/06) states:

"Quantitative Risk Analysis in road tunnels must be developed by adopting a systemic approach suitable for the specific scope of the tunnel system."

and:

"The methodology considers a gallery with its specific characteristics located on the territory and in interaction with the surrounding environment."

The systemic approach (Cafaro et al. 2013) take in account all aspects related to the tunnel system that affect the frequencies of critical events and the consequences deriving from them are considered in a more complete and coordinated way.

To be able to fully consider all the aspects, it was decided to use probabilistic quantitative models, based on the event tree, hazard flow and exodus models, which would allow us to provide a global result by evaluating the main aspects that affect the most safety of the whole system without dwelling on the details, as already analyzed in the two previous levels.

These models are collected in the computational calculation code that carries out the risk analysis related to fire events not connected to the transport of dangerous goods.

The dangerous good analysis is carried out with PIARC/OECD code named DG-QRAM. The choice is also justified by the fact that this code uses specific models for the explosion and release of toxic substances. More refined models (like CFD-3D) in case of dangerous goods fire don't give any advantages in terms of prediction of fatalities due to the high power developed and to the uncertainties related to the scenarios.

The output of the systemic approach is a distribution function to be used within Quantitative Analysis. This function is the combination of the functions determined for fires and for dangerous goods.

The method considers:

- event occurrence rates,
- the volume of traffic and its distribution,
- the geometric characteristics of the tunnel section (area, perimeter, etc.),
- the distance between the emergency exits and their standard deviation,

- the effectiveness of safety systems is defined as the number of successes of a given performance with respect to the total number of possible scenarios including the long-lasting effect of the performance.

The following systems are considered: detection, communication, ventilation, lighting, draining, Fixed Fire Fighting Suppression (FFFS) system as shown in Table 1

The model has been verified both during its tuning phase and in a specific way for the specific tunnel with CFD simulations as previously mentioned. The purpose of the model is to evaluate the weight of all factors that most affect the level of risk. The simplifications introduced, that allows the simulation of a high number of cases (over 10.000), has proven to be suitable for the desired level of analysis.

A 3D-CFD detailed analysis can be carried out for critical scenarios to validate simplified model results.

The effectiveness values of the systems are based on the best practice of fire events risk analysis (Sun, Luo, 2014) and on the analyst's experience, considering systems reliability and uncertainties. The risk analysis method adopted is compliant with annex 3 of D. Lgs. 264/06 and follows ANAS Guidelines for Safety Design in Road Tunnels (2009 version; ANAS is the Italian State Road owner).

The methodology is based on the following milestones:

1. systemic approach and probabilistic analysis of the frequency of occurrence and of the evolution of initiating events,
2. simulation of the hazard flow and exodus with different levels of detail according to the needs of the specific tunnel or tunnel portion,
3. analysis of events related to dangerous goods carried out with the PIARC model DG-QRAM.

Regarding the systemic approach it intended the characterization of the system in terms of construction characteristics, active and passive systems and variables that characterize the level of risk and which can be either deterministic (e.g. tunnel length, cross section area etc.) or of statistical type (weather conditions, traffic, intervention times) and that characterize the tunnel's response to the dangerous events.

The use of the event tree to consider some initiating events and the malfunctioning of the safety systems introduces a valid simplification to the need to analyze all the possible scenarios, but it can introduce considerable uncertainties when the case is limited to a few scenarios.

The need to set reference values to characterize an event could lead to the designer's discretion to use a series of "worst case" or a series of cases not in favor of safety.

Table 1. Effectiveness and performances of safety systems.

	Drainage	Detection	Communication	FFFS	Lighting	Ventilation
Effectiveness	0,85	0,9 Gauss distribution	0,9 Gauss distribution	0,85 Gauss distribution	0,9 Gauss distribution	0,85 Gauss distribution
Performance	15 l/min	Response 3 min Gauss	Response 2 min Gauss	100 MW mitigation in 10 min. Gauss	Lighting 2 lux Gauss	Sizing 50 MW Speed 3 m/s Gauss

To avoid these limitations, the analysis must consider the uncertainties. Therefore, the choice of the designer has been to adopt a calculation method based on the application of the Monte Carlo method to each branch of the event tree to obtain a probability distribution as continuous as possible.

The choice of such a methodology leads to the need to limit the calculation time by using simplified models of evacuation and of the hazard flow; the adoption of 3D models in fact would lead to a limited number of scenarios that, as already said, at the discretion of analyst, may be too precautionary or to the detriment of safety.

However, there are tunnels for which it is necessary to deepen certain scenarios with three-dimensional models in order to: validate the simplified models, refine the result or deepen some areas with complex geometry. It is also possible to represent some groups of scenarios with the resolution of a few representative scenarios in terms of frequency and number of fatalities obtaining points to represent on the risk side. These scenarios solved with advanced 3D fire and exodus models can provide a completion of the risk values calculated with the Monte Carlo method and simplified 1D models.

Regarding dangerous goods, has been chosen to adopt the DG-QRAM as a model that is also recognized at regulatory level in most European countries. The DG-QRAM is part of the simplified models for the fire and for the evacuation for which it presents a purely analytical and pseudo steady-state modeling. It also requires the definition of the ventilation regime to be carried out with other models not included in the package and at the analyst's choice.

On the other hand, DG-QRAM models are particularly effective for explosions and partially for the release of toxic substances. The design choice, given the complexity of the work which could have tunnels with different sections, clearing in and out of the tunnel, adopts the multi-scale approach. The hazard flow calculation models used will be:

- simplified 1D numerical model capable of simulating homogeneous tunnel sections,
- Low Mach Number 1D numerical model for possible deepening of the transitory regime on homogeneous tunnel sections,
- 1-D numerical model (IDA tunnel or SES type) for non-homogeneous tunnel sections and interconnected systems,
- CFD 3D numerical model for any geometry,
- 1D-3D CFD numerical model directly coupled.

The following figures show the hazard flow calculated with different techniques.

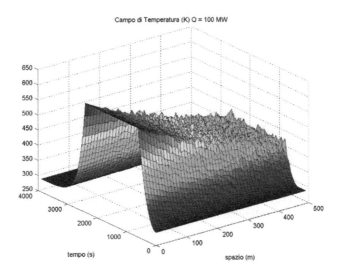

Figure 1.   Thermodynamic 1-D model.

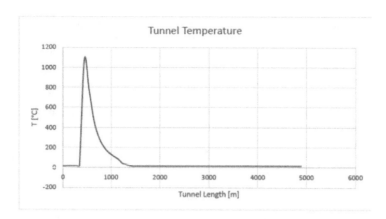

Figure 2. SES 1D model-temperature vs tunnel position.

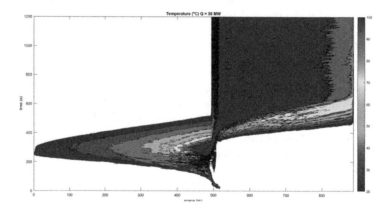

Figure 3. 3D model results - Contours of Visibility at 1.5 m from floor level.

It is specified that the aforementioned simulations are not intended to determine the performance of the ventilation system but to determine and verify the evacuation conditions. The effectiveness of the ventilation systems is verified in detail in phase 1, that is, in the ventilation system design. In particular, the effectiveness of the system is verified ensuring critical velocity considering different fire HRR, i.e. it will be evaluated the scenarios part that cannot be managed with the ventilation system.

The composition of the scenarios analyzed aims to provide a complete picture of all the problems and to identify a risk curve for each defined segment.

The adopted approach allows feedback on the design of ventilation systems too and on the definition of operational strategies and on redundancies verification aimed to ensuring continuity of operation even in case of multiple failures without incurring in oversizing the system. The methods and models adopted for risk analysis are in fact similar to those used for the design and verification phase of the system.

In addition, the methodology adopted also allows for the best management of design modifications required by the Owners during the construction phase, having set up scalable models.

For very complex cases, preliminary calculations may be performed to verify the variation of the risk as a function of the fire heat release rate for the sizing of the system in order to con-firm the initial figure.

The incident scenarios considered by the risk analysis that will be carried out are, as required by D. Lgs. 264/06: fires, collisions with fire, spills of flammable substances, releases of toxic and harmful substances.

With regard to the fire the following events are considered in a statistical way:

- car fire (5-8 MW),
- bus-van fire (15–30 MW),
- heavy vehicle fire at medium fire load (30–50MW),
- heavy fire vehicle with a high fire load (50–100 MW).

As far as collisions with fire are concerned, they are derived as subcases of the fire.

About dangerous goods, the following events are considered, including the previously mentioned puddle fire, for which a scenario analysis will be carried out:

- fire of liquid fuel puddle (150–200 MW),
- release of liquid or gaseous toxic substances.

To be congruent with PIARC recommendations, that are adopted by most of the EU Nations, DG-QRAM software will be used with regard to the dangerous goods only. Two F-N curves are derived and cumulated in a single final curve as requested by D. Lgs 264/06: fire curve for non-dangerous goods, events curve related to of dangerous goods transportation.

The Figure 4 summarizes the risk analysis process that combines the three types of models: 3D; 1D with Monte Carlo, DG-QRAM.

2.3 *LEVEL 3: scenario analysis*

At this level a quantitative analysis is carried out for selected and representative scenarios to verify, on the basis of the performance defined at LEVEL 1, the possibility to rescue the users through the modeling of the fire and evacuation event with in-depth methods and with high levels of detail.

The purpose of the analysis is twofold:

- verify that users, based on the characteristics of the tunnel, can save themselves;
- provide reference results to be used to support probabilistic models envisaged at LEVEL 4.

The analysis carried out is quantitative native and provides the assessment of fatalities number associated with specific scenarios chosen by the analyst, that must be defined accordingly with safety designer.

Each scenario is described in terms of input data and the results are given in terms of the characteristics of the danger flow (air speed, height and smoke temperatures, visibility, toxic concentrations) and of exodus process (times to reach the exits, critical zones).

Regarding the input data, the characteristics of the outbreak are defined (power, calorific value and emission rates of the species for the specific fuel), the length of the simulated domain, the approximations introduced (multiscale method), the conditions environmental and boundary conditions.

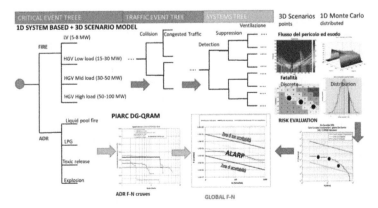

Figure 4.   Risk analysis process.

Figure 5. Coupled 3D Fire and evacuation model.

Users survival is expressed as FED (as in NFPA Standards) defined on the basis of carbon monoxide and maximum temperatures to which it is possible to survive.

2.4 *LEVEL 4: Quantitative Risk Evaluation*

The quantitative risk evaluation involves the determination of risk indicators and the comparison with risk acceptance criteria as defined by the standard and possibly as defined by the analyst-designer, as well as the subsequent cost-benefit analysis.

As far as the checks are concerned, reference is generally made to the criteria and indicators defined at point 3 of Annex 3 of D. Lgs. 264/2006:

> *"The risk associated with a tunnel is defined as the expected value of the damage or as the disbursement of the probability of exceeding predetermined loss thresholds (Cumulative Complementary Distributions reported on the so-called F - N plan).*
>
> – *The risk as the expected value of the damage is obtained as the sum of the products be-tween the probabilities of the individual critical initiator events and the corresponding summation of the probabilities of the terminal events of the individual branches of the events multiplied by the corresponding damage indicators expressed in number of victims normalized to the year.*

Table 2. Cost-benefits analysis.

Tunnel	Example			
Length [m]	5000			
RV [Risk Value]	0.095			
Period [years]	50			
COST-BENEFIT ANALYSIS				
	Emergency exits from 300m to 200m	Fixed Fire suppression System	Ventilation sizing from 100MW to 200MW	Public Address system
Effectiveness	0.85	0.75	0.4	0.1
Estimated unit cost [€/m]	€ 1,200.00	€ 800.00	€ 500.00	€ 75.00
Construction cost [€]	€ 6,000,000.00	€ 4,000,000.00	€ 2,500,000.00	€ 375,000.00
Monetary risk, reduction of RV to 0.02 fat7year	€ 5,625,000.00	€ 5,625,000.00	€ 5,625,000.00	€ 5,625,000.00
Monetary risk reduction	€ 4,781,250.00	€ 4,218,750.00	€ 2,250,000.00	€ 562,500.00
Monetary risk reduction of 0.01 fat/year	€ 637,500.00	€ 562,500.00	€ 300,000.00	€ 75,000.00
Cost-Benefit	1.255	0.948	1.111	0.667

– *The risk as a distribution of the probability of exceeding predetermined damage thresholds is shown graphically on the F-N plane (where F indicates the probability of exceeding the threshold and N the number of fatalities) from the cumulative complementary distribution (probability of exceeding the damage thresholds) obtained in correspondence with the values of the damage indicators (damage thresholds) associated with the terminal events of the individual branches of the event tree."*

Table 2 reports a cost-benefit analysis for a sample tunnel with the comparison of different safety improvement measures.

3 DYNAMIC RISK ANALYSIS

The risk analysis process could be simplified and fastened to implement in a real time software tool that allows the definition of the instantaneous and projected risk level of a single tunnel, multiple tunnel and a complete road.

The calculated risk allows the implementation of operational measures such as: maintenance times, Fire Brigade and rescue team intervention time, traffic management.

The performances of dynamic risk analysis in terms of potential risk reduction must be evaluated by data analytics on real cases on a well-defined period (at least one year). The performances are evaluated in terms of effective risk with respect to the estimated risk performed by design risk analysis. The possibility to monitor risk and to demonstrate that risk levels are maintained under the design risk value allows to define an average risk reduction.

This tool will improve with the time and we expect advantages in terms of operation improvement year by year.

The main considered hazards are: collisions, fires, dangerous goods events.

The base relation for the dynamic risk R_i related to the i-th hazard is:

$$R_i = p_i(n)k_iD_if^a\varepsilon_iN_iV/V_{ri} \tag{1}$$

where
$p_i(n)$ = probability of occurrence after n days without accidents
k_i = constant depending on tunnel characteristics
D_i = data analytics bias
f = traffic flow
a = exponent depending on the type of hazard (Collision, Fire, Dangerous Goods)
ε_i = efficiency of safety systems related to i-th hazard
N_i = Number of potential fatalities for i-th accident type
V = Average vehicle speed
V_{ri} = Reference vehicle speed
The safety level S_i related to every single hazard is defined in a scale derived directly from the risk as follows:

$$S_i = -log_{10}(R_i) \tag{2}$$

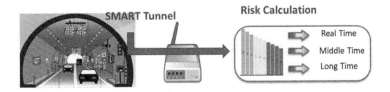

Figure 6.   Smart tunnel and dynamic risk calculation.

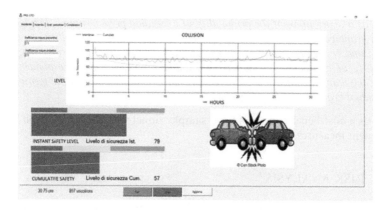

Figure 7. The PRE-STO application for dynamic risk calculation.

The safety level can be evaluated in real time or forecasted in the middle period (week, month) and in a long-term period (year) and it is compared with different predefined threshold values in order to define different actions: the intervention of safety officers, reduction of speed limits, maintenance and if necessary, the renovation of some safety systems.

4 CONCLUSIONS

The paper has shown a complete process of tunnel safety design that uses all the advanced analysis techniques and allows the general definition of the main tunnel characteristics that affects safety of users with optimal costs and suitable safety margins.

The method allows to evaluate properly different design alternatives by comparing safety levels and cost benefit.

In a second phase the tool allows a better understanding of the details of the tunnel functionalities to define or verify management procedures, the CFD scenario analysis help in solving critical scenarios while global risk analysis gives a balance of the influence of the different systems.

This method allows tunnel managers to have a virtual touch of the safety level of a new or existing tunnel to be refurbished supporting their decisions that could orient the design process.

The designer has a complete tool which results can be synthesized to let the manager understand the design process ad solutions.

Dynamic risk analysis is a condensed risk analysis that allows the road manager to maintain the risk level along the time at predetermined values by acting with operational measures.

REFERENCES

Cafaro E. & Saba F. 2013. Approccio Ingegneristico all'Analisi di rischio in galleria. In *Strade & Autostrade* n. 98: 28–35. Milan, Italy.
Directive 2004/54/EC of the European Parliament and of the Council of 29 April 2004 on minimum safety requirements for tunnels in the Trans-European Road Network.
D. Lgs. n. 264 of 5 October 2006. Attuazione della direttiva 2004/54/CE in materia di sicurezza per le gallerie della rete stradale transeuropea.
Report 2008R02EN. 2008. Risk analysis for road tunnels. PIARC
Report 2012R20EN. 2012. Assessing and improving safety in existing road tunnels. PIARC
Report 2012R23EN. 2012. Risk evaluation, current practice for risk evaluation for road tunnels. PIARC
Report 2016R35EN. 2016. Experience with Significant Incidents in Road Tunnels. PIARC
Sun X. & Luo M. 2014. Fire Risk Assessment for Super High-rise Buildings. In *Procedia Engineering* 71: 492–4501. Elsevier

Tunnels and Underground Cities: Engineering and Innovation meet Archaeology,
Architecture and Art, Volume 9: Safety in underground
construction – Peila, Viggiani & Celestino (Eds)
© 2020 Taylor & Francis Group, London, ISBN 978-0-367-46874-3

Single-channel blowing-in longitudinal ventilation theory and its applicability analysis in road tunnel

C. Guo, Z.-G. Yan & H.-H. Zhu
Department of Civil Engineering, Tongji University, Shanghai, China

ABSTRACT: Nowadays, a majority of extra-long road tunnels adopt longitudinal ventilation methods and combine the vertical or inclined shafts and jet fans. However, such methods have the problems of expensive civil construction costs and high operating costs. The single-channel blowing-in longitudinal method emerges in recent years. Only one ventilation channel is set between the left and right main tunnels, and rich fresh air at the entrance of the downhill tunnel is sent to the exit of uphill tunnel through the ventilation channel. The authors compared the theory and applicable conditions of single-channel method and single shaft method with MATLAB. It can be concluded that the single-channel method can reduce the operation energy consumption and has a good application prospect.

1 INTRODUCTION

The seventies and eighties of the last century witnessed the worldwide mainstream of road tunnel ventilation types varying from horizontal ventilation (such as Frejus Road Tunnel [Mos et al., 2009] which connects France and Italy) and semi-horizontal ventilation (for example, Cross Harbor Tunnel [Chan et al., 1996] in Hong Kong) to longitudinal ventilation (for instance, Shanghai South Hongmei Road Tunnel [Guo et al., 2013] in China). Nowadays, most long tunnels adopt the two-hole one-way traffic design, and longitudinal ventilation method is widely utilized with the combination of full jet fans and vertical or inclined shafts. By making use of this combination, shafts will divide the tunnel into several sections so that the air volume can be controlled flexibly, also the one-way traffic flow will provide piston wind which leads to fewer jet fans. The ventilation design has been adopted by lots of traffic constructions, such as Kan-Etsu Tunnel [Bobrov et al., 2001] and Tokyo Bay tunnel in Japan [Yamada et al., 1999], as well as Qinling-zhongnan Mountain Tunnel [Yan, 2006] in China. The increasing need for longitudinal ventilation stimulates further research on this method. Up till now, for longitudinal ventilation system, many scholars have studied the arrangement of jet fans [Betta et al., 2009] as well as the evacuation of fire smoke [Carvel et al., 2001; Vauquelin et al., 2006] from a perspective of theoretical derivation, numerical analysis and model experiments. However, the construction cost of ventilation shafts is relatively high, let alone the maintenance expense of blowing fans in shafts. Considering these disadvantages, some scholars began to study a new longitudinal ventilation method, Double-Hole Complementary Ventilation Method [Berner et al., 1991]. The concept was first proposed by Burner and Day in 1991. Then Xia et al. discussed its calculation methodology, and applied the ventilation method to China Dabie Mountain Tunnel [Xia et al., 2013]. Ideally speaking, the ventilation system does not require ventilation wells, thus reducing initial investment and operation costs of tunnels.

Economic as Double-Hole Complementary Ventilation is, there is a non-negligible limitation on its application. In the cases where airflow between two tunnel lines is similar, the pollutant discharging efficiency will drastically drop down. The research showed that the Double-Hole Complementary Ventilation should be adopted where the ventilation load ratio is greater than 1.5 or the tunnel line slope is within 1.5% and 2.0% [Wang et al., 2015]. To overcome these shortcomings, Yan et al. proposes a new ventilation approach for extra-long road tunnels, Single-Channel Blowing-in Longitudinal Ventilation Method. [Yan et al., 2018] Compared with longitudinal ventilation with vertical shafts, this method can save the construction costs of shafts. In comparison with Double-Hole Complementary Ventilation Method, it surmounts the limitation in traffic flow difference, simplifies the organization of ventilation system in normal condition, and enhances the convenience of fire smoke management.

2 THEORETICAL ANALYSIS OF SINGLE-CHANNEL METHOD

2.1 *Analytical Model*

As shown in Figure 1, Line L_1 is uphill while L_2 is downhill. With the same traffic speed and traffic volume, it is not far to see that the average air pollutant concentration of Line L_1 is higher than that of Line L_2. The added single-channel L_3 is x m away from the entrance of Line L_1, and it divided Line L_1 and L_2 into channel L_{11}, channel L_{12}, channel L_{21} and channel L_{22}. After single-channel L_3 is applied, the downward Line L_2 transports a part of fresh air into tunnel L_1, therefore the pollutant concentration of channel L_{12} decreases.

2.2 *The Calculation method of Single-Channel ventilation method*

In the calculation, there are three criteria, CO concentration $Q_{ireq(CO)}$, dust density $Q_{ireq(VI)}$ and minimal air exchange frequency $Q_{ireq(ac)}$.

2.2.1 *Governing equation of Single-Channel ventilation method*
Each criterion is considered under different traffic speeds which vary from 10km/h to 80km/h. The annual required air volume of tunnel line Q_{01} and Q_{02} should be the maximum of the three indexes, as shown in *Eq. (1)* and *Eq. (2)*.

$$Q_{01} = \max\left\{Q_{1req(VI)}, Q_{1req(CO)}, Q_{1req(ac)}\right\} \tag{1}$$

$$Q_{02} = \max\left\{Q_{2req(VI)}, Q_{2req(CO)}, Q_{2req(ac)}\right\} \tag{2}$$

Judging from the features of airflow, the critical sections are (1) the exit of line L_1 (2) the exit of Line L_2 (3) the joint between Line L_1 and single-channel L_3. According to the three criterial sections, *Eq. (3)*, *Eq. (4)*, *Eq. (5)* and the designed air volume of certain section Q_1, Q_2 and Q_3 are constructed as followings.

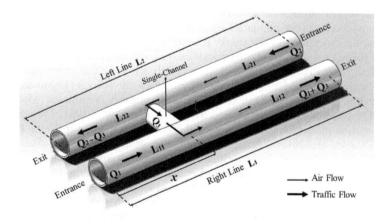

Figure 1.   Diagram of Single-Channel Blowing-in Ventilation Method.

$$Q_{01} \cdot L_2 + Q_{02} \cdot (L_2 - x) \cdot \frac{Q_3}{Q_2} = (Q_1 + Q_3) \cdot L_2 \tag{3}$$

$$Q_{02} \cdot L_2 - Q_{02} \cdot (L_2 - x) \cdot \frac{Q_3}{Q_2} = (Q_2 - Q_3) \cdot L_2 \tag{4}$$

$$Q_{01} \cdot x = Q_1 \cdot L_1 \tag{5}$$

Note that, *Eq.* (3), *Eq.* (4) and *Eq.* (5) respectively indicates that the concentrations of pollutants at the exit of the Line L_1, the exit of Line L_2, and the joint between Line L_1 and single-channel L_3 exactly reach the standard boundary.

2.2.2 *Optimal location of the single-channel in one year*
The ventilation resistance of each section ΔP_{ri}, traffic wind power ΔP_{ti} and natural wind resistance ΔP_{mi} can be calculated as the followings.

$$\Delta P_{ri} = \left(\lambda . \frac{L_i}{D_r} + \xi_i \right) \cdot \frac{\rho}{2} \cdot \left(\frac{Q_i}{A_r} \right)^2 \tag{6}$$

$$\Delta P_{ti} = \frac{A_m}{A_r} \cdot \frac{\rho}{2} \cdot \frac{N_j \cdot L_i}{3600 \cdot v_t} \cdot (v_t - \frac{Q_i}{A_r})^2 \tag{7}$$

$$\Delta P_{mi} = \lambda \cdot \frac{L_i}{D_r} + \xi_i) \cdot \frac{\rho}{2} \cdot v_n^2 \tag{8}$$

Therefore, the energy consumption of the ventilation system in a specific year can be expressed in *Eq.*(9).

$$P_j = \sum (\Delta P_{ri} + \Delta P_{mi} - \Delta P_{ti}) \cdot Q_i + \Delta P_{r3} \cdot Q_3 \tag{9}$$

2.2.3 *Optimal location of the single-channel in serving period*
In fact, traffic volume changes annually, so the optimal location of single-channel will vary every year. As *Eq.*(10) shows, P_{total} refers to the overall ventilation energy consumption for T years. As shown in *Eq.*(11), x_{fit} means the optimal location of single-channel which enables P_{total} to be minimal.

$$P_{total} = \sum_{j=1}^{T} P_j(x) \tag{10}$$

$$\min(P_{total}) = \sum_{j=1}^{T} P_j(x_{fit}) \tag{11}$$

3 APPLICABILITY ANALYSIS IN THE TUNNEL

3.1 *Design of the single-channel and Single-shaft*

A tunnel in China, adopts Two-hole One-way traffic design. The original design was to install single shaft at the right tunnel. The right line adopts shaft ventilation and introduces emergency smoke channels and the left line adopts full jet longitudinal ventilation. According to the geological prospecting data, it was found that the geological environment near the shaft was complicated. So the single-channel method was considered instead.

We compared the energy consumption and applicability of the single-channel method and single shaft method with MATLAB. One year and serving period energy consumption are calculated by changing the location of single channel and single shaft. The stability of the two methods is analyzed by changing traffic volume and the ratio of the traffic volume of right line in one year. The parameters of the tunnel and two methods are shown in Table 1, and the diagram of two methods are shown in Figure 2.

3.2 Feature and stability analysis

The P_{total} is the sum of ventilation energy consumption for 30 years. This section probes for the variation regulation of optimal location of single channel and single shaft x_{fit}, one-year ventilation energy consumption P_{one}, total ventilation energy consumption P_{total}.

Table 1. Parameters of energy consumption calculation in MATLAB.

Calculated Parameters	Value
Right Line Length $L_1/(m)$	7531
Left Line Length $L_2/(m)$	7531
Clearance area of tunnel $A_r/(m^2)$	65.18
Tunnel Equivalent Diameter $D_r/(m)$	8.29
Tunnel drag coefficient λ	0.02
Air density $\rho/(kg/m^3)$	1.29
Local resistance coefficient of tunnel portal ξ_1	0.5
Local resistance coefficient of tunnel exit ξ_2	1
Bifurcation Loss coefficient ξ_3	0.244
Confluence Loss coefficient ξ_4	0.7
Automobile equivalent resistance Area $A_m/(m^2)$	3
Single-channel length $L_3/(m)$	30
Single-channel Area $A_3/(m^2)$	9.13
Length of exhaust and blowing shaft $L_4/(m)$	340
Exhaust shaft area $A_4/(m^2)$	18.62
Blowing shaft area $A_5/(m^2)$	11.59
Design speed of automobile $v_t/(km/h)$	80
Natural air velocity $v_n/(m/s)$	2.5
Traffic volume proportion of right line L_{ratio}	0.3–0.7
Designed peak hourly traffic volume $N/(pcu/h)$	300–1500

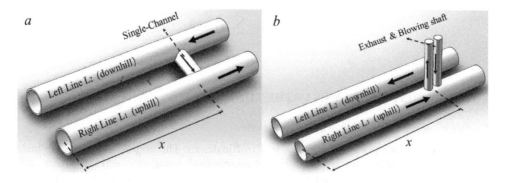

Figure 2. Diagram of Single-Channel Method and Single Shaft Method.

In the calculation, the x represents the distance between the single channel or single shaft and the entrance of right tunnel. Noted, the term, "moving right', denotes that the distance x increase, while 'moving right' means x decrease. Actually, designed hourly peak traffic volume N increases with the year. There are tremendous predicting models for traffic volume, and in this paper, a linear increasing model is adopted.

3.2.1 Variation regulations of ventilation energy consumption

To make data more different under various cases, the ratio of traffic volume of right line to that of left line is 7:3. According to Figure 3, there are four findings.

1. In the single-channel method, with x increasing and single-channel moving right, P_{one} decreases monotonously then increases slightly. All x_{fit}, corresponding to the location of single-tunnel where the minimal P_{one} is reached, is within 6000m-6500m.
2. In the single shaft method, with x increasing, P_{one} decreases first and then increases. All x_{fit}, is within 4000m-5000m.
3. It is assumed peak hourly traffic volume N has a constant taken from 300 pcu/h to 1500 pcu/h with a step of 100 pcu/h from year 2015 to 2045 in serving period. Single-channel x_{fit} is 6445m, where the corresponding minimal P_{total} is 9.36×10^6W. Single shaft x_{fit} is 4664m, where the corresponding minimal P_{total} is 8.07×10^6W, 13.8% lower than single-channel method.
4. P_{one} varies drastically if x changes a lot, and the maximal P_{one} can be 50 times of minimal P_{one}. The x_{fit} of one year is different from x_{fit} for serving period. Therefore, the final design

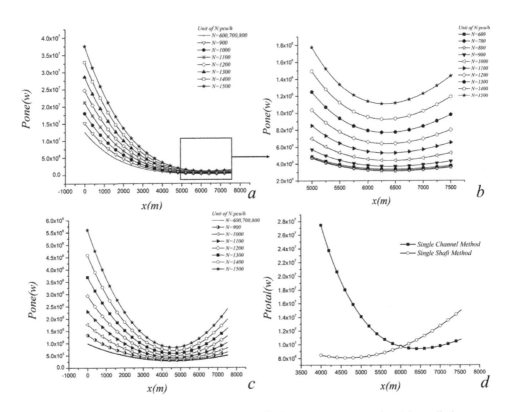

Figure 3. Diagram of the position of x versus ventilation energy consumption (a) ventilation energy consumption in one year P_{one} versus single-channel x (b) figure a. partial enlargement (c) ventilation energy consumption in one year P_{one} versus single shaft x (d) total ventilation energy consumption P_{total} versus single-channel and single shaft x.

of single-channel cannot make all of P_{one} to reach the minimum, and under some circumstances, none of P_{one} can reach the theoretical minimum.

3.2.2 *Discussion on x_{fit} and Pone in one year*

Considering that $L_{ratio} = 0.3, 0.4, 0.5, 0.6$ and 0.7, also assuming that N is a constant taken from 300 pcu/h to 1500 pcu/h with a step of 100 pcu/h, the x_{fit}-N curves and P_{one}-N curves of two methods are presented in Figure 4 and Figure 5. The two methods have the same variation trend as follows.

1. When L_{ratio} is stationary, if x_{fit} increases, single-channel or single shaft moves right and *Pone* dwindles. Thus, x_{fit} has a negative correlation with P_{one}.
2. When $L_{ratio} = 0.3$ or $N \leq 800$pcu/h, *Pone* declines with the increasing of N. Because the traffic airflow of the downhill line is fully used, both two methods can be more energy-efficient even though traffic becomes busier.
3. With the same N, if L_{ratio} descends, the downhill left line takes more traffic flow and *Pone* decreases.
4. However, when $L_{ratio} \geq 0.4$, x_{fit} and P_{one} show linear or quadric growth after N reaches a critical value. This feature can be explained by the regulation of required air volume Q_{01} and Q_{02}. Before N achieves the critical point, the required air volume is defined by minimal air exchange frequency which is a constant. If N exceeds the threshold and the uphill line

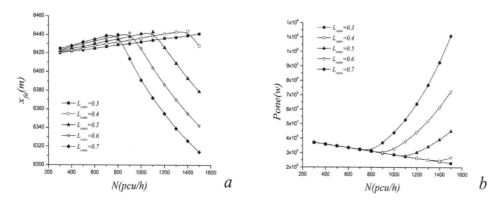

Figure 4. Diagram of (a) single-channel method optimal location x_{fit} and (b) one-year ventilation energy consumption P_{one} versus traffic volume N.

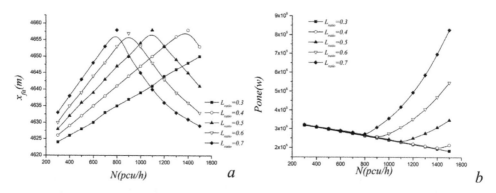

Figure 5. Diagram of (a) single shaft method optimal location x_{fit} and (a) one-year ventilation energy consumption P_{one} versus traffic volume N.

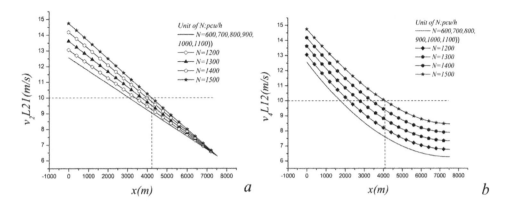

Figure 6.   Diagram of air velocity (a) v_{21} and (b) v_{12} versus x in the single-channel method.

shares a large proportion of traffic, Q_{01} is defined by dust density which arises sharply when N increases, and *Pone* ascends in the same manner.

Both two method shows a better energy-efficiency property and stability of single-channel and single shaft position.

3.2.3 *Discussion on air velocity in tunnel*

According to Guidelines for Design of Ventilation of Highway Tunnels, v_i, the wind speed of Line i, should be less than 10m/s, and it can be calculated as *Eq.*(12) presents.

$$v_i = \frac{Q_i}{A} \tag{12}$$

With the increase of traffic volume, the air velocity of each section in the single shaft method meets the requirements regardless of the exhaust and blowing shaft location. However, in the single-channel method, the distance from single-channel to the entrance of right line must be greater than 4227m, so that all sections in the single-channel method can meet the requirements of wind speed. Therefore, the location of a single-channel takes into account not only energy consumption and geological conditions, but also the limitations of tunnel wind speed.

4 CONCLUSION

This paper compares the theory and applicable conditions of single-channel method and single shaft method with **MATLAB**, and proves effectiveness of the single-channel method. The two methods have the same variation regulations. It can be concluded that optimal position x_{fit} is relevantly stationary even traffic volume N or ratio of right line traffic volume L_{ratio} changes drastically. Besides, x_{fit} and one-year ventilation energy consumption P_{one} versus traffic volume N of the two methods have a similar variation trend.

In the case of linear growth of traffic volume, the two methods achieve the minimum energy consumption at different locations and the energy consumption is close. The single-channel method can be used as a substitute for shaft ventilation method in practical engineering, especially in the mountain tunnel where it is difficult to build shafts.

During the operation period, the energy consumption of the tunnel is proportional to the square of the length of the tunnel and is proportional to the cube of the ventilation quantity. The energy saving principle of single-channel is to increase the ventilation energy consumption of the shorter tunnel sections and reduce most of the energy consumption of longer tunnel sections. So that the whole tunnel achieves the goal of operating energy-saving.

ACKNOWLEDGEMENT

The research was supported by China major science and technology special project in Guizhou province (Research and demonstration on key technologies of efficient construction and intelligent operation and maintenance of ultra-long and large-span highway tunnel in Karst area).

REFERENCES

Berner, M. A., Day, J. R., 1991. A New Concept for Ventilation Long Twin-tube Tunnels. in: *7th International Symposium on the Aerodynamics and Ventilation of Vehicle Tunnels*, Cranfield UK, pp. 811–820.

Betta, V., Cascetta, F., Musto, M., Rotondo, G., 2009. Numerical study of the optimization of the pitch angle of an alternative jet fan in a longitudinal tunnel ventilation system. *Tunnelling and Underground Space Technology* 24(2),164–172.

Betta, V., Cascetta, F., Musto M, Rotondo, G., 2010. Fluid dynamic performances of traditional and alternative jet fans in tunnel longitudinal ventilation systems. *Tunnelling & Underground Space Technology* 25(4),415–422.

Bobrov, O. P., Khonik, V. A., Zhelezny, V. S., 2001. Operation and adequacy of the new ventilation system of the Kan-etsu Road Tunnel. *Journal of Non-Crystalline Solids* 223(3),241–249.

Carvel, R. O., Beard, A. N., Jowitt, P. W., 2001. The influence of longitudinal ventilation systems on fires in tunnels. *Tunnelling and Underground Space Technology* 16(1),3–21.

Chan, L. Y., Zeng, L., Qin, Y., Lee, S.C., 1996. CO concentration inside the Cross Harbor Tunnel in Hong Kong. *Environment International* 22(4),405–409.

Guo, Q. C., Yan, Z. G., Zhu, H. H., Shen, Y., 2013. Numerical simulation on fire characteristics and smoke control of a longitudinal ventilation road tunnel. In: *World Conference of Acuus: Advances in Underground Space Development*, pp:232–240.

Industry Standard of the People's Republic of China. JTG/T D70/2-02—2014. Guidelines for Design of Ventilation of Highway Tunnels. Beijing, China Communication Press.

Mos, A., Brousse, B., Chabert, A., Giovannelli, G., 2009. Modification of a transverse ventilation system: A probability and CFD-based assessment of the improvement in fire safety. In: *13th International Symposium on Aerodynamics and Ventilation of Vehicle Tunnels*, France, pp. 239–253.

Vauquelin, O., Wu, Y., 2006. Influence of tunnel width on longitudinal smoke control. *Fire Safety Journal* 41(6):420–426.

Wang, Y. Q., Jiang, X. M., Wu, Y. K., Xie, Y. K., 2015. Analysis on Applicability of Twin-tube Complementary Ventilation in Road Tunnel. *Modern Tunnelling Technology* 52(3),14–21.

Xia, F. Y., Wang, Y. Q., Jiang, X. M., 2013. Analysis and Field Test about Operation Ventilation Effect of Dabieshan Highway Tunnel. *Applied Mechanics & Materials* 438–439, 1000–1003.

Yamada, N., Ota, Y., 1999. Safety systems for the Trans-Tokyo Bay Highway Tunnel project. *Tunnelling & Underground Space Technology* 14(1),3–12.

Yan, Z. G., 2006. Experimental study of shaft ventilation modes for road tunnels in case of fire. *China Civil Engineering Journal* 39(11),101–106.

Yan, Z. G., 2018. Single-channel blowing-in longitudinal ventilation theory and its application in road tunnel. *Tunnelling & Underground Space Technology*.

Tunnels and Underground Cities: Engineering and Innovation meet Archaeology,
Architecture and Art, Volume 9: Safety in underground
construction – Peila, Viggiani & Celestino (Eds)
© 2020 Taylor & Francis Group, London, ISBN 978-0-367-46874-3

Virtual reality for hazard assessment and risk mitigation in tunneling

E. Isleyen, S. Duzgun & P. Nelson
Colorado School of Mines, Golden, CO, USA

ABSTRACT: Virtual reality (VR) simulations provide realistic training environments for safety training and decision-making enhancement. A typical example in tunneling is hazard assessment after blasting. Ground support systems must be installed based on the stand-up time to ensure the safe access of personnel and equipment to the working area. This study presents VR models developed for hazard assessment training in tunneling. The VR simulations involve decision-making tasks to help users learn how to create a safe working environment. First, they are asked to identify the potential hazard around the tunnel face. If the hazard identification is successful, users are requested to install support systems by specifying certain attributes. Finally, participants evaluate the roof fall hazard by examining the displacements around the opening. The initial results indicate that VR is a useful tool for safety training, and have great potential for enhancing decision-making process.

1 INTRODUCTION

Virtual reality (VR) has shown up in many industries as a technology with financial and operational benefits. VR technologies bring a "creative destruction" in labor training by replacing traditional methods with digital techniques that provide cost-effectiveness. As VR hardware becomes more available and affordable, the number of companies using VR as a training environment has been increasing significantly. Pioneering industries of VR-based training are the military, healthcare and aviation, and various other high-hazard industries including tunneling. Tunneling is an industry that requires frequent operational training. However, the working environment introduces various hazards to workers. Therefore, having a safe environment where the employees can experience the most dangerous conditions without any real danger has been seen as the cornerstone of effective labor training in tunneling. Improvements in VR technology offer a solution to this problem by providing immersive, realistic and safe training environments.

Implementation of the VR simulations in various fields for safety training and hazard identification has become popular in recent years. In the mining industry, VR-based training has been in use for some time, in specific operational areas to overcome problems related to safety and to create an overall safe work environment. Lucas and Thabet (2008) developed a VR simulation to train young miners on how to interact with a conveyor belt in mining operations. Grabowski and Jankowski (2015) developed a VR simulation for training workers dealing with blasting operations in mining. They tested different VR equipment and concluded that head-mounted displays with a wide field of view are the optimum VR system. In the context of mining, the VR systems are still considered to be in early developments. Therefore, a systematic evaluation methodology is needed to evaluate the effectiveness of VR-based training simulations objectively (Tichon & Burgess-Limerick 2011). Fire safety training is another field that could benefit from the potential of VR. Cha et al. (2012) introduced a real-time dynamic data conversion methodology for smoke and flames in VR for fire safety training.

Xu et al. (2014) developed a fire training simulator in VR which requires participants to assess the hazard based on smoke levels and identify the safest evacuation path. The growing potential for VR applications in safety training and hazard assessment has also been investigated by the construction industry. VR-based training improves construction safety by maintaining participant attention and preserving the effects of training over longer periods of time than conventional training methods (Sacks et al. 2013). A significant number of accidents happening each year in this industry indicates a need for innovative approaches to safety training (Bhoir & Esmaeili 2015). Li et al. (2018) provided an overview of VR-based construction safety research, trends in state-of-the-art hazard assessment and safety training tools, and future directions. VR-based safety training has been implemented in other fields such as oil and gas, aviation, vehicle driving, chemical processing and maintenance works (Manca et al. 2013, Gavish et al. 2015, Muhlberger et al. 2015, Buttussi & Chittaro 2018). Although there are various examples of VR-based training, methodological approaches for developing and implementing VR-based training in tunneling is not sufficiently established.

This paper introduces a methodology to implement VR in hazard assessment training in tunneling operations. The methodology is applied in a virtual tunnel that includes predefined hazards. The participants are asked to perform hazard identification and hazard mitigation by installing different support systems. They have an opportunity to carry out self-assessment by viewing the outcome of their support decision. The targeted participants are defined as foreman and/or safety engineers, who will be responsible to make tunnel inspections, in order to detect any unexpected potential failures. The developed VR environment is designed for improving the decision-making skills of the participants who are expected to make decisions on failure types and support measures in real-world under several constraints like time, visibility, and proximity.

2 METHODOLOGY

The methodology presented in this study consists of five steps (Figure 1). The first step is to develop a scenario that accurately mimics the real-world tunneling hazard. The second step is to create a 3D model of a tunnel in which the scenario will be implemented. The third step is the real-time rendering of the 3D models and interactions in a game engine. The fourth step is the VR implementation for the selected scenario. The last step is testing the simulation by participants and receiving feedback about potential improvements.

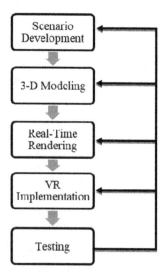

Figure 1. Methodology.

2.1 Scenario Development

The purpose of the simulation is to train people and to support their judgment on hazard identification and risk mitigation in tunneling operations. In this study, roof fall is selected as the hazard condition and the following scenario is proposed. Participants are taken into a virtual tunnel where the blasting operation has been completed recently. Participants' role is to conduct tunnel inspection and detect the potential hazards. After the correct hazard identification, they are asked to install rock bolts as the support measure by selecting the design parameters (e.g. bolt length and spacing). Displacement amounts are shown to the participants around the tunnel and they're updated with each installation of different bolt systems. Finally, participants select the most suitable bolt pattern for the underground opening.

2.2 3-D Modeling

A 3D tunnel model was developed which is consistent with the conditions defined in the scenario, using Blender software. The 3D tunnel model is based on the point cloud of the Edgar Experimental Mine of Colorado School of Mines (CSM) which is located in Idaho Springs, Colorado. However, only a small part of the original point cloud is used for the purpose of this study. Figure 2 shows the steps of the 3D modeling phase.

The point cloud data is converted to the quadrangulated mesh and the solid model is created (Figure 2a and Figure 2b). We generated texture coordinates and applied sample material properties in Blender in order to test the final rendered model before exporting it to a game engine (Figure 2c). The final form of the 3D model is exported as OSGB (OpenSceneGraph Binary) file.

2.3 Real-Time Rendering

The Vizard game engine was used for real-time rendering. Interactions between the user and the environment are defined in the Vizard scripts. Vizard is mostly used for VR applications and it's fully compatible with the available VR infrastructure. Rock-like material properties with seamless rock texture images were prepared to create a virtual environment with a realistic and continuous appearance. Headlights are the only light source in the scene to implement visibility constraint in the tunnel environment. Standard shading effects such as normal, occlusion and specular maps are applied to improve lighting effects.

2.4 VR Implementation

The virtual tunnel simulation runs on the walking VR system in the VR laboratory of the Mining Engineering Department at CSM. The system makes it possible to walk and interact with the virtual environment inside the tracking space by high precision motion tracking. The motion tracking system used is the Worldviz PPT Precision Position Tracker. The

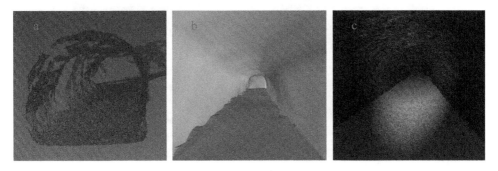

Figure 2. 3-D modeling steps: (a) point cloud, (b) solid model, (c) rendered model.

participant's head and hand motion are tracked by the infrared (IR) cameras through the sensors attached to the

headset and the controller. The motion tracking in the virtual environment is at the real scale of the physical environment. Moving at scale enables the participants to feel the full immersion while minimizing the motion sickness which is the most frequently experienced problem with VR systems. The components of the walking VR system are shown in Figure 3.

The VR lab is equipped with six IR cameras (Figure 3a) in a 4 m × 4 m room, controllers with LED sensors (Figure 3b) and Oculus Rift headsets with LED sensors (Figure 3c). The camera tracks the motion of the LED sensors. Thereby, the participant's head and hand motions are conveyed to the virtual world with millimetric accuracy. Participants interact with the virtual environment by using the buttons on the controller. They can move around in the virtual environment by walking or by using the joystick buttons (Figure 4).

At the initial step of the VR simulation, a graphical user interface asks participants to inspect the tunnel and determine if there is a rock fall hazard or not. The participants are expected to walk around in the tunnel to find and identify the potential hazard (Figure 5).

If the hazard identification is correct, participants proceed to the risk mitigation stage. A new interface appears and informs the users that they are expected to use rock bolts as the

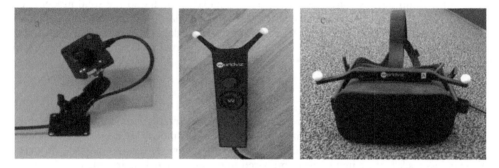

Figure 3. Components of the walking VR system: a) IR camera, b) controller with LED sensors, c) headset with LED sensors.

Figure 4. A user participating in the VR simulation.

Figure 5. Starting interface for hazard identification and the rockfall hazard in the tunnel.

support system and asks them to select a suitable rock bolt length. Additionally, the dimensions of the tunnel and the displacement amounts at certain points around the tunnel are shown. After selecting the rock bolt length, the participants are asked to select the spacing between the rock bolts (Figure 6). Then, the rock bolt pattern is shown to the user around the tunnel and the displacement amounts are updated depending on the selected bolt pattern. Also, tunnel geometry changes based on the new displacement amounts after the bolt installation.

The users have the option to cycle through all the rock bolt pattern combinations to see their effects on the tunnel displacements. Moreover, they can move outside the tunnel, to inspect if the rock bolts are effective to hold the hanging rock blocks. Figure 7 shows an ineffective rock bolt pattern for holding the large blocks. VR simulation gives the users an opportunity to visually distinguish the effects of different bolt lengths and change the design parameters if they decide the current pattern is insufficient. In addition, the VR simulation also acts as a medium for visualizing numerical analysis results for displacements and stress fields around the tunnel (Figure 8).

2.5 Testing

After the successful implementation of the scenario in the VR environment, several faculty members in the CSM Mining Engineering Department were invited for an initial trial of the VR simulation. Volunteers commented on the usefulness and the immersion of the simulation. The purpose of the testing phase is to validate the resemblance of the VR simulation to the actual field conditions by expert opinion, before using it for training purposes.

The experts participated in the VR indicated the promising potential of the simulation in terms of training engineers for roof hazard assessment. They noted the advantages of investigating the

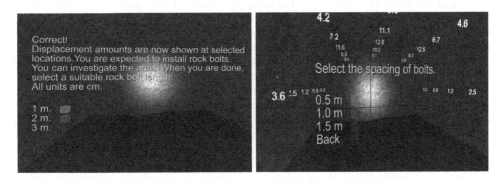

Figure 6. Selection of rock bolt length and the spacing between the bolts.

Figure 7. Viewing rock bolts from outside of the tunnel.

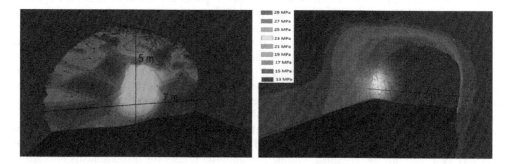

Figure 8. Visualization of numerical analysis results for displacement vectors and stress distributions.

effects of different support systems in a fully immersive simulation. Understanding how different support measures change the displacements around the tunnel is the key element to improve decision-making skills in the virtual tunnel. Also, the trainees have the opportunity to judge the results of their own hazard mitigation decisions and gain experience through a self-assessment.

3 CONCLUSION

This study introduces a methodology for developing VR-based simulations for hazard identification and risk mitigation training in tunneling operations. The objective of the VR simulation is to train personnel for improved decision-making skills. The participants are asked to identify the potential failure in the virtual tunnel. After the correct hazard identification, different rock bolt lengths and spacings are offered as the possible hazard mitigation measures. The VR infrastructure incorporates high precision motion tracking that maximizes the feeling of immersion while minimizing the motion sickness.

VR provides a safe training environment to improve the efficiency of visual detection of all kinds of ground control problems, geological structures, and instabilities. Since these features may be unfamiliar to the personnel in various underground works, having a training system to enhance their situational awareness is an important safety measure. In addition, implementing this type of simulations in engineering education is likely to have positive effects on student's understanding of tunneling operations and geological features.

The proposed methodological approach in this study is suitable for developing various training modules. Implementation of this methodology demonstrates its versatility to be adopted for developing alternative training exercises. The feedback loop based on testing provides opportunities for continuous improvement needed for developing advanced training modules.

REFERENCES

Bhoir, S. and Esmaeili, B. (2015). State-of-the-art review of virtual reality environment applications in construction safety. *AEI 2015 Milwaukee, Wisconsin, March 24–27,2015.*

Buttussi, F. and Chittaro, L. 2018. Effects of different types of virtual reality display on presence and learning in a safety training scenario. *IEEE Transactions on Visualization and Computer Graphics* 24 (2): 1063–1076.

Cha, M., Han, S., Lee, J. and Choi, B. 2012. A virtual reality based fire training simulator integrated with fire dynamics data. *Fire Safety Journal* 50: 12–24.

Gavish, N., Gutierrez, T., Webel, S., Rodriguez, J., Peveri, M., Bockholt, U. and Tecchia, F. 2015. Evaluating virtual reality and augmented reality training for industrial maintenance and assembly tasks. *Interactive Learning Environments* 23 (6): 778–798.

Grabowski, A. and Jankowski, J. 2015. Virtual reality-based pilot training for underground coal miners. *Safety Science* 72: 310–314.

Li, X., Yi, W., Chi, H., Wang, W. and Chan, A. 2018. A critical review of virtual and augmented reality (vr/ar) applications in construction safety. *Automation in Construction* 86: 150–162.

Lucas, J. and Thabet, W. 2008. Implementation and evaluation of a vr task-based training tool for conveyor belt safety training. *ITcon* 13, Special Issue Virtual and Augmented Reality in Design and Construction: 637–659.

Manca, D., Brambilla, S. and Colombo, S. 2013. Bridging between virtual reality and accident simulation for training of process-industry operators. *Advances in Engineering Software* 55: 1–9.

Muhlberger, A., Kinateder, M., Brutting, J., Eder, S., Muller, M., Gromer, D. and Pauli, P. 2015. Influence of information and instructions on human behavior in tunnel accidents: a virtual reality study. *Journal of Virtual Reality and Broadcasting* 12 (3).

Sacks, R., Perlman, A. and Barak, R. (2013). Construction safety training using immersive virtual reality. *Construction Management and Economics* 31 (9): 1005–1017.

Tichon, J. and Burgess-Limerick, R. 2011. A review of virtual reality as a medium for safety related training in mining. *Journal of Health & Safety Research & Practice* 3 (1): 33–40.

Xu, Z., Lu, X. Z., Guan, H., Chen, C. and Ren, A. Z. 2014. A virtual reality based fire training simulator with smoke hazard assessment capacity. *Advances in Engineering Software* 68:1–8.

Tunnels and Underground Cities: Engineering and Innovation meet Archaeology,
Architecture and Art, Volume 9: Safety in underground
construction – Peila, Viggiani & Celestino (Eds)
© 2020 Taylor & Francis Group, London, ISBN 978-0-367-46874-3

The convergence of all communication and data requirements onto a single IP backbone

D. Kent & D. Edmonds
MST Global, Australia

ABSTRACT: This paper discusses the application of high bandwidth networks to converge all voice and data requirements onto a single communication backbone. Reference will be made to the application of Wi-Fi & Internet Protocol (IP) networks in the Singapore T302 and the West-Connex/NorthConnex projects in Sydney. The industrial world has been converging all their communication & data requirements onto a single IP backbone to ensure the capacity & reliability of the system is optimum. Tunnelling is no exception, where the potential to optimize machine performance & personnel performance by ready access to high quality data to monitor, report & proactively manage processes delivers significant cost benefits. The application of purpose built, open protocol IP networks, allows this convergence of data to be available to the underground industry in a form factor that makes it practical to deploy and use for the life of a project.

1 INTRODUCTION

The need for greater bandwidth and quality of service (QoS) from underground communication systems has been growing over the last two decades. Initially this increasing bandwidth was met with just adding more data cables, including optic fibre and attempts at Ethernet over leaky feeder. But the ability to extend and maintain several data networks has become increasingly onerous to the point where the attitude of "this has worked for us to now" has reached its limit and network planning is becoming key consideration for tunnelling projects wanting to leverage modern IT and data solutions.

An initial driver was the development of remote operating and guidance systems and other high bandwidth applications, such as video.

The network, together with the process control and production networks, all now fall under what is termed the Operational Technologies (OT) category. The significant difference between the newer OT and the previously recognised term of Information Technologies (IT) is that OT encompasses hardware and software that affects the productivity and profitability of a tunnelling or mining operation. It also generally infers that the systems and hardware are deployed in rugged, harsh and "environmentally unfriendly" locations of an operation.

Both the OT and IT networks are critical to the success of a modern tunnelling operation and demand careful and considered design consultation to enable the delivery of a holistic technologies platform for a data ready project.

This paper will discuss the journey MST has been through to develop a modern convergent OT platform and to explain the what, where and how of the practical elements in the underground network puzzle.

2 WHAT HIGH BANDWIDTH NETWORKS ENABLE

Once a pervasive network in is in place, value is realised by using this core platform to enable a myriad of applications and devices to be used to better plan, monitor, analyse and proactively manage underground tasks and processes.

Personal Communications

A fundamental requirement for any communication network, the digital, Wi-Fi networks uses Voice over Internet Protocol (VoIP) telephony for vice and text communications. But being Wi-Fi based, the system offers several advantages over more traditional radio systems:

– License free band used
– Individual, private dialling
– 24 Push to Talk (PTT) channels
– Emergency "man-down" button
– Text messaging capability

Data Requirements

The increasing ability to monitor, analyse and manage underground activities has seen an increasing requirement for digital networks more generally in underground projects.

Currently there is a lot of buzzwords around "big data", but the fact is that modern machines, plus intensified environmental monitoring systems and video surveillance, does can produce vast amounts of data. To be anywhere near useful, a first step is the accurate transmission of that data to the various software applications and systems that can utilise that data and allow productive value to be extracted from it.

So again, the data and communication network plays a core role in accessing the data and being able to proactively use it to optimise underground tasks and processes.

Guidance and Automation Systems

TBM and roadheader guidance systems require good data connectivity to enable remote monitoring. Often done with dedicated fibre or other data cables, this connectivity is achieved using the IMPACT network backbone, wired or wirelessly.

That is, rather than requiring separately installed and maintained data links, the guidance systems can be supported by the single, converged network installed for other applications and devices.

Tracking Equipment and People

Tracking the movement of people, machines and equipment in a tunnelling project provides a powerful safety and management tool. Examples in WestConnex and NorthConnex include:

– Personnel safety and emergency preparedness
– Logging machine use and availability
– Ensuring safety modules and sleds are in correct provisions and within required distances to the tunnelling face.
– Monitor truck and loader cycle times for production management and deviation alerts.
– Easy remote programing and software updates on tunnelling machines.

Short Interval Control

In many projects, exampled in mining with groups like Anglo American and Glencore, there is growing efforts to proactively manage operations to minimise the impact of any deviation from plan. Tunnelling is no different.

This means that tasks are planned and then monitored using tools such as tracking/location systems, machine diagnostics systems, and operator communications. By setting business rules for tasks their progress can be monitored to plan and should a deviation occur (e.g. as simple as delivering segments out of order, a truck or loader taking the wrong route, or delivering equipment to the wrong location) alerts can be raised in the system and key personnel advised so the situation can be remedied and corrective action taken.

The chain of events from deviation to corrective action being taken is controlled by the various tools and applications a good quality network allows. The key being minimising the time between deviations to corrective actions – short interval control.

Basically, it is the proactive management of tunnelling tasks and processes underground by having sufficient reliable data (such as machine location and status), analysis and reporting tools, and the means for communication of corrective action.

Paperless Reporting

Many doubt the "paperless office" concept, as modern computing and printing has made the generation of reams of paperwork much easier. But open protocol connectivity like Wi-Fi enables many common business tools to be used underground. An example is the use of tablets to replace paper reports, such as pre-start check reports for vehicles. The connected tablet logs the location and time of the report, which is automatically uploaded into the project database for historical recording.

Rather than a sheet of paper that is filled in, then possibly countersigned and logged, a tablet running a pre-start application is much more efficient and effective.

Connecting Devices

IoT, or whatever other buzz word you want to use, is nothing new. Underground sensors and equipment monitoring has been around a long time. Having broad Wi-Fi coverage provides an ideal system to allow easy connection to these various sensors and devices to relay their data accurately to where it can be used.

The opportunity this opens up is the ability to remotely monitor in detail many machine and environmental parameters. So along with the associated increase in computational power, be able to analyse greater and greater amounts of data and apply sophisticated analytic tools to better understand and anticipate events.

3 THE TECHNOLOGY

The evolution of IT infrastructure to a rugged, fit for purpose OT infrastructure for use in the underground environment, has enabled all the features and functionality that we have grown accustomed to having at our disposal in our comfortable surface office to be available throughout the tunnel right up to the tunnelling face. Through the use of firewalls and VLAN segregation, what was once considered and protected to be only corporate IT data can now be available underground across the single piece of infrastructure delivering OT data applications such as Tunnel Boring Machine (TBM) guidance systems, production reporting & management, and automation.

A typical high bandwidth network topology is shown in Figure 1, which shows the logical connections between the various IT and OT applications which can be utilised in an underground tunnel.

The core infrastructure components that comprise the MST communications platform are:

- Wireless Network Switches
- Wireless Access Points
- Composite Cable
- Wireless Repeater Node (WRN)
- Vehicle Intelligence Platform (VIP) data logger and Wi-Fi bridge
- ImPact Communications Appliance (ICA) – MineDash software

A more detailed overview of some of these key components follows.

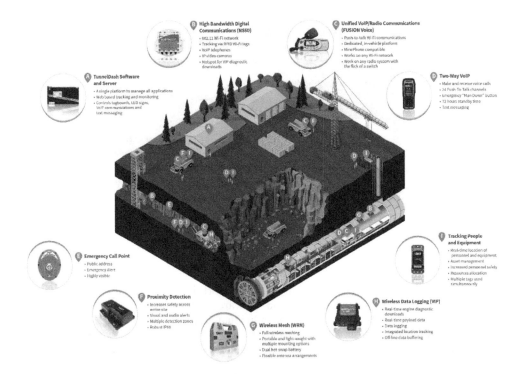

Figure 1. Key Wi-Fi Network Components and Applications.

3.1 Wireless Network Switch – NS50

The "workhorse" of the wireless infrastructure platform in MST's case, is the Wireless Network Switch, known as the NS50. The NS50 is the third generation field device and consists of four managed, fibre optic Ethernet switches, two 802.11 b/g wireless access points, and four Power over Ethernet (PoE) ports providing scalable wired and wireless network access. Its high bandwidth and low latency design enables multi-service applications to share the network infrastructure. This multi-service aspect means on top of the data transport role to support automation systems the switch also facilitates functionality such as VoIP, IP video streaming, remote PLC programming, mobile data acquisition, real-time vehicle diagnostics and asset/ personnel tracking.

3.2 Composite Cable – Optic Fibre & Power

Though power is often not an issue in tunnelling operations, soemthimes it distribtuion can be difficult and expensive. Traditional enterprise networks have a star topology which requires power at every network node, which is not a cost effective solution in the tunnelling environment. The challenge of limited power availability is overcome using Composite Fibre Cable, which carries both power and the single mode optical fibre cores.

3.3 Wireless Repeater Node (WRN)

The WRN is the most recent addition to the "network family" and was developed to simplify connectivity in the most dynamic areas of the tunnel, that is the tunnelling face, and to do so wirelessly. The WRN provides up to 120 hours of operation and is considerate of the 24-hour reserve required for emergency communications in certain safety jurisdictions.

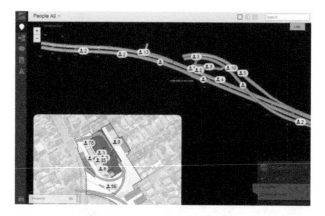

Figure 2.   The ICA prvides the user intrerafce and visualisation of tracking and alarm data.

3.4   IMPACT Communication Appliance (ICA)

The ability to remotely manage and monitor the network is critical to any network deployment. The ICA is a server based hardware and software platform that constantly monitors the network and provides tools for managing the individual network components.

It also runs the TunnelDash software, which is the operators' interface, where data can be viewed and reported on, including tracking, voice communications, machine health data, system alerts, etc.

The key elements outlined above make up the single network to support a range of key functions in the tunnel additional to any machine guidance or automation specific requirements.

3.5   The Devices

Once the IMPACT network is in place it enables the use of any Wi-Fi or Ethernet enable device.

These can be IP Video cameras, tablets and laptops, etc.

However a number of key devices are used for communication and tracking are specifically supplied by MST to meet the environmental and use requirements, these include:

3.5.1   RFID Tags

A range if Wi-Fi based RFID tags are sued for personnel and equipment to enable the tracking of people and equipment throughout the tunnel and surface areas. The Tags are active, in that transmit a Wi-Fi signal that is detected and logged by the NS50 APs.

Figure 3.   RFID tags to suit a variety of applications.

3.5.2 Voice Communication

Using Voice over Internet Protocol (VoIP) technology two-way voice is provided with a range of mobile and fixed telephone options that operate using the Wi-Fi signals set up with the IMPACT network.

Figure 4.   VoIP Telephones are sued for voice communications and apps.

3.5.3 Vehicle Intelligence Platform (VIP)

A key part of VIP system is the VIP Data Logger/Wi-Fi Bridge module. This VIP module I fitted to key assets, such as trucks and loaders, and performs a number of functions:

– Logs and stores machine data (such as load, pressures, temperatures, etc).
– Reads and logs fixed RFID Tags to allow accurate logging of location (e.g. TBM, cross over switches, dumps, shaft bottom). This is sometimes referred to as reverse tracking.
– Above data is time stamped when read and stored.

Uploads the time stamped data whenever in range of Wi-Fi signals. Hence this could be streaming in real-time when Wi-Fi coverage is present in all areas, or uploaded when in signal range (store and forward functionality).

Figure 5.   VIP Datalogger/Wi-Fi Bridge.

4 CASE STUDIES

A number of recent tunnel projects in Australia and Singapore have deployed high bandwidth networks to support a number of applications required by the tunnel contractors.

4.1 *T302 Project, Singapore*

MST collaborated with DJS Solutions to provide a total communication and electrical solution for the main contractor, China Rail First Group. MST Global provided a complete communication solution and DJS Solutions provided the electrical requirements.

China Rail First Group (CRFG) required data to the TBM for navigation and monitoring, telephone communications, camera system for monitoring and also fire alarm points in the 4 tunnels as they were being excavated.

MST Global designed an integrated system for the T302 metro project. Based on the IMPACT data network system connected via optic fibre from the office to shaft bottom and through the tunnel to the TBM control cabin.

The MST Headend which includes an ICA, optic fibre and Ethernet switches is positioned in the office next to the control room. A PC workstation is provided for MSTs TunnelDash and a big screen provided for monitoring of the video surveillance system. Wireless Access Points (WAPs) are used in the office and on the surface to provide Wi-Fi coverage. A fibre optic break out enclosure and outdoor UPS is positioned at shaft top which distributes the fibre and power into the 4 tunnels.

Underground, the system utilises the NS50s to provide Wi-Fi and Ethernet break out points throughout the tunnel. The NS50s unique design allows MST to distribute the fibre and the power together in a plug and play solution. The Fire Alarm Call Points (FACP) utilise the PoE Ethernet ports on the NS50 for connection, the FACPs are configured to allow an incoming and outgoing Ethernet port allowing daisy chain connection every 100m between the NS50s. Cameras also utilise the NS50s Ethernet ports for network connection throughout the tunnel.

In summary, key elements are:

IMPACT: The MST network solution is IMPACT, ruggedized IP65 hardware suitable to the tunnelling environment. With "plug and play" composite cables and connectors the fibre backbone and MST network switches provided at intervals along the bored tunnel provides access to the Ethernet TCP/IP network (PoE) and also Wi-Fi propagation throughout the tunnel.

The ImPact system provides access to data for 3rd party equipment such as gas detection, cameras, ventilation and pump control.

Voice communications: MST utilises a digital PBX VoIP system connected over Ethernet and fibre. VoIP telephones are utilised in the office as deskphones, outdoors as weatherproof phones and telephones are integrated as part of the FACPs. Seamless interconnectivity is achieved with the MST MinePhone so that the mobile units can be used throughout the tunnel or on the move on site surface.

Fire Alarm Call Points: MST provided and integrated Fire Alarm Call Points at the required 100m intervals for IP VoIP telephone connectivity throughout the tunnel, the surface, Main Contractor and LTA site offices and portable MinePhone radio units. The Fire Alarm Call Points also feature an emergency break glass alarm activation point that triggers a 120dB alarm and shows the location of the triggered unit in the tunnel through MSTs Tunnel-Dash system. The system provides for all the LTA emergency voice communication requirements.

Figure 6.   Fire Alarm Call Point Module.

TBM data connection: MSTs fibre backbone provides a dedicated fibre line to the TBM. The TBM also houses 2 Fire Alarm Call Points and an NS50 network switch for PoE connection. This provides the required data connection for real-time monitoring and reporting navigation connection to the surface and retains the telephone and fire alarm network through to the TBMs manlock.

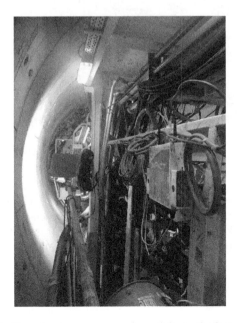

Figure 7.   NS50 on TBM trailing structure ensy-ures cinnectivity to the face.

4.2   *NorthConnex/WestConnex, Australia*

The number of projects made up the large NorthConnex and WestConnex road scheme in Sydney.

Rather than a few large TBM drives, the project consisted of almost 40 roadheaders working in different sections of the massive project.

This added larger communication and management challenges for the project. Managing the large number of people and pieces of equipment to support all the concurrent excavation presented a challenge to tunnelling contractors.

MST was chosen to provide a converged communication and tracking system to enable the multiple sites to manage and to ensure the whereabouts of all personnel were known to better manage safety should any incident occur. In particular, as sections broke through and the number of entry and exit points increased, then ensuring people were in a safe location should an emergency occur, meant active electronic tagging was the only option of tracking people on the move through the tunnel complexes.

Figure 8.   NS50 providing Wi-Fi signal around a roadheaader at North Connex.

A number of constructions joint ventures are delivering four main projects within the overall WestConnex and NorthConnex complex. MST was chosen by each direct JV to provide the networking and tracking system. Something that has made the overall delivery more streamlined as different tunnel sections merge and a seamless integration of systems was undertaken.

In all to cover the 42km of tunnelling over 200 x NS50 AP/Switches, 150 x WAPs, 5000x Wi-Fi RFID Tags, 900 x VoIP Phones and various emergency call points, power supplies etc made up the various networks.

Figure 9. All areas at WestConnex required Wi-Fi Connectivity.

5 CONCLUSION

Modern tunnelling operations are no different from other major construction and mining projects in that they require high quality connectivity to ensure safe and productive operations.

The convergence of data and communication onto a single IP network is leveraged to deliver this connectivity underground.

MST has focused on customizing open standard Wi-Fi and IP networks to have their topology suit the underground environment, so that delivering this connectivity and subsequent benefits can be done in a practical and cost effective way.

Tunnels and Underground Cities: Engineering and Innovation meet Archaeology,
Architecture and Art, Volume 9: Safety in underground
construction – Peila, Viggiani & Celestino (Eds)
© 2020 Taylor & Francis Group, London, ISBN 978-0-367-46874-3

Safety in construction of underground structures in complex geology of Tehri Pump Storage Project – successful implementation of safe working practices – a case study

R.K. Khali
Hindustan Construction Co Limited, Mumbai, India

ABSTRACT: Safety, central to the construction of any mega project, is nothing but awareness of executing projects without any mishap or loss of life. The construction of underground tunnels, shafts, chambers, and passageways is indispensable yet fraught with dangers. Reduced light conditions, difficult or limited access and egress, or possibility of being exposed to air contaminants and the hazards of fire and explosion are the threats an underground construction worker may face. To ensure the safety and health of underground workers, the Occupational Safety and Health Administration has been made operational at the prestigious Tehri pump storage project. Safety operations are not delimited only to the permanent structures and persons working on it, they extend to the infrastructure and its users as well. The technology incorporated for safety in construction during the course of the project has been brought to focus in the present paper. Examples illustrating the efficacy of safety measures in averting accidents have been cited. Some photographs to illustrate the methodology have also been appended. The paper seeks to establish that while it is difficult to maintain safety at construction site, it is even more challenging to maintain safety in the entire gamut of infrastructural framework which supports the construction of the main work. It is felt that more attention needs to be paid to frame the safety guidelines for infrastructural works per se and ensure zero accident at project sites.

1 INTRODUCTION

The Tehri PSP is located on the left bank of the Bhagirathi river in the district of Tehri, about 1.5 km downstream of its confluence with the Bhilangana river (now forming a part of Tehri Reservoir), falling between 78°30' and 79°00'E longitudes and corresponding 30°30' and 33° 30'N latitudes. The nearest railhead is Rishikesh (Uttarakhand), located approximately at about 82 Km south of the project site. The construction of the Tehri PSP by HCC Ltd under progress, envisages an underground Machine Hall on the left bank of river Bhagirathi, housing 4 reversible pump turbine units, each of 250 MW capacity. The reservoir of the Tehri dam will operate as the upper reservoir and the Koteshwar reservoir as the lower reservoir for this project. Two completed Head Race Tunnels (HRTs) (3 and 4) will be augmented by respective Surge Shafts at the end of each HRT. These HRTs will bifurcate into two steel lined penstocks at the base of the Surge Shaft to feed two turbines each. The water from the turbine units will discharge into two TRT's which, in turn, will carry the water of all the four units into the downstream reservoir. The availability of water for the Tehri Pump Storage Project shall be governed by the mode of operation of the Tehri Power Complex. During non-peak hours, water from the lower reservoir will be pumped back to the upper reservoir by utilizing the surplus available power in the grid.

Interestingly, the underground Tehri PSP, which is in the advance stage of construction, is located in the same hillocks where the Tehri Hydro Power Project (HPP) is. The major project components of the PSP are: (a) 25.4m (W) X57.3m (H) X203m (L) size Machine Hall to accommodate 4 of turbines of 250 MW each, (b) 22m dia. 02 nos. Upstream Surge Shafts with Surge Chambers, (c) 77m (L) X24m (H) X10m (W) size BVC, (d) 81m (L) X20m (H) X13m (W) PAC and (e) 18m dia 02 nos. downstream surge chambers. (f) 02 nos. of 1081m & 1176m TRTs.

For such a huge crucial project which is fully underground, ensuring safety is a challenging task. The challenge is further compounded by the complicated Himalayan geology which surprises us at every stage of construction.

2 PROJECT COLONY

Large projects, more so, hydropower projects, are located in remote areas where large infrastructure consisting of several components is required. The Project Colony including residential and non-residential buildings, is an integral part of such an infrastructure. The location of a colony has to be particularly chosen with utmost care to ensure that the area is not prone to slips, floods and other natural risk. In case such a location is not available, which usually is the case, necessary measures have to be taken to ensure safety of the colony. The slopes cut to provide benches for the colony have to be stabilized. For this gabion walls or other structural and/or non-structural measures need to be carried out along with the construction of buildings. (Figure 1)

If it be a seismic zone, the buildings should be earthquake resistance. The engineering history reveals that buildings constructed with thick mud walls have collapsed and caused loss of human lives; while the conventional Assam type buildings were least damaged. If the area is prone to flash floods, proper warning measures have to be ensured. Some lead time provided with such warnings has proved to be of immense help in minimizing loss of life and to some extent of the property as well. A suitable flood warning site can usually be ratified through interaction with Central Agencies viz. Central Water Commission and the State Agencies viz. Irrigation Department and the District Administration. For a effective use of the warning system, the residents must also be educated how to respond to the warning system. Adoption of such a system instills a sense of security among workers and their families. The system gets further tested through periodic drills. Our past experience shows that in the absence of proper training, the panic stricken populous starts running in the wrong directions.

Fire is another potentially dangerous threat. Even though the buildings in the colony are of a temporary nature, foolproof electric installations must be put in place as per specifications. Installation of Earth Leakage Circuit Breakers (ELCBs) has proved to be an asset, ensuring the safety of human life in case of faults occurring in the insulation. The installation has meticulously to be fixed and monitored periodically to prevent tripping resulting from human contect with faulty appliances.

Sometimes even well-designed buildings succumb to minor accidents. In a project, the CGI roof sheets were blown off the buildings during a storm for want of wind ties being fixed. The residents were injured and those surviving had to undergo nightmares until the buildings were repaired.

Figure 1.   View of colony constructed for the working staff at Tehri Project.

No borrow pit should be allowed to exists in the vicinity of a colony. They can be a potential health hazard due to collection and stagnation of water. In an extreme case a borrow pit left undressed had resulted in the death of an infant by drowning in the pit. Therefore, the filling up of the site after completion of the job should be made mandatory.

3   ROADS

Most of the dam sites are located in the interiors of the hills where the river valleys are very narrow. There is very little place available for the construction of roads. Hence, these roads have to be constructed by cutting benches and providing Zigs. The width, gradient, line of sight and stability of the road go a long way towards ensuring the safety of vehicles and human lives. Majority of lives lost on these projects are on account of accidents on these hill roads. More often than not, sufficient land is not available from the owner which also leads to construction of improper roads. Parapets on the outer side would prevent rolling down of vehicles. (Figure 2). In case it is not possible to have a road on which two vehicles can cross, refuges should be provided at regular intervals. Proper road signs, preferably luminous, must be installed at the required places. More signs rather than less should be wished for.

4   PLANT & MACHINERY

All plant and machinery on the Project have to be kept in very good working order to ensure that no accidents occur on account of their failure. The daily checking of plants before commencement of shift should be meticulously followed. Slings, wire ropes and hoist gears should be given special attention.

Accidents occur on account of faulty roads and human and mechanical errors. Moving away from road safety already discussed, let the vehicles plying on the road be repaired with the help of preventive and not breakdown maintenance. These allowances must take into account the roadworthiness of steering, brakes and transmission system. Although other provisions are equally significant; but failure of brakes, steering and transmission system are fatal and have to be prevented.

Most accidents as a matter of fact, occur entirely due to the fault of the drivers who in most of the cases are inexperienced and unlicensed helpers of the persons operating the vehicles. A common misconception is that the project area does not come under the traffic regulatory authority. The ground reality is that these roads require highly skilled drivers compared with other places. The ill practice of parking vehicles only in the heavy gear, instead of applying parking brakes and/or wheel block, has resulted in rolling of vehicles into the valley. In a recent case, while a worker was attending to a vehicle from inside, it rolled off in the valley as it was parked without applying parking brakes or wheel block. This resulted in the injury to the worker. Inexperienced drivers have sometimes driven straight into the valley while turning around the vehicle. A close monitoring of the performance of drivers has to be carried out, and quick, stringent remedial measures taken to ensure safety.

Figure 2.   Condition of road at project site in river valley project-Tehri PSP.

Figure 3. View of approved quarry and crushing plant for Tehri PSP.

5 QUARRY

A quarry is the infrastructure that feeds the main construction material for the work. Usually the quarry is located at a distance from the main work; site. The progress on the main work depends on the production of construction material from the quarry, yet the working at quarry usually does not get the desired attention. The job at the quarry is hazardous and has led to several accidents. As the quarrying operation involves removal of over-burden, blasting of rock, mucking and loading of blasted material by excavators – both men and machines are used under extremely difficult situations. The labour force is preoccupied in turning out maximum output and is usually not careful in keeping an eye out for boulders which may slip or the slopes may fail. Alertness of dedicated persons has reduced the number of fatal accidents. Such a person should be indispensably posted on all quarries. The blasting operations at the quarry have to be strictly in accordance with the rules and have to be followed in letter and spirit. The person in charge of blasting should be fully aware of the rules on the subject. The inspection of misfired holes has to be carried out by a competent person whose competence ought to be monitored periodically. The quality of illumination should be very good to ensure full safety of the personnel and the plant. The timings of blasting should be well notified and also warned through powerful sirens. (Figure 3).

6 MEDICAL FACILITIES

In the first place possibility of accidents at the site should be ruled out and prevaricated. However, the medical facilities both at the first aid level and at the hospital should be adequate for cases of injury or trauma as also for treatment of diseases. The level of facilities to be provided in the project hospital should depend upon the size of the project, nature of operations and the distance from the nearest well-equipped hospital. Ambulances should be provided and kept in working condition all the time. Attention has also to be paid for the prevention of diseases like malaria and diarrhea etc. and to prevent the same, blood samples of labour and staff entering the project area should be examined. The water supply should be suitably treated and periodically checked to ensure that no water-borne disease spreads out in the area. The sanitary arrangements should be well planned and maintained.

7 HSE MANAGEMENT SYSTEM – OUR EXPERIENCE

Health, Safety and Environment play an important role for the development of any infra project. The purpose is to preventing job related accidents. The role of HSE management is to achieve construction safety and health, a purpose which is generally over sighted. This function does not supersede, override, or take precedence over the other assignments of the officers who are ultimately responsible for safety and health. The key function of HSE management, i.e., construction safety and health, has to be monitored with another standard system called Integrated Management System. This system covers the all norms of ISO 9001 and 14001.

In the Tehri Pump Storage Plant, the effectiveness of HSE is so calculated that we have maintained the Zero Incident goal. To ensure the status, each individual has been assigned the following chart of duty:-

• Enforce carefully and systematically the planning and implementation to avoid the human injury, property damage, and loss of productivity.
• Create a safety culture by sensitizing the safety and environmental awareness of the employees through the Employee Safety Training Program, taking assistance from the management and the organized labor. The program includes orientation of all new employees, regular safety meetings, pre-task planning, and ongoing safety training.
• Minimize hazards/disruptions to the traveling public by controlling access to construction areas, following established safety procedures prescribed in the 'airport operational systems', and secure work areas adjacent to those spaces.
• Establish and maintain a system that promptly identifies and corrects unsafe practices.
• Establish emergency procedures to respond within a minimum time to fire accidents and eventual call up-for an ambulance.

Individual initiative and commitment is predominantly effective in the operations of safety. A healthy, safety culture has to be inculcated among all the employees, officers or workers. Having adopted the procedures of ISO 9001, the HSE department at our project has rigorously followed the specified parameters stated below:-

7.1 Safety Induction Training and Premedical Test

Safety Induction Training and Premedical Tests are invariably carried out for all new joining officers, workers and sub contractors. The training schedule and content is dependent upon the kind of job and location the employees are likely to be given. The high indicators of the Occupational Safety and Health Ordinance (OSHO) in the company prove the degree of adherence to the norms cited in OSHO.

7.2 Tool Box Talk on daily basis

Safety meetings and tool box menu is conducted on regular basis to educate the workers and officers and also ensured through its compliance. The practice prevents workers from getting complacent and taking the safety for granted. Tool box talks are the essence and strategy to have officers and workers inspect the equipment, tools and PPE they will be working with.

The mechanism creates a safety culture and thereby protects its employees. Tool box meetings carried out regularly enhance the knowledge of safety awareness related to the job hazards. The status of Zero Accident level at our Project has been given in Figure 4.

Figure 4.   Tool Box meeting at Site on daily basis along with officers and workers.

7.3 Monthly Training for Safety Awareness

Training for safety awareness is conducted every month on safety norms related to mechanical and electrical operations. All persons engaged in the projects have to compulsorily possess adequate safety knowledge and have a high degree of safety awareness. They are given the following ready-reckoner:-

- Recognize the importance of safety and assign sufficient resources to handle it.
- Give proper consideration to safety during planning and design to eliminate/reduce prospective safety hazards.
- Avoid performing unsafe acts, creating unsafe conditions, identifying unsafe conditions and take initiative for rectification.

Under Section 6a of the Factories and Industrial Undertakings Ordinance and Section 6(c) of the Occupational Safety and Health Ordinance, it is mandatory to inform, instruct, train and supervise the workers on the operations involving safety. Our Site Safety Committee consists of 34 members, including workers & Managers.

The Project Manager chairs the meetings and HSE Manager functions as Secretary. Generating Minutes of meetings (based on all non-confirmatory issues of whole month) with their photo documentation is assigned the Secretary.

7.4 Hazard Identification, Risk Assessment and Controls

Hazards & risks are identified and feasible control measures proposed to reduce potential hazards are stated at the planning stage by the Project Manager, the Project Construction Head, DPM- Planning & Controls, Project HSE Head. The risk assessment at project level will be carried out for every activity undertaken. When the project is in progress, risks and measures for control will be recorded in the risk-control documents by the operators in the prescribed format. The section in charge, site engineer and supervisors along with the Project HSE Head will review, the risks involved and propose appropriate measures.

7.5 Procedure for Hazard Monitoring and Action

The Site Safety Department will carry out regular inspections of the work place. Reporting of safety deficiencies and/or violations will be done on the basis of the daily inspection report. The details of items requiring actions and the time limit for the same will be furnished. In a separate proforma, deficiencies' will also be reported under the head 'unsafe acts and unsafe conditions'. The reports will subsequently be distributed to all the sites concerned.

Figure 5.   Safety training meeting under the leadership of Project Manager.

7.6 Mock Drill

Sensitization of employees through mock drills is highly essential for emergency situation like fire, chemical disaster, flood, cyclone, blast etc.

7.7 Monitoring the Environment

Environmental monitoring is used in the preparation of Environmental Impact Assessment (EIA), as well as in many circumstances where human activities harm the natural environment. All the monitoring strategies and programmes are justified with facts of the current status of an environment and prescription of the parameters to be followed vis-a-viz the environment. Results of such monitorings have to be reviewed, analyzed statistically and published.

7.8 Emergency Plan by our HSE Department

- Emergency Alarming System have been provided at all sites.
- Emergency Assembly points have been declared and sign boards also displayed at their locations.
- Emergency Ambulance facility is available for 24 hours (all working days & nights).
- First aid trained Pharmacists are available on 24 hours basis at the Site Medical Dispensary.
- First aid boxes and treatment has been provided at all sites besides the one at the Medical Dispensary on 24 hours basis.
- Rescue teams have been formed.
- Half yearly mock drills are conducted to evaluate the entire Emergency System.

7.9 Celebration of Safety Week at Site

Every year the Tehri PSP team celebrates National Safety Week by conducting various safety awareness activities at site. The aim of this initiative is to inculcate a culture of safety and to seek inclusion of all the workers from all the project sites. During the week, workshops, training programmes, medical camps, screening of safety films, and seminars are held. Banners, posters at prominent places notifying the safety week theme are displayed. The theme this year was "Reinforce positive behavior at workplace to achieve safety & health goals". HCC Tehri PSP team introduced Proactive Safety Observation Programme (PSOP) in 2013 and Behaviour Based Safety Programme (BBS) in 2015.The PSOP programme analyses the cause of possible accidents which are then categorized into near misses, unsafe conditions and unsafe acts. Each of these causes is systematically reported and corresponding actions are taken to mitigate them. The BBS programme goes one step further by engaging every employee to take responsibility for his own safe or unsafe behavior. Then implementation of a data driven decision-making process is prepared. The another interesting concept was introduced at the celebration of safety week in the form of "Best Safety Practice Competition". The competition was conducted at all the projects sites. The purpose was to:-

- Engage and motivate HSE staff to reduce safety risk in their work area.
- Set a benchmark of HSE practices for all projects
- Provide an opportunity to explore the professional concepts

The celebration concluded with the valedictory function where the winners of the safety competitions were awarded prizes and certificates. (Figure 6).

Figure 6.   View of annual safety week celebration and prize distribution at Tehri PSP site.

8   CONCLUSION

Human life is the most precious thing on earth because humans are the primum mobile of all developmental drives. The pace of any enterprise depends on the efficiency of the human resource engaged on it. The leitmotiv of human efficiency lies in security of life and the wages earned. The first question of the worker instinctively asked is 'whether or not the working conditions are secured'. The question acquires highest magnitude when asked by a worker who has to work in an underground tunnel where loss of life may occur because of falls, incidents with site vehicles, collapsing materials and contact with overhead power lines. The machines themselves if ineptly handled may add to the list of the causes of death. Taking into consideration this human psychology the first job of a manager is to make a workplace thoroughly secured with the multiple option security system. Therefore we have adopted the ISO 9001: 2008, OHSAS 18001: 2007, ISO 14001: 2004, OSHS, PSOP and BBS system to ensure accidental and hygienic security of our workers. If a worker feel psychologically secure, his efficiency increases, the increased efficiency leads to speedy and increased production and the increased production paves way to the well being and the growth of the company. And a prosperous and trustworthy company makes the nation strong.

This is the basic safety doctrine of the Hindustan construction company.

Tunnels and Underground Cities: Engineering and Innovation meet Archaeology,
Architecture and Art, Volume 9: Safety in underground
construction – Peila, Viggiani & Celestino (Eds)
© 2020 Taylor & Francis Group, London, ISBN 978-0-367-46874-3

Study on strength and deformation characteristics of early age shotcrete in tunnel cutting face

N. Kikkawa & N. Hiraoka
National Institute of Occupational Safety and Health, Tokyo, Japan

K. Itoh
Tokyo City University, Tokyo, Japan

R.P. Orense
University of Auckland, Auckland, New Zealand

ABSTRACT: When inserting detonators into boreholes and when mounting steel arch supports inside tunnels, workers approach a cutting face about 10 minutes after concrete is sprayed. It is still not clear whether such early-age shotcrete could support what size of rock and how shotcrete would deform just before rockfall. In this study, a new punching test equipment is developed which could apply a load of rock mass to the base concrete. Based on the test results, the base concrete with thickness of 40–50 mm and compressive strength of around 0.3 N/mm^2 easily fails with a few hundred newtons of resistance and several micrometers of extrusion displacement. However, cracks on the surface of concrete could be captured during post-failure behaviour at around few millimeters of extrusion displacement; thus, if measurement instrument could detect the crack generation and growth, and then it would be possible to warn against rockfall.

1 INTRODUCTION

Recently, accidents still happen due to rockfalls that kill or seriously injure workers at cutting face during conventional tunnel construction. In Japan, for example, a large-scale columnar rock mass 6.6 m high, 3.6 m long, and 1.5 m wide collapsed in 2015 while workers were inserting detonators and blasts into the cutting face; as a result, a worker was squashed and killed by the falling of the rock mass. In 2016, when workers approached the face in order to check if the steel arch support could be mounted on the circumference of the excavated rough surface, a rock mass on the surface fell and a worker was killed. In 2017, when a worker in a drill-jumbo's man-cage was about to start inserting the detonators and blasts into the face, three-quarters of the face area collapsed, and the worker was buried with the man-cage and died. In addition, during the same year, when a worker approached the side of the face while preparing to mount a steel arch support on the side, a rock mass fell from the side and hit the worker, who suffered severe pelvic fracture.

Based on these incidents, there is a tendency for workers to suffer serious damages due to rock falls or face collapses during those operations, such as inserting detonators and blasts, checking the excavated rough surface and mounting steel arch support when workers need to approach the face. Considering these accidents and the severities of rockfall and face collapse to date, the Labour Standards Bureau of the Ministry of Health, Labour and Welfare decided to issue guidelines concerning countermeasures against rockfall at tunnel cutting face on 26th December 2016 and then revised the guidelines on 18th January 2018. In principle, the guidelines prohibit workers from entering the face. This is because the ground characteristic curve (ground response curve) and the Fenner-Pacher curve (e.g. Möller 2006) clearly demonstrated that the excavated surface is definitely still unstable until the support structure exerts its function. As the stress acting on the excavated surface decreases, the stress rearranges at the contacts between the rocks on the surface, and the plastic deformation region expands and the inner diameter of the tunnel shrinks with elapsed time; in other words, the tunnel cutting face and the circumference are in the state of mechanical disequilibrium stress. In principle, allowing workers near the face with such stress imbalance should be avoided.

In reality, however, the present situation requires that workers must enter the face to perform various operations, such as charging, checking an excavated surface and mounting a support. Therefore, at almost all construction sites, after sprayed concrete is applied to the face, workers approach the face under the supervision of the face monitoring supervisor. There is also a situation when workers approach the face in order to mount steel arch support within 10 minutes after the primary shotcrete is applied to the face. The strength of the sprayed concrete at this time is indirectly discussed from the compressive strength estimated from the MEYCO pin penetration test, pneumatic pin penetration test, pull out test and so on; however, there has been no discussion about the direct strength in case where an unstable rock punches into the shotcrete and the rock falls. Considering the curing time, researches on strength and deformation characteristics of sprayed concrete at early-age of several hours to several days have been done (Barrett & McCreath 1995); however, there is no previous research on punching strength deformation characteristics of sprayed concrete at curing time of around 10 minutes.

Obviously, it is very difficult to support large-scale rock masses only with sprayed concrete, but it is very significant to estimate its strength, at least when workers approach the face in order to clarify the limit of application. It is also considered significant to develop and promote the use of measurement instruments by clarifying the accuracy and the interval of the instruments for capturing an indication of rockfall. In particular, evaluating the deformation characteristics of spray concrete, such as crack initiation, development and expansion, can also be considered in order to capture the initiation of rockfall. In other words, it is very important not only to try to prevent rockfall accidents with sprayed concrete, but also to capture the initiation of rockfall in combination with measuring instruments in order to warn or evacuate workers appropriately and to prevent accidents in advance.

In order to contribute to the prevention measures against rockfall, elastic wave velocity measurement tests, unconfined compression tests and punching tests were carried out in this study on the base concrete with 6 hours curing time and compressive strength of 0.3 N/mm^2, which is equivalent to the strength of sprayed concrete at 10 minutes curing time. From the results, the strength and deformation properties of the base concrete were evaluated.

2 MIX PROPORTIONS AND PHYSICAL PROPERTIES OF CONCRETE

2.1 *Mix proportions of concrete*

The mix proportions of concrete in this study was referred from the design code (road edition) of the Kanto Regional Development Bureau of the Ministry of Land, Infrastructure, Transport and Tourism. According to the design code, the additive rate of accelerator to the cement is 5.5%. Since it is very difficult to complete various experimental tests within 10 minutes, we selected the mix proportions of concrete without the accelerator. The mix proportion of the concrete adopted in this study is shown in Table 1.

The fine aggregate used was a mixture of crushed sand (absolute dry density: 2.58 g/cm^3, saturated surface-dry density: 2.61 g/cm^3) and sand (absolute dry density: 2.53 g/cm^3, saturated surface-dry density: 2.57 g/cm^3) at a mass ratio of 7:3. For fine aggregate, we adjusted the surface moisture ratio at 2% and 5% in crushed sand and sand, respectively, prior to mixing them. For the coarse aggregate, crushed stone for concrete was used, with maximum grain size of 15 mm, absolute dry density of 2.66 g/cm^3, and saturated surface-dry density of 2.68 g/cm^3. The coarse aggregate used for mixing was given moisture to such an extent that the surface was almost evenly wetted, and it was considered as the saturated surface-dry state.

Table 1. Mix proportion of concrete used in this study.

Compressive strength N/mm^2	Slump test cm	W/C %	Maximum grain size of coarse aggregate mm	Cement kg/m^3	Fine aggregate kg/m^3	Coarse aggregate kg/m^3
$\sigma_{28}= 18$	10 ± 2	56	15	360	1086	675

After mixing fine and coarse aggregates based on the mix proportion shown in Table 1, the particle size distribution was measured according to JIS A 1204: 2009 "Method for testing grain size of soil". The particle size distribution is shown in Figure 1. The average particle size was 1.5 mm and the aggregate contained 45.4% gravel, 52.6% sand, 2.0% silt and 2.0% clay. Also, the minimum and maximum densities were determined according to JGS 0162-2009 "Method for testing minimum and maximum densities of gravel". The minimum and maximum densities were 1804 kg/m^3 and 2155 kg/m^3, respectively.

Table 2 shows the density of the aggregate, the density of the aggregate and cement, and the density of the aggregate, cement and water in the concrete prepared by the mix proportion specified in Table 1. It is seen from the table that the density of the aggregate is 1761 kg/m^3, which is slightly smaller than the minimum density of 1804 kg/m^3. Therefore, in the concrete specimen prepared according to the specified mix proportion, some of aggregates are supposed to be suspended in the cement hydrates and some of the grains are in contact with each other with minimum coordination number.

2.2 Elastic wave velocity of concrete

Prior to the experimental punching test, bender & extender element tests (Lings & Greening 2001; Kikkawa et al. 2013) were performed on a concrete specimen whose dimension was around 50 mm in diameter and 100 mm in height and was prepared with the mix proportion specified in Table 1 with about 5.5 hours curing time. The elastic wave velocities were measured as shown in Figure 2. Figure 3 shows the pulse wave transmitted and the wave propagated and received inside the specimen by the tests. Two measurement results with different phases are

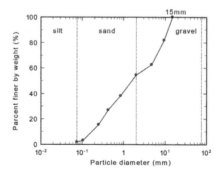

Figure 1. Particle size distribution of aggregates used in this study.

Table 2. Densities of aggregate, cement and water in the concrete specimen.

Density of aggregate kg/m^3	Density of aggregate and cement kg/m^3	Density of aggregate, cement and water kg/m^3
1761	2121	2323

Figure 2.   Image of bender-extender element tests and unconfined compression test.

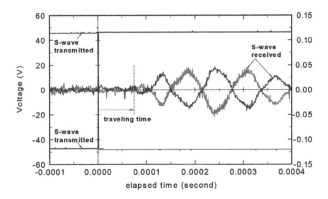

Figure 3.   S-wave transmitted and received in the bender-extender element tests.

shown together in the figure for both the wave transmitted and received. The propagation distance (almost equal to the diameter of specimen) was measured by vernier calipers prior to the tests and the propagation time was decided from the time differences between the rising points of waves transmitted and received in the figure and then the elastic wave velocities were calculated by dividing the propagation distance by the propagation time. As a result, P-wave velocity, V_p= 1093 m/sec, and S-wave velocity, V_s= 656 m/sec, were measured. Ye et al. (2001) measured the ultrasonic velocity of early-age concrete considering the water-cement ratio and curing temperature as parameters. The ultrasonic velocity (P-wave velocity) was about 800 to 1000 m/sec in the concrete with a particular mix proportion (cement content: 350 kg/m^3, aggregate content: 1792 kg/m^3, water-cement ratio: 55 %) which was almost the same as that in this study, at temperature of 20°C with about 5.5 hours curing time. That V_p was almost the same as the velocity measured in this study, showing the validity of our measurement.

2.3   Unconfined compression strength of concrete

Unconfined compression test was also carried out using the concrete specimen whose dimension was around 50 mm in diameter and 100 mm in height and was prepared with the mix proportion shown in Table 1. This specimen was totally the same as the concrete used for the experimental punching test. Because it was carried out after the punching test, the curing time was about 6.5 hours instead of 6 hours. From this test, unconfined compressive strength, q_u = 0.354 N/mm^2, was measured as shown in Figure 4. The European standard on sprayed concrete, BS EN 14487-1: 2005 "Sprayed concrete - Part 1: Definitions, specifications and conformity," provides a compressive strength of 0.3 N/mm^2 for general sprayed concrete considered as J2-class

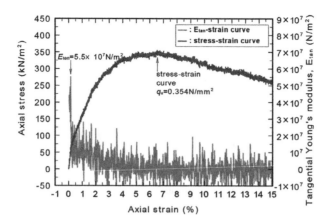

Figure 4.   Stress-strain relationship in the unconfined compression test.

with curing time of 10 minutes. The compressive strength based on European standard was the same as that of the base concrete specimen prepared in this study.

Therefore, we considered that the base concrete without accelerator at curing time of about 6 hours and the sprayed concrete with accelerator at curing time of 10 minutes were almost the same material, after which the following experimental tests were performed.

In addition, the relationship between the tangential Young's modulus and the axial strain is shown in Figure 4. From this figure, the tangential Young's modulus was read as 55 N/mm^2 (= 5.5 × 10^7 N/m^2) during the initial stage of loading. Shiozaki et al. (1989) showed the relationship between the elastic modulus, E, and the compressive strength, q_u, as follows:

$$E = 63.82 q_u^{1.139} \tag{1}$$

where, q_u: compressive strength (kgf/cm^2).

Substituting q_u = 0.3 N/mm^2 = 3.1 kgf/cm^2 to the above equation results in E = 228 kgf/cm^2 = 22 N/mm^2, and therefore the tangential Young's modulus of the base concrete with curing time of 6 hours seems to be not so different from the elastic modulus of sprayed concrete with curing time of 10 minutes.

3   OUTLINE OF EXPERIMENTAL PUNCHING TESTS

3.1   *Experimental equipment for punching test*

The experimental equipment for punching test used in this study is shown in Figure 5. The equipment has the structure where the concrete was placed inside a mould with specified internal dimensions (600 mm × 600 mm × 110 mm) and the concrete was punched by a loading plate from the bottom to the top. The diameter of the loading plate, D, was 100 mm. The load and the displacement were measured during test by load cells whose capacity was 5 kN and 100 kN, and by a displacement transducer with capacity of 50 mm.

The experimental punching test is also described in the "European Specification for Sprayed Concrete" issued by EFNARC. In the specification, concrete should be placed into a mould of given internal dimensions (600 mm × 600 mm × 100 mm) and should be punched out with a square loading plate of 100 mm × 100 mm from the top towards the bottom. The displacement rate of the square loading plate is 1.5 mm/min. In the case of a square loading plate, there is concern that local stresses would concentrate at the apexes of the four sides of the square, and so a loading pedestal of circular shape was adopted in this study. In addition, since we handled early-age concrete, in order to prevent concrete drooping from the mould due to gravity, we adopted a method of loading from the bottom upward in this study. Other dimensions are nearly equal to those specified in the EFNARC's European standard.

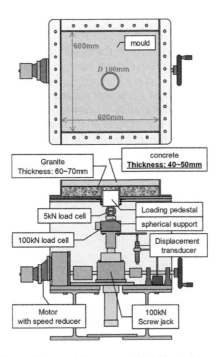

Figure 5. Experimental equipment for punching test used in this study.

3.2 *Experimental procedure for punching test*

The concrete used in the experimental punching test was the same as that used for the unconfined compression test and was prepared based on the mix proportion shown in Table 1. Prior to the experimental test, granite (thickness of about 60 to 70 mm) was mounted on the mould (600 mm × 600 mm × 110 mm), which simulated a tunnel cutting face. Granite with a diameter, D, of 100 mm and a thickness of about $10 \sim 20$ mm was also applied on the loading pedestal. Then, after kneading the concrete according to the specified mix proportion, we confirmed that it passed the slump test, and we placed the concrete into the mould. A straight edge with a length of 600 mm was used to flatten the upper end surface of the concrete, after which the concrete was cured for about 6 hours. The thickness of the concrete was about 40 to 50 mm, while the cross-sections of the mould, granite and concrete are as shown in Figure 6. In the punching test, the granite pasted on the loading pedestal was raised at a constant displacement rate of 2 ± 0.2 mm/min with a screw jack, and concrete was pushed out with rock. This simulated a rockfall from a cutting face. It should be noted that the direction of gravity differed from the rockfall at an actual tunnel cutting face by about 90°.

Kneading, curing and punching test of concrete were carried out inside a constant temperature and humidity chamber all the time. The temperature was controlled at 20 ± 0.2°C and the humidity was over 50 %.

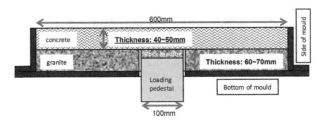

Figure 6. Cross-sectional sketch of the mould, granite and concrete.

3.3 *Three-dimensional measurement of surface shape*

The surface shapes of the granite mounted on the mould and the concrete applied to the granite were measured with a three-dimensional laser scanner. As shown in Figure 7, the scanner was installed at a distance of 1.5 m or more away from the concrete surface, since a minimum distance between targets and scanner of 1.5 m was needed for the scanner.

The measured surface shape of the granite is shown in Figure 8. It can be seen from the figure that the surface of the granite was around 40 to 50 mm below the top end of the mould. For this reason, when the concrete was placed on the granite and the surface of the concrete was flattened with the straight edge, then concrete with thickness of around 40 to 50 mm could be prepared. The laser scanner was able to measure three-dimensional coordinates with 5 mm pitch in a grid pattern on the xy-plane and so the average of all z-coordinates was - 41 mm. In other words, the mean thickness of the base concrete was 41 mm.

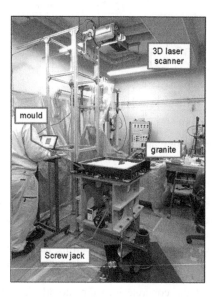

Figure 7.   Setting up all equipment for experimental punching test.

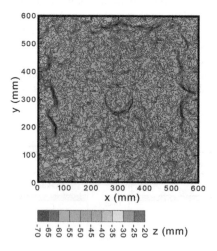

Figure 8.   Three-dimensional measurement of surface of granite.

In this way, not only the surface shape of the granite, but also that of the concrete was measured by the scanner once every 30 seconds during the punching test. As described above, since the vertical displacement rate of the loading pedestal was 2 ± 0.2 mm/min, in terms of vertical displacement by punching, it was scanned once per vertical displacement of 1 mm.

4 RESULTS OF EXPERIMENTAL TEST

Figure 9 shows the load against the vertical displacement during the punching test. From the figure, the maximum load during the experiment was 385 N, and the vertical displacement at that time was 0.34 mm. Therefore, in consideration of the actual tunnel cutting face, it is understood that a rock mass of less than 40 kg could only be supported for 10 minutes time after applying the sprayed concrete with accelerator. Also, since the displacement at maximum resistance was 0.34 mm, we need a very accurate measuring instrument in order to capture the initiation of rockfall occurrence.

The cross-sectional shapes of the concrete and granite during the punching test are summarized in Figure 10 at several vertical displacements. As can be seen from the figure, for the surface shape corresponding to a vertical displacement of 8.4 mm, a straight line could connect the left and right ends of the floating points of the concrete surface with the ends of the rock loading; the straight lines seem to indicate failure lines of 14° and 21°, respectively. In this way, the early-age concrete failed at an angle less than 45° when the diameter of the rock loading was 100 mm and the thickness of the concrete was 40 to 50 mm with curing time of 6 hours.

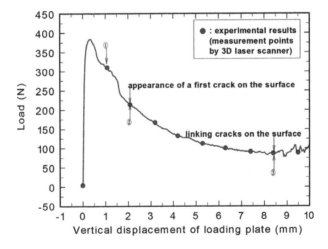

Figure 9.   Load against vertical displacement of loading pedestal in experimental punching test.

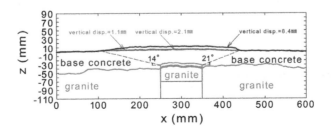

Figure 10.   Cross-sections of base concrete, granite and failure lines.

On the other hand, Figure 11 shows the digital images and the three-dimensional measurement results of the concrete surface during the punching test. From the digital image, cracks could not be seen yet on the surface of the concrete for the vertical displacement of 1.1 mm just after the load showed the maximum value. As the experimenter pointed out with the digital image, the first time that a crack was visually confirmed on the surface was when the vertical displacement was 2.1 mm. After that, cracks progressively grew and expanded circumferentially and the surface cracks were completely closed circles in shape when the vertical displacement was 8.4 mm. In three-dimensional measurement during this time, it can be seen that the deformation region extended to around 300 mm in diameter, which was three times larger than the diameter, $D = 100$ mm, of the rock loading. It was also found that the load value showed a constant value at this time, as illustrated in Figure 9.

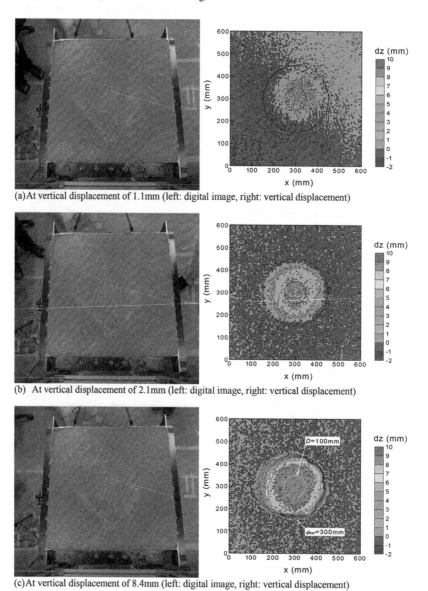

(a) At vertical displacement of 1.1mm (left: digital image, right: vertical displacement)

(b) At vertical displacement of 2.1mm (left: digital image, right: vertical displacement)

(c) At vertical displacement of 8.4mm (left: digital image, right: vertical displacement)

Figure 11. Digital images of concrete surface and three-dimensional measurement of vertical displacement in experimental punching test.

5 CONCLUSIONS

During tunnel construction, around 10 minutes after the primary shotcrete is applied, workers approach a tunnel cutting face to mount steel arch support on the circumference. In order to understand the strength deformation characteristics of sprayed concrete at that time, in this study, a simple punching test was conducted on early-age concrete. If a sprayed concrete with accelerator would be used for experiments, then it would be very difficult to completely prepare the specimen and testing it within approximately 10 minutes after the shotcrete is applied. Therefore, the base concrete without accelerator was used with 6 hours curing time, with compressive strength of around 0.3 N/mm², equivalent to that of the sprayed concrete with accelerator at 10 minutes curing time. The major observations from this study are as follows:

1. From the punching test results, the base concrete with curing time of 6 hours had a punching resistance of 385 N only, and the extrusion displacement at that time was 0.34 mm, which was very small. For this reason, in actual tunnel construction, measurement instruments with high precision are required to capture an indication of rockfall occurrence from the extrusion of sprayed concrete.
2. Also, focusing on the cracks on the surface of concrete, cracks were generated on the surface at the extrusion displacement of about 2.1 mm, and the cracks formed closed circles at extrusion displacement of about 8.4 mm; hence, there is a possibility that we can capture an indication of rockfall occurrence with measurement instruments that can detect the generation, progression and expansion of the cracks. In other words, if the crack loops closed, the unstable rock mass would fall due to gravity, so it is necessary to detect those initiations of the cracks before closing.
3. It is important to note that the directions of gravity and of the rock loading during the test were different from those in the actual tunnel cutting face. We still need to examine the punching resistance, its displacement and the crack growths of concrete in the state where the direction of gravity and the face are parallel to each other.

REFERENCES

Barrett, S. V. L. & McCreath, D. R. 1995. Shotcrete Support Design in Blocky Ground: Towards A Deterministic Approach. *Tunneling and Underground Space Technology*. Vol. 10, No. 1: 79–89.
BS EN 14487-1:2005. 2005. Sprayed concrete –Part 1: Definitions, specifications and conformity.
EFNARC. 1996. European Specification for Sprayed Concrete. ISBN 0 9522483 1 X.
Kikkawa, N., Hori, T., Itoh, K. and Mitachi, T. 2013. Study on decision maker of parameters for discrete element method in bonded granular materials, *Journal of Japanese Geotechnical Society*. Vol. 8, No.2: 221–237 (in Japanese).
Lings, M. L. & Greening, P. D. 2001. A novel bender/extender element for soil testing. *Gèotechnique*. Vol. 51, No. 8: 713–717.
Möller, S. 2006. Tunnel induced settlements and structural forces in linings. *Doctoral Thesis*. Institute of Geotechnical Engineering, University of Stuttgart. 149p.
Sezaki, M., Kibe, T., Ichikawa, Y. and Kawamoto, T. 1989. An experimental study on the mechanical properties of shotcrete. *Materials*. Vol. 38, No. 434: 1336–1340 (in Japanese).
Ye, G., Breugel, K. V. and Fraaij, A. L. A. 2001. Experimental study on ultrasonic pulse velocity evaluation of the microstructure of cementitious material at early age. *HERON*. Vol. 46, No. 3: 161–167.

Tunnels and Underground Cities: Engineering and Innovation meet Archaeology,
Architecture and Art, Volume 9: Safety in underground
construction – Peila, Viggiani & Celestino (Eds)
© 2020 Taylor & Francis Group, London, ISBN 978-0-367-46874-3

Assessment of health condition and management system for road tunnel by inspection data in Japan

S. Kimura
Kanazawa Institute of Technology, Ishikawa, Japan

M. Nomura
CTI Engineering Co., Ltd., Tokyo, Japan

T. Yasuda
Pacific Consultants Co., Ltd., Tokyo, Japan

ABSTRACT: There are approximately 11,000 road tunnels in Japan. As with bridges, the majority of them were constructed during the period of rapid economic growth, in other words, tunnels reaching 30–50 in-service years are increasing. Road tunnels, by aging progresses, there is an increasing risk of an accident. In 2010, an accident to collapse the ceiling plate to be installed in Sasago road tunnel has occurred. This triggered the law to oblige the inspection of road tunnels. Based on the deterioration state and management system of the infrastructure in Japan, JSCE, as a third party institution, has also decided to assess the soundness of the current infrastructure. This paper describes the discussion that JSCE independently assesses the health status of the road tunnels and its management system using inspection data of the road tunnels.

1 INTRODUCTION

There are approximately 11,000 road tunnels and 725,000 bridges (over 2m length) in Japan. As with bridges, the majority of tunnels were constructed during the period of rapid economic growth (1960–1990), in other words, tunnels reaching 30–50 in-service years are increasing. The law of Road in Japan required all of road tunnels to be inspected and assessed at least once every five years was enacted in 2014. As of 2018, the inspection data for three years of the road tunnels has been published.

On the other hand, based on the deterioration state and management system of the infrastructure in Japan, Japan Society of Civil Engineers (JSCE), as a third party institution, has also decided to assess the soundness of the current infrastructure.

This infrastructure health assessment is carried out by means of a method developed independently by JSCE, that of collecting released published data and surveys of inspection results and maintenance system information of the facilities. By accessing data provided by each administrator, the national average is expressed as an index.

This paper describes the results of examining the health assessment of the road tunnels and the management system using the inspection result of the road tunnels.

2 PRESCRIBED RULES ON MAINTENANCE IN JAPAN

2.1 *Number of existing road tunnels and implementation of inspection*

Management areas for the maintenance of the existing road in Japan are divided into 9 local blocks (Figure 1). Currently in 2017, there are approximately 11,000 road tunnels. Regarding

Figure 1. Divisional areas to manage road maintenance.

the number of tunnels by administrator, approximately 32% are managed nationally or by Expressway Companies, approximately 46% are managed by the prefectures or government ordinance-designated cities and approximately 22% are managed by municipalities (Figure 2).

In Japan there are many mountainous areas, and the urban transportation system is over-crowded, so the number of tunnels present in each area is different.

Inspection results of the road tunnels have been obtained corresponding to 47% of all of road tunnel by the inspections which began three years from 2014 (Figure 3) (MLIT 2015–2017).

2.2 *Inspection methods and frequency of inspection execution*

Inspection of tunnel facility is implemented as the objects of the lining and the appendages of lighting, ventilation fan and other facilities those have installed in the road tunnels. Regular inspection methods are obliged to visually observe in close proximity the state of deterioration and damage in close proximity and to ascertain whether or not the deteriorated concrete portion falls due to hammer impacting test. The inspection of road tunnels is legally mandated to

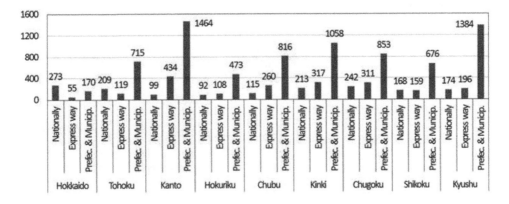

Figure 2. The number of existing road tunnels in Japan (MLIT 2017).

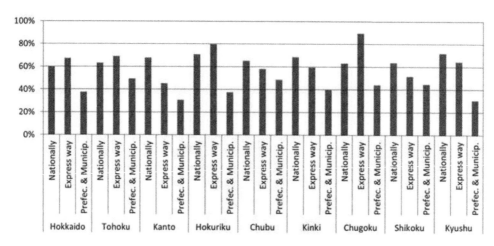

Figure 3. Percentage of inspection has been performed (MLIT 2017).

Table 1. Types of deterioration and damage in tunnel lining.

Types of deterioration and damage	Typical phenomena of deterioration and damage		
1 Cracks in lining	Uunbalanced pressure	Concrete shrinkage	Cracks with leakage
2 Floating and peeling of lining concrete	Expansion earth pressure	Unfilled concrete	Aged deterioration
3 Deformation, movement, settlement of lining	Bottom swelling	Ground uplift	Subsidence of ground
4 Exposure and corrosion of steel materials	Aged deterioration of concrete materials		
5 Lack and decrease of lining thickness	Missing cross section of lining caused by freeze-thaw action		
6 Degradation of lining by leakage	Ejection of water leak	Flow of water leak	Formation of icicles

be performed regularly at a frequency of once every five years. Regular inspection of the road tunnel emphasizes the reliable execution of the maintenance cycle consisting of inspection, diagnosis, measures and records.

2.3 *Inspection results of tunnel lining*

This section is described inspection data in view of the tunnel lining, also subsequent chapters are discussed on it. The types of deterioration and damage caused in the tunnel lining are as shown in Table 1. Table 2 defines the state of deterioration and damage occurring in the tunnel lining and the necessity of countermeasures against them I, II, III and IV shown in Table 2 are indicators to evaluate the soundness of the tunnel lining respectively.

Figure 4 shows the results of soundness evaluation of tunnel lining by each administrator. Figure 4 shows the ratio of the number of tunnels requiring countermeasures immediately or as soon as d eterioration and damage progress (III and IV), and that's average value is about 45%.

Table 2. Indicators of soundness and state of deterioration and damage.

	Clacification of status	Condition of deterioration and damage
I	Stage of sound	There is no obstacle to the function required by the tunnel.
II	Stage of preventive maintenance	There is no obstacle to the function required by the tunnel, but it is desirable to take measures for preventive maintenance.
III	Stage requiring measures at an early stage	There is a possibility that trouble may occur in the function required by the tunnel, and it is necessary to take countermeasures at an early stage.
IV	Stage requiring urgent measures	Since the possibility that a function required by a tunnel is obstructed or a trouble occurs is extremely high, it is necessary to take countermeasures urgently.

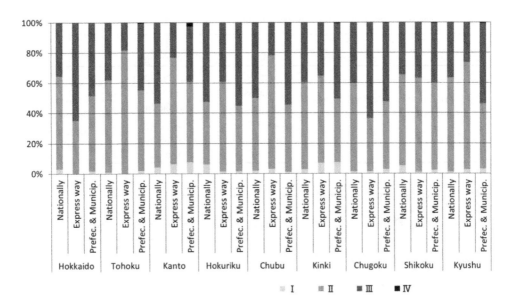

Figure 4.   Results of inspection (Soundness of tunnel linings) (MLIT 2015–2017).

3 HEALTH REPORT FOR THE ROAD TUNNEL TRIAL VERSION BY JSCE

3.1 *Health assessment index*

Health assessment is carried out by means of a method developed independently by JSCE, that of collecting released published data and surveys of inspection results and maintenance system information of the road facilities (JSCE 2016–2018). By accessing data provided by each administrator, the national average is expressed as an index (see Table 3 and Table 4). Health assessment index of the road tunnel is evaluated from the inspection result shown in the previous chapter by the ratio of the number of tunnels whose level is determined to be IV, III and II. Evaluation of the management system of the road tunnel is determined by the validity of the inspection plan conducted by the management organization, the feasibility of the inspection, and the implementation status of those information disclosure.

3.2 *Assessment of health status by administrators*

In Japan, most existing road tunnels were constructed with mountain tunneling. Existing road tunnels were constructed by Timbering Support Method until 1980 and NATM after the 1980s.

It is known that the phenomenon of deterioration and damage occurring in the tunnel lining varies depending on the construction method in which the tunnel was constructed. Figure 5 shows the concept of tunnel lining constructed by Timbering Support Method and New Austrian Tunneling Method. Most of the causes of deterioration and damage occurring in the lining were due to poor construction and material deterioration. In particular, many phenomena of deterioration and dam-age occurred at the joint of lining concrete. That is, since Timbering Support Method and NATM have different positions of joints remaining in lining, the positions of occurrence of deterioration and damage are different. Furthermore, in Timbering Support Method, water leakage phenomenon occurred frequently as waterproof sheet was not installed.

The details of the health condition of the lining were evaluated according to circumstances such as the service period of the road tunnel, occurrence of deterioration and damage, and regulations on closing the road.

Table 3. Health assessment Index of the road facilities.

A Sound	B Satisfactory	C Caution	D Warning	E Critical
No deterioration is seen in most facilities	Deterioration is seen in some facilities	Deterioration is progressing in quite a few facilities, requiring early repairs	Deterioration is obvious in many facilities, requiring repairs and reinforcements	Deterioration is serious overall, requiring urgent measures

Table 4. Assessment Index for management system of the road facilities.

→	⇒	↘
The state in which, if the present management system continues, the health condition will likely progress toward improvement.	The state in which, if the present management system continues, the current health condition will continue.	The state in which, unless there is an improvement in the present management system, there is a possibility that the health condition will deteriorate.

As a result of the evaluation, the condition of health of many tunnel lining was in a state where deterioration or damage had occurred and repair was necessary (Table 5). As for the health condition, tunnels managed nationally, those managed by Expressway Companies, those managed by the prefectures, those managed by municipalities were in good health condition, in that order. Although they are not in a serious condition, it is thought that there are a great deal of tunnels requiring early repairs. Assessment results of the soundness of the intracity tunnel were good, because the service period of those tunnels was short. Only 3 years have elapsed since systematic inspection was carried out, and there is a possibility that some administrators may have started inspecting tunnels with a prior knowledge of an inferior health condition.

On the other hand, the health status result of the tunnel was mainly assessed for its lining inspection data. The in the tunnel are present, such as lighting fixtures, also have occurred an accident due to their fall. Therefore, it is necessary to sufficiently consider the risk of falling, such as lighting fixtures.

3.3 Assessment of management system of tunnel facilities

In a recent three years, the trend of improvement of the health management system was not as good (see Table 5). Although the information relating to inspection results has been disclosed, the implementation of planned inspection is delayed for the initial schedule. And the validity of the inspection plan over 5 years is ambiguous.

Regarding the system of managing the health of tunnels, from the perspective of the storage of facility register management, inspection records and repair records, tunnels managed by the Expressway Companies are good and those managed by the prefecture are basically good. However, tunnels managed by municipalities require future effort. Each administrator has proposed a long-term inspection plan and therefore, improvements in the health condition of tunnels are anticipated, although planning and improvements in the securing of the number of inspection engineers and certification of skills are necessary.

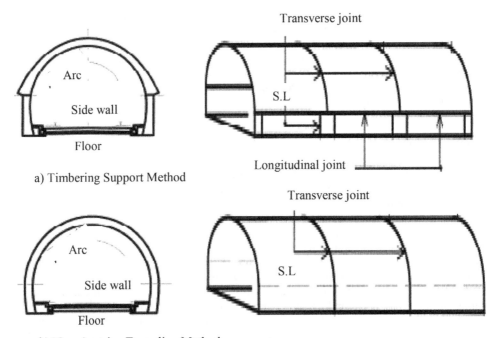

a) Timbering Support Method

b) New Austrian Tunneling Method

Figure 5. The concept of tunneling methods.

Table 5. Assessment of health status and management system by administrator (JSCE 2018).

Nationally-managed roads	Intracity roads managed by Expressway Companies	Intercity roads managed by Expressway Companies	Roads managed by the prefectures	Roads managed by municipalities
C	A	C	D	D
➡	➡	➡	⬊	⬊

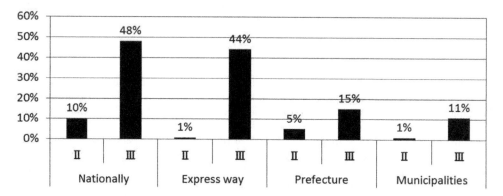

Figure 6. Implementation rate of repair measures (MLIT 2017).

3.4 *Correspondence to the tunnel judged to require measures*

In 2017, data on the implementation of repairing existing tunnels with deterioration and damage was released. Subject to tunnel repairs, from the results of inspections were carried out two years 2015 2014, a tunnel judged to require large repairs deterioration or damage. Figure 6 shows the status of implementation of repairs. The tunnel managed by municipalities, corresponding is delayed to have been determined to require early repair for the tunnels (III). Further, in all the administrators, the corresponding is delayed from the point of view of preventive maintenance to tunnel judged necessary repairs (II).

4 CONCLUSIONS

As of July in 2018, the inspection data for three years of the road tunnels had been published. Over the next two years, it is scheduled to inspect the end of the first round for all of the road tunnels. According to the inspection results by 2017, it is expected that about half of number of existing road tunnels will need to be repaired.

In addition to this, appendages such as lighting, ventilation fan, information system and others are installed in the road tunnel. Figure 7 shows inspection target portions in the road tunnel facilities. Falling of the appendage in road tunnels directly affects users. Therefore, the maintenance of these appendages is also important (Suzuki et al. 2015).

Currently, the budget necessary for repairing the road tunnel is not secured. In addition, training of engineers who conduct inspection and diagnosis of road tunnels has also been delayed. On the other hand, in 2014, ISO55000s, the international standard for asset management systems, was issued. In Japan, a management system of the road based on the Road Law is established, but a rational management system based on ISO55000s is required.

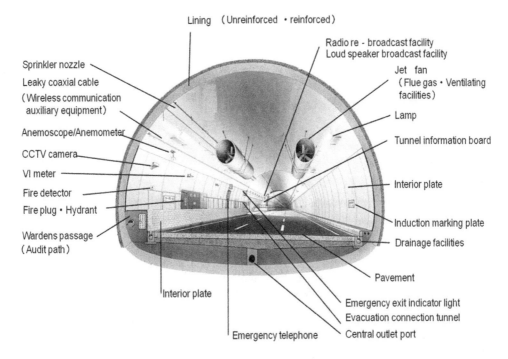

Lining　(Unreinforced ・reinforced)

Radio re - broadcast facility
Loud speaker broadcast facility

Jet　fan
(Flue gas ・ Ventilating facilities)

Sprinkler nozzle

Leaky coaxial cable
(Wireless communication auxiliary equipment)

Lamp

Anemoscope/Anemometer

Tunnel information board

CCTV camera

VI meter

Interior plate

Fire detector

Fire plug ・ Hydrant

Induction marking plate

Wardens passage
(Audit path)

Drainage facilities

Interior plate

Pavement

Emergency exit indicator light
Evacuation connection tunnel

Emergency telephone　Central outlet port

Figure 7.　Iinspection target portions in the road tunnel facilities.

Inspected until end of all road tunnels, rational and efficient management system based on the asset management system is desired to be developed.

REFERENCES

JSCE committee on health assessment of infrastructure. 2016. Infrastructure health report - Road sector trial version. Tokyo, Japan: JSCE.

JSCE committee on health assessment of infrastructure. 2017. Infrastructure health report - Road sector trial version. Tokyo, Japan: JSCE.

JSCE committee on health assessment of infrastructure. 2018. Infrastructure health report - Road sector trial version. Tokyo, Japan: JSCE.

MLIT. 2015. Annual report of maintenance for road facilities in Japan.

MLIT. 2016. Annual report of maintenance for road facilities in Japan.

MLIT. 2017. Annual report of maintenance for road facilities in Japan.

Suzuki, T., Kimura, S., Moriyama, M. & Hira, T. 2015. A risk exposure method of tunnel management for expressways. Extended abstract: World Tunnel Congress 2015: 22–28.

Tunnels and Underground Cities: Engineering and Innovation meet Archaeology,
Architecture and Art, Volume 9: Safety in underground
construction – Peila, Viggiani & Celestino (Eds)
© 2020 Taylor & Francis Group, London, ISBN 978-0-367-46874-3

Constructing the Arncliffe ventilation connection – a case study from the WestConnex New M5 project

A. Lippett
Aurecon, Australia

D.A.F. Oliveira
Jacobs Engineering Group and University of Wollongong, Australia

ABSTRACT: The WestConnex New M5 provides a duplication of the M5 East corridor through the design and construction of twin motorway tunnels, nine kilometers long, from Kingsgrove to St Peters. The Arncliffe ventilation connection, located midway along the alignment, comprises of a complex series of ventilation adits, 4 overhead shafts, and 6 large span tunnel junctions each with unique geometry and construction sequencing. Furthermore, unfavorable geology, including a 40 m deep palaeochannel and the intersection of the inferred Woolloomooloo fault zone placed further constraints on construction. These aspects compound with an ambitious construction sequence which has resulted in a challenging scope of works for the construction team to deliver. The design of the Arncliffe ventilation connection was previously presented by the author. This paper provides an update on the construction phase, highlighting the progress, challenges, and lessons learnt during this phase of the project.

1 OVERVIEW

1.1 *WestConnex and the New M5 Project*

Sydney's $16.8 billion WestConnex project will provide 33 kilometres of motorway to link western and south-western Sydney with the city, Kingsford Smith Airport and Port Botany precincts. It will largely be constructed in the M4 and M5 corridors and will comprise approximately 14 kilometres of road above ground and approximately 19 kilometres of tunnels, including a new tunnel linking the two corridors. The WestConnex program of works is being delivered as a series of projects, each of which is subject to a stand-alone planning assessment and approvals process.

The New M5 project will be open to traffic in 2020 and is the second stage of the WestConnex works. The $4.3 Billion project contains twin 9 km long motorway tunnels extending from Kingsgrove to a multi-level interchange at St-Peters. The twin tunnels cater for two lanes in each direction, with future provision for an additional lane. The project will also provide future connections for the planned F6 project and WestConnex Stage 3, resulting in approximately 1.3km of 4-lane tunnel in each direction, and 29 m span Y-junction caverns.

1.2 *The Arncliffe Ventilation Connections*

The focus of this paper is an underground ventilation complex located midway along the New M5 alignment, which provides exhaust and supply ventilation connections for both the New M5 tunnel and the future F6 tunnel. The area extends over approximately 235 metres of main-line tunnel, and comprises of four overhead rock shafts which tie into ventilation adits connecting to the New M5 tunnels, and stub adits which will later be developed as part of the future F6 tunnel. The exhaust shafts are tapered off southwest to allow for the complete

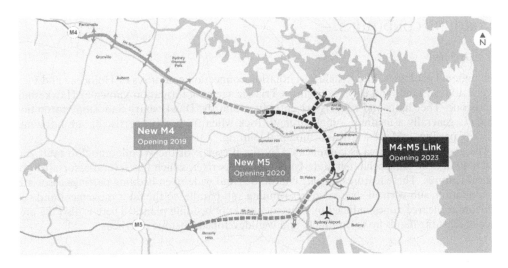

Figure 1. WestConnex Overview.

development of the tunnels offline from the shaft development, whereas the supply shafts directly overlie the tunnel. Ventilation adits in most instances overlie the mainline tunnels, with decking above the roadway to compartmentalize the airflow. The complete layout is shown in Figure 2 below:

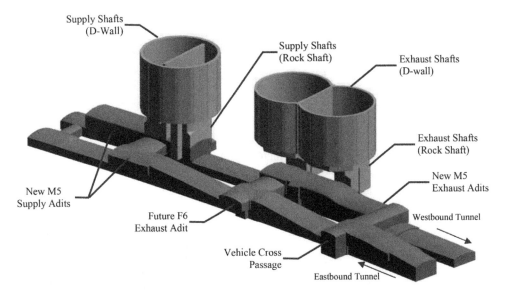

Figure 2. Geometric arrangement of the ventilation connections. Note some modifications to the construction lay-out have allowed for the removal of two ramping sections shown above.

2 GROUND CONDITIONS

2.1 *Overview*

The regional geology of the Arncliffe ventilation connection comprises of Holocene and Pleistocene alluvial sands and clays overlying a Triassic age rock formation known as Hawkesbury Sandstone. The shafts and tunnels are entirely within the Hawkesbury Sandstone formation, which is generally a medium – high strength rock when fresh, with persistent sub-horizontal bedding partings and sub-vertical orthogonal jointing.

The geology in the area is complicated by the presence of the Woolloomooloo fault zone and a 40 m deep palaeochannel underlying the Cooks river, which has several associated geological features. These features include continuous and wide-open bedding partings associated with dilation, sub-vertical sheared zones running sub-parallel to the fault movement and sub-horizontal sheared zones induced by movements along bedding planes. There is also the presence of siltstone bands up to 3m thick, and well-developed siltstone interbeds.

2.2 *As-Built Geology*

Ground conditions were generally better than what was anticipated in design. The position of critical sections of the fault zone were encountered 40 m down chainage from the expected location, which meant that the critical area was encountered away from the interface of the

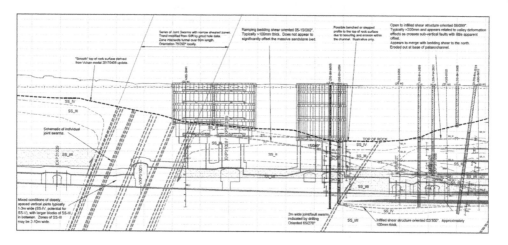

Figure 3. Expected geology along the alignment (Golder Associates, 2017).

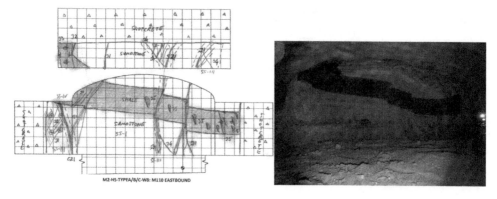

Figure 4. As-built geology typical of the fault zone (Golder Associates, 2017).

exhaust shafts, and only in portions of the vehicle cross passage. Moreover, sub-vertical shear zones were more widely-spaced than expected, and while sub-vertical jointing was closely spaced as anticipated, the lack of cross-jointing meant that the profile was generally maintained during cutting.

Elsewhere, the weathering profile of the rock was generally deeper than anticipated, and intact rock was generally medium strength in areas where sub-horizontal shear zones were encountered. Persistent sub-vertical jointing and minor joint swarms were encountered in adjacent areas of the faulting. Areas of massive sandstone were encountered, although generally rock strength was still below 20 MPa verified by point load samples.

3 GROUND SUPPORT

3.1 Design Overview

As discussed in Lippett et al (2017), the geology was separated into ground types GT-H-1 to GT-H-4 based on the varying geotechnical properties of the rock mass, and each ground type had a corresponding ground support type which was divided again based on different geometric parameters. Support for ground types GT-H-1 to GT-H-3 were designed as per typical Sydney tunnel design – a flat roof with rock bolts being the primary means of support for global stability, with thin shotcrete to contain smaller wedges and areas of spalling which may form between bolts. Support for the ground type GT-H-4 (representative of expected geotechnical conditions in the fault zone) was designed as an arched profile whereby both the rock bolts and shotcrete arch each contribute to the global stability by providing additional confinement to the rock mass.

3.2 Ground Support Feedback

Given the favourable ground conditions encountered, the support type used was better than expected. This also meant a profile change to an arched roof was avoided. Occasional areas where sub-vertical seams widened and the ground type was classified beyond the lower bound limits, additional bolting to locally reinforce the feature was favoured, rather than changing the profile. The additional bolting involved longer, and more robust bolts to effectively hang the midspan beyond the typical ground support zone, thereby acting as a third abutment and subdividing the span. This approach is commonly used in mining, for example in hanging wall design for large stopes (Hutchinson & Diederichs, 1996), and was also used elsewhere in the project (Aurecon Jacobs Joint Venture, 2017).

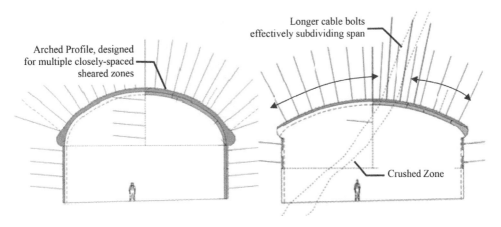

Figure 5. Comparison of altered design concept based on improvement in ground conditions LEFT: Previous design concept RIGHT: Modified design concept during construction.

Additionally, the lack of cross jointing allowed the advance length to be increased to 3.75m for most areas in the fault zone, even for a 14 m wide full-face excavation sequence.

3.3 Tunnel Sidewalls

Sydney tunnels excavated in Hawkesbury sandstone are well known for their vertical sidewalls with little to no rock support, even in tunnels greater than 10 m in height (Pells, 2002). During benching excavation of some of the 16 m high ventilation tunnels, ground movements observed were larger than expected for the given amount of excavation, with an amber trigger being breached after the first bench removal as shown below in Figure 6.

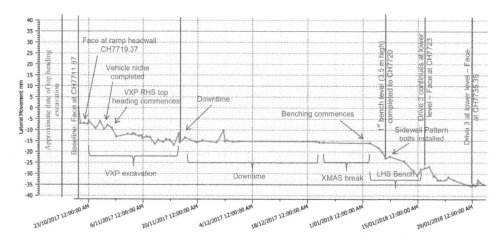

Figure 6.   Annotated monitoring history for sidewall convergence in ventilation tunnels.

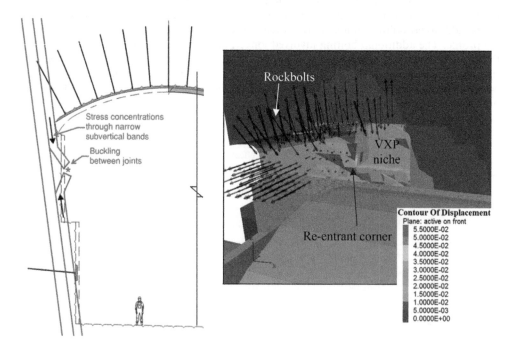

Figure 7.   LEFT: Buckling mechanism in sidewalls RIGHT: Predicted displacements based on updated observations and additional support.

It was inferred that sub-vertical jointing in the fault, which was running subparallel to the excavation, was creating thin wedges which were not readily identifiable in the geological mapping. Furthermore, the narrow sub-vertical bands were subject to stress concentrations, even when a kinematic wedge was not formed, which was potentially resulting in buckling of the thin beams formed between closely spaced joints, as shown in Figure 7 below. Additionally, ground movements increased further due to geometry of the vehicle cross passage openings, with re-entrant corners exacerbating the sidewall convergence. 5 m long pattern CT bolts were then specified for these areas of the tunnel and were installed immediately. 3D DEM modelling using 3DEC was undertaken to model the behaviours and verify the proposed additional support, predicting maximum sidewall movement of 40 mm, and 50 mm predicted at some re-entrant corners of the vehicle cross passage. Since implementing the additional support and excavating the full height of the walls, sidewall movement is observed to be between 35–40 mm in areas where jointing is prevalent.

4 EXHAUST SHAFT BREAKTHROUGH

Excavation in some sections of overhead ventilation tunnels were carried out by drill and blast to allow for simultaneous excavation in multiple drives. Despite drill and blast works being carried out with better end-profile in other areas of the project, strongly-bedded siltstone laminations and clasts encountered in this area gave rise to somewhat large areas of delamination in the crown. The issue was compounded by the need to use a drilling rig with no electronic guidance due to access requirements, and ultimately resulted in overbreaks as much as 1.6 m. Given that the pillar between the exhaust tunnels and shafts was only 2.6 m in the first instance, the pillar was deemed too narrow to transfer the stresses as per the design.

Additional bolting was considered for remedial works; however, the area was already heavily congested with bolts from the tunnel crown and the exhaust shaft headwall. Moreover, supplementing with longer bolts was not possible given the proximity to the tapered exhaust shafts, where even existing tunnel bolts had clashed with exhaust shaft excavations (Refer to Figure 9 below). The final solution involved reinstating the original profile with shotcrete, bringing the pillar back to the design thickness, and thereby allowing stresses to arch over the adit and providing additional confinement to the rock mass.

Figure 8.   LEFT: Well-developed siltstone laminations generated overbreak during drill and blast works RIGHT: As-built of tunnel overbreak, with the effective reduction in rock pillar shown.

Figure 9. LEFT: Tunnel bolts were encountered during exhaust shaft excavation due to tunnel over-break RIGHT: Successful breakthrough between shaft and tunnel.

5 CONSTRUCTION SEQUENCING

During the design phase a challenging construction sequence was already in place, due to the need to excavate most of the ventilation tunnels prior to shaft development. However, an overhaul of the construction sequence followed on from the design phase due to a second tunnel access point which was introduced later. This led to some unusual construction methodologies being implemented to accelerate the construction.

5.1 Ventilation Adit Crossover

With the second access point developed, the construction team identified an opportunity to link the two access points for improved emergency egress and allowing faster spoil transport out of the tunnel. To link the two areas as soon as possible, the lower east-bound drive was prioritized above the overhead ventilation crossover adits. This involved excavating under the crossover section with a narrowed heading width, backfilling the tunnel to provide temporary access above, and excavating the crossover tunnel through drill and blast.

Conceptually speaking, the completed geometry remains unchanged when compared to the previous sequence; however, there exists a temporary case during crossover where the pillar between the two tunnels is too narrow to transfer stresses. Additional reinforcement was provided to stitch a larger portion of rock together, and hence enable the rock to redistribute stresses without loosening or yielding the rock. Figure 11 below provides a summary of the stress behaviour.

Construction on this area is currently underway, with the crossover being developed at the time of writing this paper. Current crown convergence is less than 12mm in the lower east-bound tunnel.

5.2 Supply Shaft Breakthrough

It was recognised that the supply shafts would now take more time to develop than the ventilation tunnels. This meant that a key driver for the construction team was to build as much of the tunnels prior to shaft excavation. This led to a solution whereby the tunnels were

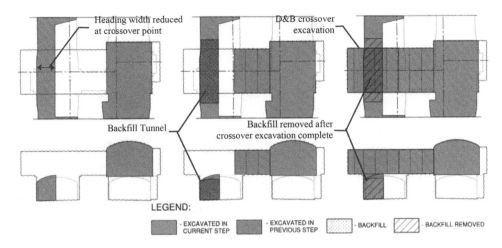

Figure 10.   Sequence of works for ventilation adit crossover.

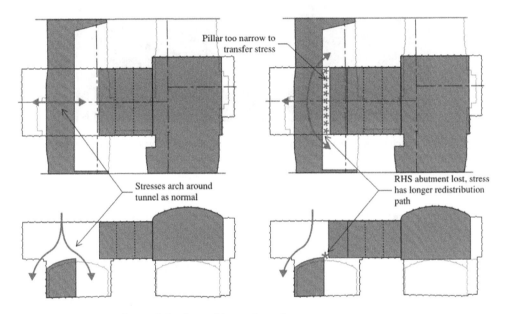

Figure 11.   Summary of stress behaviour mid-way through crossover sequence.

excavated under the footprint of the supply shafts, and the shaft team would then excavate the final portion by blasting the remaining rock plug between the tunnels.

The "plug" thickness (i.e. the remaining portion to be blasted), needed to be large enough for the tunnel to remain supported, while also supporting construction loads from works in the shaft. At the same time, a smaller plug thickness was desirable in order to limit the size of the blast, thereby reducing vibrations and simplifying the blast design. Furthermore, due to a slender rock pillar between the two supply shafts, the original design specified pillar support which could now only be installed down to the top of the plug, and in doing so, would leave the blasted portion of the shaft unsupported.

As shown in Figure 13, the design plug thickness was specified to be 6 m nominally from the apex of the crown. 9 m long cable bolts were installed from the tunnel at the boundaries of the exhaust shaft, which were intended to act as temporary replacement for the pillar support until blast spoil could be removed and the permanent pillar support solution could be

Figure 12. Split bench in the vehicle cross passage to allow simultaneous development of multiple headings. Note the local narrowing in the tunnel further ahead, to allow for the crossover sequence.

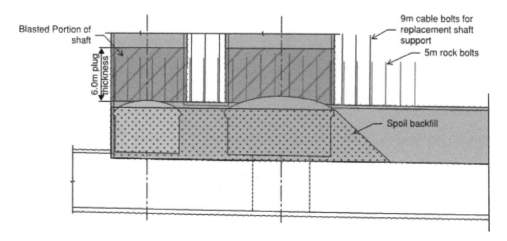

Figure 13. Schematic of supply shaft breakthrough.

installed. The tunnel below was backfilled with spoil to allow access to supply shaft such that the final support could be installed on the supply shaft walls.

At the time of writing, the supply shaft has commenced excavation in rock. The top heading of the ventilation tunnels under the supply shaft are nearing completion, after which, the tunnels will be backfilled, ready for the final.

6 CONCLUSIONS AND LESSONS LEARNT

The multiple shafts and large span tunnel junctions that comprise the Arncliffe ventilation connection have created many technical challenges for the New M5 construction team. Since the original design, there have been several key changes to expedite the construction programme, which have resulted in the need to deliver a technical construction sequence. Opportunities to optimise the ground support design by responding to the as-built ground conditions have also been undertaken to enable faster excavation. Outlined below are several lessons learnt:

- Complex packages of work require close collaboration between the designer and the contractor, requiring both parties to maintain the appropriate level of practicality and flexibility to create a safe and constructible design.
- It is recommended to create a detailed set of criteria for assessing the sidewalls and appropriate support required, especially where sidewalls are >8m high, or where sub-vertical jointing is more prevalent – this can be independent of ground type; hence a separate assessment would be appropriate.
- Profile changes associated with different ground conditions are potentially incompatible with the observational method, as it can be difficult to respond in time to change the profile prior to encountering an adverse feature, even with the benefit of geotechnical probing. Given that many of these adverse features are only encountered for a short length, it is preferred to reinforce locally using additional bolts, which may be longer and more robust than the typical bolts used in the crown (i.e. long cable bolts), rather than alter the tunnel profile.
- It can difficult to control profile in drill and blast where defects are running unfavourably to the profile geometry, even using trimming techniques. In these instances where structure is unfavourable and geometry has stringent requirements, the blast holes should be taken in from the final profile, which generally result in zones of underbreak to be trimmed afterwards. Excavation through this strategy can be slow, but may be necessary where tolerances to the geometry are more stringent.
- Intricate construction sequences in development of tunnel intersections require due consideration to the practical issues such as access, exclusion zones, and limitations on advances and support installation. The examples shown in this paper are intended to provide some insight on how to approach such tasks and develop a practical design concept.

ACKNOWLEDGEMENTS

The Arncliffe ventilation connection is a significant package of works, one which requires extensive collaboration between all parties to enable a successful delivery. With that said, the authors are immensely grateful to CPB-Dragados-Samsung-JV for their close working relationship, open-mindedness, and allowing the author to publish this paper.

The authors would also like to thank Golder Associates for their contribution to building the geological models for which us designers rely on.

The authors' colleagues who assisted in the design are also gratefully acknowledged, namely, Brad Sandford, Tony Peglas, Harry Asche, and the many others who contributed to this complex and multi-disciplinary design effort.

REFERENCES

Aurecon Jacobs Joint Venture 2017. *Design Report - Ground Support (Primary and Secondary) – St Peters Caverns*. Sydney.

Golder Associates 2017a. *Arncliffe Secondary Grout Program - Structural Model*. Sydney.

Golder Associates 2017b. *Geological Mapping.* Sydney.

Hutchinson, D.J. & Diederichs, M.S. 1996. *Cablebolting in underground mines.* Richmond, B.C.: Bitech Publishers.

Lippett, A.L. Sandford, B. Peglas, T. & Rae, D. 2017. WestConnex New M5 - Design of the Arncliffe Ventilation Connection. *16th Australasian Tunnelling Conference.*

Pells, P.J. 2002. Developments in the design of tunnels and caverns in the Triassic rocks of the Sydney region. *International Journal of Rock Mechanics and Mining Sciences, 39*(5), 569–287.

Tunnels and Underground Cities: Engineering and Innovation meet Archaeology,
Architecture and Art, Volume 9: Safety in underground
construction – Peila, Viggiani & Celestino (Eds)
© 2020 Taylor & Francis Group, London, ISBN 978-0-367-46874-3

The first huge pressurized ventilation system in the world, realized in Brenner, with heavy duty centrifugal fans

N. Lucatelli
CBI S.p.A. – Milano, Italy

M. Bringiotti & F. Serra
GeoTunnel S.r.l. – Genoa, Italy

A. Cicolani
Astaldi S.p.A – Roma, Italy

E.R. Vitale
Ghella S.p.A. – Roma, Italy

ABSTRACT: For the first time in Europe, the installation of centrifugal fans for renewal of sanitary air makes its debut in an underground construction site, seeing the use of new heavy-duty machines.

When very high ventilation system performances are required, centrifugal fan steal the show to the more traditional solutions obtained by means of single and multiple stage axial fans. This is the case of the BTC Mules 2-3 worksite where the new ventilation system is required to cope with sanitary ventilation of the underground construction site with a total flow rate of 420 m³/s and 6,000 Pa of total pressure and with total electric power of about 3.2 MW.

The system is required to guarantee air renewal for pollutants dilution produced by almost 10,000 kW of total power expected during underground operations and to guarantee the salubriousness of the air for the workers involved (200 people), along ca. 40 km of tunnels.

1 INTRODUCTION

Since the beginning of 2016 the BTC consortium, formed by Companies Astaldi, Ghella, PAC, Cogeis and Oberosler, has been involved in the construction of the longest underground railway link in the world: the Brenner Base Tunnel, which forms the central part of the Munich-Verona railway corridor.

The whole project consists of a straight railway tunnel, which reaches a length of about 55 km and connects Fortezza (Italy) to Innsbruck (Austria); next to Innsbruck, the tunnel will interconnect with the existing railway bypass and will therefore reach a total extension of about 64 km.

The tunnel configuration includes two main single-track tubes, which run parallel with a 70m span between each other through most of the track, and linked every 333 m by cross passages (Figure. 1).

Between the two main tunnels and driven 12 meters below, an exploratory tunnel will be excavated first. Its main purpose during the construction phase is to provide detailed information about the rock mass Furthermore, its location allows important logistic support during the construction of the main tunnel, for transportation of excavated material as well as that of construction material. During the operations, it will be essentially used for the drainage of the main tunnel.

The excavation process is divided into 2 blocks; the first one will be bored by n. 3 TBMs (n. 2 for the main tunnels and n. 1 for the exploratory tunnel, toward North). The second

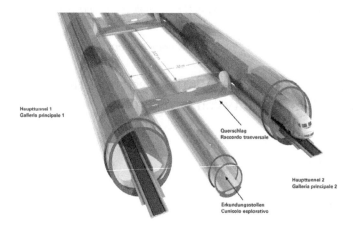

Figure 1.   Brenner Basie Tunnel.

one will be bored with traditional method, including mainly drill & blast in the competent material and special drilling techniques in the faulty zones, toward South and in some areas toward North.

Company CIPA S.p.A. is currently handling most of the tunnel excavations activities in the lot named "Mules 2-3" with the French drilling partner Robodrill SA. These mainly consist of:

− Excavation and lining of the ADIT Tunnel at the Trens Emergency Stop and the Central Tunnel, with a total length of approx. 4,500 m;
− Excavation and lining of the Exploratory Tunnel by traditional method, with a total length of approx. 830 m;
− Excavation and lining of the Main Tunnel toward South, East tube and West tube in single track section, with a total length of approx. 7,320 m;
− Excavation and lining of the Main Tunnel toward South East tube and West tube in double track section, with a total length of approx. 2,590 m;
− Excavation and lining of 19 Connecting Side Tunnels linking the two main tubes, with a total length of approx. 900 m.

All the material excavated in traditional method by the subcontractors Cipa, Europea92 and LSI, runs from the central cavern up to the surface, sized by a crusher, by means of Marti Technick belt conveyors. The material is also divided in 3 different geological classes and moved according to needs in 2 different depony areas.

1.1   Mules 2-3

As already said the Mules 2-3 construction lot is the main part of the BBT track on the Italian side; in particular it is between the State border, to the North (km 32.0 + 88 East Reed) and the adjacent lot "The Isarco River underpass", to the South (km 54.0 + 15 East Bound Track). The works for this construction lot, the biggest in the Brenner base tunnel, has started on September 2016. The lot Mules 2-3, which value is about for 993 million euros, has been assigned to joint venture called BTC S.C.A.R.L. constituted by Astaldi S.p.A., Ghella S.p.A., Oberosler Cav Pietro S.R.L., Cogeis S.p.A. and PAC S.p.A. General lay out of Mules 2-3 lot is shown in Figure 2.

Complex excavation activities has to be conducted including: about 39.8 km for the line tunnels and about 14.8 km of exploratory tunnel, the emergency station in Trens together with the related adit tunnel, and the cross passages tunnels, which will connect the main tunnel every 333 meters. By the end of 2023, 65 kilometers of tunnels within this construction lot will be excavated. Once finished "Mules 2-3", all the Italian side underground works will be completed.

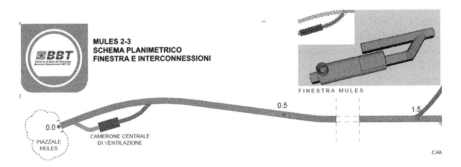

Figure 2.   Mules 2-3 plan view.

2   GEOLOGICAL CONDITIONS

As anticipated, the Brenner Base Tunnel is the high-speed rail link between Italy and Austria and therefore establishes a connection with North East Europe. It consists in a system of tunnels, which include two single-track tunnels, a service/exploratory tunnel that runs 12 m below them and mostly parallel to the two main tunnels, bypasses between the two main tunnels placed every 333 m, and 3 emergency stop stations located roughly 20 km apart from each other. The bypasses and the emergency stops are the heart of the safety system for the operational phase of this line.

Average Overburden is, between 900 and 1.000 m, with the highest one about 1.800 m at the border between Italy and Austria.

The excavation will be driven through various geological formations forming the eastern Alpine Area. Most of these are metamorphic rocks, consisting of Phyllites (22%), Schist (Carbonate Schist and Phyllite Schist, 41%) and Gneiss of various origin (14%). In addition, there are important amounts of plutonic rock (Brixen Granite and Tonalite, 14%) and rocks with various degrees of metamorphism, such as marble (9%).

Among the tectonic structures in Italy, we find the above mentioned Periadriatic Fault. As mentioned, the unknown characteristics of the rock masses along this stretch determined the necessity of excavating the exploratory tunnel long before the main tunnels, in order to allow the identification of lithological sequences of the various types of rock mass within this heavily tectonized area, as well as a detailed study of their responses to excavation. These analysis were subsequently used to adjust the consolidation and support measures both for the exploratory and for the main tunnels. The excavation of the exploratory tunnel inside the Periadriatic Faults allowed to determine the actual sequence of lithologies within this area.

A short geological description is necessary in order to understand the chosen excavation process, the mucking out system and the subsequent fresh air need. This is divided in mechanized tunnelling for the good and competent rock types along the exploratory and the main tunnels toward North, D&B for a limited distance toward South (2-3 km) and in some minor

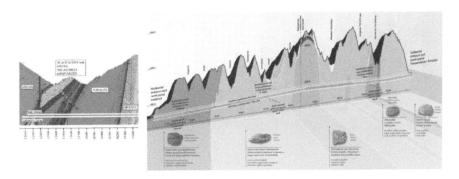

Figure 3.   Geological distribution.

short tunnels as well as several bypasses and traditional method (mainly self drilling anchors protection umbrellas and hammer excavation) in the faulty areas. It is also interesting to understand difficulties related to the excavation in the hard granite as well the related problems linked to the abnormal wear and tear due to high abrasivity.

A summary of the encountered rock sequences is illustrated with Figure 2.

3 THE BELT IDEA

The conveyor belt system allows a straightforward and efficient mean of transportation for material going both in and out from the underground rock crusher, and in and out from the underground concrete batching plants. It evolves while the project is carrying out and is gradually implemented in accordance to the work's progress.

This system has been chosen not only for its efficiency in general but also, and may be mainly, to lower down fresh air need along the full project.

The conveyor system has been rationalized during the tender phase maintaining its high potential, flexibility and capacity. The differentiation of the two belts in Aica (belt 1 and belt 2) allow each single band to be allocated with a certain type of material; therefore belt 1 is for material A (good material for concrete) and belt 2 is for material B+C (semi-good and not concrete-accepted material). The differentiated configuration avoids alternating different material types with temporary stocking buffer on the same conveyor. However, the system provides the possibility to switch the material types in Aica according to necessities. This kind of realization allows an easier management of the whole system.

Such works may be identified in three major phases. This short report is aimed to describe the choices taken within the Construction Project and to list them congruently to the time-related phases.

Without taking into consideration the specific calculations regarding the Executive Work Program's Space-Time Diagram, the 4 major phases may be described as follows:

– Phase 1: excavation of the Exploratory Tunnel (CE) with the conventional excavation method and the TBM assembly chamber; realization of the excavation of the first part of the North Line Tunnels (GLN) by traditional system and excavation of the South Line Tunnels always by traditional method (done by Cipa S.p.A. with its Drilling Partner Robodrill SA);
– Phase 2: continuation of the activities following the previous phase; realization of "in cavern" assembly areas for the two TBMs, which will excavate the Line Tunnels (toward North), excavation prosecution of the South Line Tunnels with drill & blast system (GLS) and realization of the mechanized excavation of the CE Northwards.

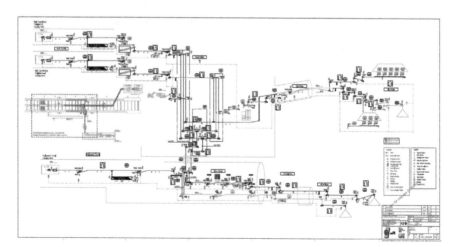

Figure 4. Belt lay out with cavern zoom (center left).

Figure 5. Mules configuration and A22 highway transfer belt to Genauen 2 intermediate waste dump.

- Phase 3: prosecution of the precedent activities, excavation of the North Line by mechanized tunnelling (GLN) and subsequent finishing job site works.
- Phase 4: main TBM tunnels secondary final lining. This also represents a complex activity, which will take more than 1 year. A part of this job will be anticipated during TBM excavation, deviating logistic and traffic from one tube to the parallel one. The phases are integrated and shown in the complex lay out indicated in Figure 3.

All belts generally are connected to the five excavation fronts with the Logistic Knot (N). The system is flexible, but actually in the current phase belts are directly used only for the North bound tunnels (3 TBMs). For the South bound (conventional excavation) the material is transported by dump truck up to the logistic Knot, where after resizing (by the crushing plant housed underground within the cavern) it is transferred and transported by belt conveyor system. After that, from Knot N:

- Material type A is carried towards the jobsites Mules and Genauen 2 (depony area reached by a 180m long bridge belt conveyor – 300 ton/h – which crosses the A22 highway, the Isarco river, the National road S.S. 12 and the existing railway line, Figure 5), it is stocked and/or crushed outside and/or transported back to the underground batching plant, in order to provide D&B tunnels' primary and secondary lining.
- Material type A/B+C is transported to Hinterrigger jobsite and it is used to produce the precast elements and pea-gravel for the TBM's back filling. Concrete segments and pea gravel will be then transported back into the tunnel with a dedicated rolling stock through the Aica Tunnel.

4 THE VENTILATION IDEA USING HEAVY DUTY & HIGH PERFORMANCE CENTRIFUGAL FANS

For the first time in Europe it makes its debut in an underground construction site for the construction of a railway infrastructure, the installation of centrifugal fans for the supply of fresh air to the excavation fronts. The debut could not be more striking of such a new heavy-duty machine, designed and built by CBI Industrie S.p.A., for the key lot of the new Brenner Base Tunnel.

When a very high-performance ventilation scheme is required the centrifugal fan can steal the spotlights to the most conventional solutions such as the single and multiple stage axial fans. This is the case of the mules 2-3 jobsite, where the new ventilation system is required to cope with the renewal of the underground breathing air through a total flow of 420 m^3/s at a pressure of 6,000 Pa with and total installed power of approx. 3.2 MW.

4.1 *Main Ventilation System – Supply of fresh air to the underground working areas*

The ventilation system requires, during the operation, to assure the safest environmental conditions, clearly indicated by the HS authority, and in particular to dilute the concentration of harmful substances in the entire underground area.

For the Mules 2-3 construction lot, the ventilation system is required to guarantee air renewal and the dilution of the pollutants produced by the almost 10,000 kW of power produced by the underground machineries thus in order to guarantee the breathiness of the air for the workers involved (approx. 200 units).

As per the following table a list of tunnel machineries foreseen to be used in the project.

Id.	Type of Machinery	Type of Engine (Diesel/Electric)	Installed power (kW)
1	Waterproofing Gantry	E	20
2	Re-bar Gantry	E	20
3	Service Car 4WD	D	70
4	Concrete batching Plant	E	150
5	Concrete batching Plant	E	150
6	Concrete Pump	D	70
7	Jumbo Drilling Rig (BIG)	D	70
8	Jumbo Drilling Rig (SMALL)	D	75
9	Stone Crusher	E	200
10	Dumptruck (BIG)	D	300
11	Concrete Truck Mixer	D	240
12	Belt Conveyor	E	800
13	Forkliftt	D	80
14	Service Car 4WD	D	100
15	Gantry Crane (BIG)	D	80
16	Gantry Crane (SMALL)	D	60
17	Minibus	D	80
18	Grout Injection Plant	E	10
19	Overhead Gantry	E	50
20	Exploratory Tunnel - Service Gantry	E	1350
21	Northbound Tunnel – Service Gantry	E	1500
22	Southbound Tunnel – Service Gantry	E	1500
23	Wheel Loader (BIG)	E	200
24	Wheel Loader (SMALL)	E	120
25	Tunnel Formwork	E	50
26	Tunnel Invert Formwork	E	20
27	Shotcrete	D	80
28	Exploratory Tunnel - TBM	E	4100
29	Northbound Tunnel – TBM	E	4600
29	Southbound Tunnel – TBM	E	4600
30	Dump Truck	D	130
31	Crawler Excavator	D	130
32	Underground Workshop	E	200
33	Service Train - Cross Passages (Finishing)	D	155
34	Service Train - Cross Passages (Excavation)	D	155
35	Service Train – Muck Transport	D	155
36	Service Train – Segments' Transport	D	273
37	Service Train – Segments' Transport	E	273
39	Service Train – Personnel Transport	E	50
40	Dump Truck	E	50
41	Booster	D	273

The air produced by the Main Ventilation Unit is provided to the working areas by means of a series of flexible ventilation ducts installed along the main adit tunnel to the exchange cavern (Figure 6) and from there branch out into six header feeding the six excavation fronts.

The design criteria for the air demand calculation are:

– supply unitary flow rate for workers: ≥ 3 m^3/min/person
– supply unitary flow rate for pollutants dilution: ≥ 4 m^3/min/kW
– air velocity inside tunnel: ≥ 0.30 m/s (maximum admissible value 5.0 m/s)
– air velocity inside tunnel (in presence of potential CH_4): ≥ 0.50 m/s

Figure 6.   Ventilation ducts in the Mules adit.

4.2   *The ventilation station and the service plenum*

The ventilation station is installed in a dedicated recess/room realized along the adit tunnel, where no. 4 centrifugal fans are installed sucking fresh air provide by a ventilation shaft ended up into the chamber.

The need to interface the no. 4 centrifugal fans to the no. 3 flexible ventilation ducts blowing air along the adit tunnel, has determined the realization of a "tunnel plenum" (Figure 7). The plenum has been realized installing bulkheads, along the excavated chamber, between the ventilation units and the intersection with the adit tunnel.

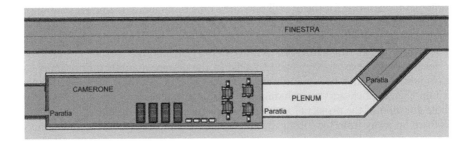

Figure 7.   Plan view of the ventilation chamber.

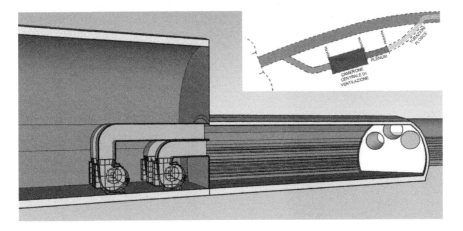

Figure 8.   Typical installation lay-out of ventilation ducts interfaced to a ventilation plenum.

On the bulkhead interfacing the fans no. 4 diffusers are installed together with no. 4 ON/ OFF dampers; on the opposite bulkhead no. 3 Dapò dampers are installed thus in order to interface the no. 3 flexible ventilation ducts (Figure 8).

4.3 Ventilation Fans

The ventilation system is therefore provided with no. 4 high-performance centrifugal fans of the CH series with a 10-blade aerodynamic impeller able to guarantee overall the nominal performances required by the project (Figure 9). The fans are supplied @ 400V-3ph-50Hz, equipped with a power supply/control unit complete with VFD.

The provided solution allows to achieve:

– The modularity of the loads and the ventilation capacities.
– The control of the machines performances during the flow throttling phases (considered for about 90% of the operating time).
– The system availability assuring continuous supply of fresh air to the excavation fronts of even in case of machines failure or machines stopped for maintenance. In comparison to a single machine linked to the related ventilation ducts, the presence of a ventilation plenum assure simultaneous distribution of the flow rate through the three ventilation duct lines.

The use of centrifugal fans therefore allows the following benefits:

– Maximizing the operating efficiency.
– Minimizing the installed power.
– Optimize energy consumption.

Only the use of this type of machine, in fact, allows to obtain, with the nominal performances required, always exceeding 84% with maximum peaks close to 90% (89% in the conditions of partial loading), thus assuring for the overall operation time to minimize the power consumption and maximize the energy requirements therefore the direct operating costs.

Table reported in Figure 10 is just a saving estimation related to only 3 working phases on ca. 20 on which the job site is foreseen to move in the years along.

The reliability and experience gained by CBI in the design, manufacturing, supply and servicing (Figure 11) of such heavy duty fans has allowed BTC S.C.A.R.L. to sign an agreement for a comprehensive field service including the guarantee extension of the supplied machines for the whole duration of the project (7 years!). The presence of a well-organized field service at less than 300 km far away from the jobsite, has undoubtedly comforted the Customer in his choice.

4.4 Adjusting the flow rate to the excavation fronts

Each of the three ventilation ducts installed along the adit tunnel (Figure 12), once reached the logistic cavern, branches off to the provide fresh air the 6 excavation fronts, by means of different sized flex ducts and steel distribution boxes.

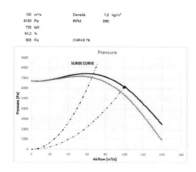

Figure 9.   Curve performance example and FAT of the first Centrifugal Fan CH4S 76 (SIS7).

		PROJECT DATA				PROJECT HYPOTHESIS			SOLUTION WITH 4 CENTRIFUGAL FANS					SAVING	
Machine	Phase duration	Total working hours (3 shifts, 24 h)	Total air flow	Project prevalence Ventilation (total pressure)	Fan total efficiency	Electric energy (as per project = 0,82)	Electric energy cost (calculated at 0,15 €/kWh)	Fans flow	Fans pressure	Effective efficiency in the working point	Electric energy consumption	Electric energy consumption cost (calculated at 0,15 €/kWh)	Saving with the solution offered	Saving in money (calculated at 0,15 €/kWh)
Phases Description	GG	[h]	[m3/s]	[Pa]	[-]	[kWh]	[€]	[m3/s]	[Pa]	[-]	[kWh]	[€]	[kWh]	[€]
Phase 10 — GLN and GLES traditional excavation, GLN lining, TBM CE and GLOS excavation	120,0	2.880	312,7	5.821	0,82	7.102.492	1.065.373,8	314	5.966	0,886	6.765.956	1.014.893,4	336.536	50.480,4
Phase 15 — GA traditional excavation, GA lining, TBM CE and GLN excavation	189,9	4.056	351,3	5.357	0,82	10.342.925	1.561.438,7	361	5.491	0,869	10.272.055	1.540.808,2	70.870	10.630,5
Phase 20 — GLN, CE, CT of FdE lining	168,0	4.032	299,6	4.198	0,82	6.870.516	1.030.577,4	305	4.203	0,863	6.650.305	997.545,8	220.211	33.031,6

	TOTALE	627.617	94.143

Figure 10.   Energy saving estimation in function of 3 working scenario phases.

Figure 11.   Fans installation phase.

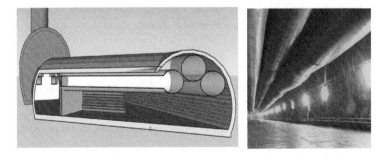

Figure 12.   Flex ducts lay out after the plenum.

The flow rate adjustment the no. 6 ventilation branches is provided by motorized regulating dampers (2,000 x 2,000 mm and 2,500 x 2,500 mm) equipped with of flow/pressure control units.

In order to guarantee the full functionality of the system, the regulation system is integrated by a series of differential pressure sensors thus to measure the static and total pressure of each

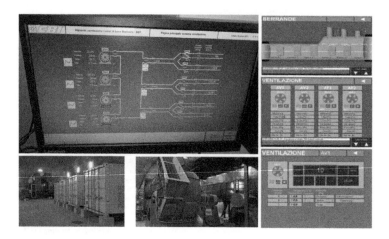

Figure 13. PLC integrated system with the ventilation plant.

fan downstream of each Dapò damper and the additional six pressure sensors upstream of each regulating damper. The complex centralized control system therefore provides a comprehensive operator interface able to constantly monitor the operating parameters of the fans.

Of course an integrated PLC system is controlling the full system, launching data and orders from a control cabinet (Figure 13).

5 CONCLUSION

The installation of the powerful Heavy Duty Fans has allowed ventilation of a very complex construction site ensuring ventilation modularity (adaptability to the process requirements) and combining high performances and operating efficiency.

Together with the continuity and reliability of operation, the HDF system ensures consistent energy saving without giving up the safety and the healthiness of the work areas.

ACKNOWLEDGMENT

All the BTC Team must be acknowledged for the hard and professional work done together, starting from feasibility study to the contractual and supply set-up: G. Frattini, S. Citarei, M. Ferroni, M. Secondulfo, A. Caffaro, R. Paolini, G. Vozza.

REFERENCES

Bringiotti, M., Duchateau, JB, Nicastro, D. & Scherwey P.A. 2009. *Sistemi di smarino via nastro trasportatore - La Marti Technik in Italia e nel progetto del Brennero*, Convegno "Le gallerie stradali ed autostradali - Innovazione e tradizione", SIG, Società Italiana Gallerie, Bolzano, Viatec,

Bringiotti, M., Parodi, G.P. & Nicastro D. 2010. *Sistemi di smarino via nastro trasportatore*, Strade & Autostrade, Edicem, Milano, February

Fuoco, S., Zurlo, R. & Lanconelli, M. 2017. *Tunnel deformation limits and interaction with cavity support: The experience inside the exploratory tunnel of the Brenner Base Tunnel*, Proceedings of the World Tunnel Congress– Surface challenges – Underground solutions. Bergen, Norway

Rehbock, M., Radončić, N., Crapp, R. & Insam, R. 2017. *The Brenner Base Tunnel, Overview and TBM Specifications at the Austrian Side*, Proceedings of the World Tunnel Congress – Surface challenges – Underground solutions. Bergen, Norway

Tunnels and Underground Cities: Engineering and Innovation meet Archaeology,
Architecture and Art, Volume 9: Safety in underground
construction – Peila, Viggiani & Celestino (Eds)
© 2020 Taylor & Francis Group, London, ISBN 978-0-367-46874-3

Life safety applied in full face excavation

P. Lunardi
Lunardi Geo-Engineering, Milan, Italy

G. Cassani, M. Gatti, A. Bellocchio & C.L. Zenti
Rocksoil S.p.A., Milan, Italy

ABSTRACT: The management of the risks associated with tunnelling work follow a systematic process, which involves identifying hazards, assessing and controlling risks. Tunnel design is different to surface structure design, because it is difficult to accurately predetermine geological and stress conditions, properties and variability along the tunnel. ADECO-RS (Analysis of Controlled Deformation in Rock and Soil) defines a method for design and construction of tunnels. The paper summarizes the main issue of this method, which is related to the design procedure and approach that ensure maximum safety level, accurate scaling and job site organization. During the construction stage (operational phase) the work process is to be verified by monitoring the Real Deformation Response. Control is developed by perfecting the design, balanced on the basis of the results of the Analysis and on the field of the possible variables that have already been planned in the design stage. The management of the different work phases is a fundamental safety aspect during tunnel construction. Limiting the number of the operators at the face and their occupancy time in that area significantly increases safety level of the excavation process.

1 INTRODUCTION

The construction of underground tunnels, shafts, chambers, and passageways are essential yet dangerous activities. Working under reduced light conditions, difficult or limited access and egress, with the potential for exposure to air contaminants and the hazards of fire and explosion, underground construction workers face many dangers.

Many aspects of tunnel design influencing whether a tunnel is constructed safely, like tunnelling in rock or soft ground, are decided in the concept design stage. During this phase duty holders should consult and plan to manage risks which may occur when constructing, using and maintaining the tunnel. Consultation between duty holders will give a better understanding of the time needed for geotechnical investigations, tender preparation, construction and potential delays. The principal contractor must ensure a work health and safety management plan is prepared for the tunnelling work before tunnelling work starts. The client should include this requirement in contract documents. Underground construction is an inherently dangerous undertaking, and hazardous conditions are given mainly by - but not limited to - falls from height, falling objects, hit by machinery, handling of heavy objects.

2 THE DESIGN AND CONSTRUCTION OF TUNNELS

Estimates of uncertainties and risks of the construction process are essential information for decision-making in infrastructure projects. The construction process is affected by different types of uncertainties. We can distinguish between the common variability of the construction process and the uncertainty on occurrence of extraordinary events, also denoted as failures of the construction process or of the geological model. In tunnel construction, a significant part of the uncertainty results from the unknown geotechnical conditions. The construction performance is further influenced by human and organizational factors, whose effect is not known in advance. All these uncertainties should be taken into account when modelling the uncertainty and risk of the tunnel construction.

Underground structures are man made objects constructed in heterogeneous and complex natural environment. For planning and designing of the structures it is thus crucial to describe the behaviour of the geological environment by parameters, which can be used in the structural analysis and for planning and monitoring of the construction process.

For reliable predictions, it is essential to realistically estimate the parameters of the probabilistic model. This is so true especially for the definition of geological and geotechnical model which often is characterized by uncertainty due to:

– low number and poor quality of investigations
– wrong sampling
– wrong technological choice for in situ and laboratory tests
– unfair geotechnical characterization

It is fundamental try to solve this possible initial lack of information because geological and geotechnical knowledge represent the basis of tunnel design.

In any case there are practical limits to the detail in which an investigation can describe the ground to be encountered by a tunnel. The most widely accepted and successful way to deal with the uncertainties inherent in dealing with geological materials came to be known as the observational method (used since 1969).

The European Standard EN 1997-1: Eurocode 7. Geotechnical design - Part 1: General rules in paragraph "2.7 Observational method" includes the following definition:

(1) When prediction of geotechnical behaviour is difficult, it can be appropriate to apply the
(2) approach known as "the observational method", in which the design is reviewed during construction.
(3) P During construction, the monitoring shall be carried out as planned.
(4) P The results of the monitoring shall be assessed at appropriate stages and the planned contingency actions shall be put into operation if the limits of behaviour are exceeded.
(5) P Monitoring equipment shall either be replaced or extended if it fails to supply reliable data

of appropriate type or in sufficient quantity.

The application of observational method should lead to a fine tuning of the design during the construction works.

2.1 The design and construction stages

According to the A.DE.CO-RS Approach for full face tunnel excavation to frame the design and the construction of underground works it is necessary to divide them into two chronologically separate moments (Figure 1): a design stage and a construction stage (Lunardi, 2008). The design stage consisting of:

– survey phase: referred to the geological, geomechanical and hydrogeological knowledge of the ground and to the analysis of the existing natural equilibriums;
– diagnosis phase: referred to the analysis and the theoretical forecasting of the behavior of the ground in terms of Deformation Response, in the absence of stabilizing operations, according to the stability conditions of the core-face;
– therapy phase: referred, firstly, to the definition of the methods of excavating and stabilizing the ground to control the Deformation Response; and subsequently, to the numerical evaluation of the effectiveness of the solutions chosen; in this phase the section types are composed and the possible variability depending on the actual deformation behaviour of the tunnel in the excavation phase, which will be measured during the operating phase

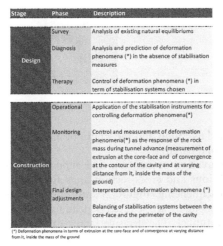

Figure 1. Design and Construction Stage.

The construction stage consisting of:

– operational phase: referring to the actual construction of the tunnel, in which the application of the stabilizing instruments for controlling the Deformation Response is implemented.
– monitoring and final design adjustment phase: during the course of the work, referring to the measurement and experimental interpretation of the actual behaviour of the ground to excavation in terms of Deformation Response, for the finalization and the balancing of the stabilizing systems implemented between the core-face and the excavation perimeter, and for checking the chosen solutions by means of comparing actually measured deformations with the ones that are expected theoretically.

The behaviour of the face is then influenced by these factors:

– Geotechnical parameters of the soil (strength and deformability)
– Overburden of the tunnel (geostatic stresses)
– Size of the tunnel (diameter and shape)
– Excavation system
– Constructional procedures

The construction of tunnels consists in two main phases: tunnel excavation (including construction of the tunnel support) and equipment of the tunnel with final installations (ventilation system, lighting and safety systems etc.). The latter is not discussed in this paper. Three main tunneling technologies are commonly utilized in present practice (mechanized, conventional and cut&cover) but the paper paid a special attention to the conventional tunneling excavation method.

3 TUNNEL OPERATIONAL CYCLE

According to definition of International Tunnelling Association (ITA-AITES, 2009), the conventional tunneling technology is construction of underground openings of any shape with a cyclic construction process of:

– excavation, by using the drill and blast methods or mechanical excavators (road headers, excavators with shovels, rippers, hydraulic breakers etc.)
– mucking
– placement of the primary support elements such as: steel ribs or lattice girders, soil or rock bolts, shotcrete, not reinforced or reinforced with wire mesh or fibres.

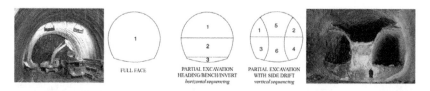

Figure 2. Typical excavation sequencing types in conventional tunneling.

One cycle of the construction process is denoted as round and the length of the tunnel segment constructed within one round is denoted as round length.

The excavation method, round length, excavation sequencing and support measures are selected depending on the geotechnical conditions and cross-section area of the tunnel. The excavation method, round length, excavation sequencing (Figure 2) and support measures (in sum denoted as the construction method in this paper) are selected depending on the geotechnical conditions and cross-section area of the tunnel. The decisive factor for the selection is the stand-up time of the unsupported opening. To give an example, a tunnel constructed in very good ground conditions with long stability of unsupported opening can be excavated full face with round length of several meters and it requires only simple support. On the contrary, in difficult ground conditions, shorter round length and demanding support measures must be applied. In poor ground conditions, auxiliary construction measures can be used. These are for example jet grouting, ground freezing, pipe umbrellas or face bolts.

The conventional tunneling technology has many modifications depending on the local experience and geological specifics.

3.1 The conventional tunnel excavation common risk

The risk is inherent all engineering applications. The common practice area of mining and civil engineers, tunnel construction, is also prone to several hazards originating from different sources.

In the early design and tender and contract negotiation phases, certain risks may be transferred, either contractually or through insurance, others may be retained and some risks can be eliminated and/or mitigated. In the construction phase, possibilities of risk transfer are minimal and the most advantageous strategy for both owner and contractor is to reduce the severity of as many risks as possible through the planning and implementation of risk eliminating and/or risk mitigating initiatives (Eskesen S.D., 2004).

Underground workers are at risk for serious and often fatal injuries. Some hazards are the same as those of construction on the surface, but they are amplified by working in a confined environment. Other hazards are unique to underground work. These include being struck by specialized machinery or being electrocuted, being buried by roof falls or cave-ins and being asphyxiated or injured by fires or explosions. Tunnelling operations may encounter unexpected impoundments of water, resulting in floods and drowning.

Control measures should be identified to eliminate or minimise, so far as is reasonably practicable, risks associated with tunneling work. These mostly arise from working underground and can be identified during consultation and the risk assessment process.

The risks related to the conventional tunneling excavation methods can be listed as follow:

- Rock Fall from crown and face
- Small working spaces (Figure 3.a)
- Numbers of Workers on face
- Small size equipment (Figure 3.b)
- High water and mud inflow
- Gas inrush
- Falls from height
- Loss of lighting
- Manual tasks like handling air tools, drill rods, supports, cutters
- Heat stress

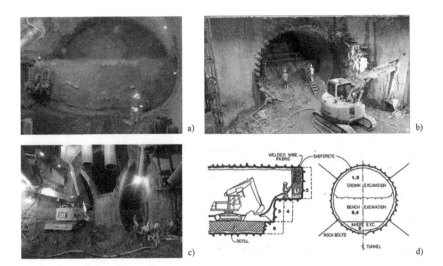

Figure 3. Risks related to the conventional tunneling excavation methods: a) Small working spaces; b) Small size equipment; c) e d) synchronized actions.

but the most critical issue is the synchronized actions highlighted by the Figure 3c and Figure 3d (ITA-AITES, 2009). Each work phase is characterized by specific risk and the overlap of different actions cause an exponential increase of the risks.

4 SAFETY & HAZARD MITIGATION

Underground construction is inherently a dangerous undertaking. Work progresses in a noisy environment in close quarters with moving heavy machinery. Careful attention must be paid to the layout of the worksite and workers must be protected at all times.

In order to excavate the opening required for the tunnel, the natural properties of the ground are disturbed. The ground is rarely a homogeneous mass but has been subjected to massive natural forces and has been substantially altered. Once the opening has been excavated it must be supported in order for the workers to be protected from falling material, collapse or other deterioration of the tunnel roof or crown. The excavation can be carried out in different methods depending on the type of material to be excavated.

4.1 The hazards control in underground construction

The best practice establishes a process and determines controls for achieving reduction in risks using the following hierarchy:

– Hazard elimination: Avoiding risks and adapting work to workers, (integrating health safety and ergonomics when planning new construction site)
– Substitution: Replacing the dangerous with the less or non-dangerous (increasing technology)
– Engineering controls: implementing collective protective measures
– Administrative controls: Giving appropriate instructions to workers
– Personal protective equipment (PPE): Providing PPE and instructions for PPE use/ maintenance

Every step of the underground excavation should be planned with safety in mind. The normal surface safety concerns are also appropriate and often amplified for underground construction including: workers must be safeguarded from falling from the work platforms used in the mining process, from being struck by the moving equipment and from electrocution amongst many hazards. However there are also many additional hazards that workers must be protected from and guarded against.

It is typically the job of the Construction Engineer to plan on making the tunnel opening stable to allow workers to move freely and without the concern of falling material.

The contractor should ensure that suitable and sufficient tunneling equipment for the type of work to be carried out is provided and is operated and maintained in accordance with manufacturers' instructions. Furthermore, the contractor should reduce the risk to workers underground through the elimination or control of hazardous materials and processes.

The full-face excavation requires a specifically risk analysis for the construction process applied. The risks are well known and should be managed in a properly way.

4.2   Standard & Code

Safety Issue has been considered an important topic since a long time. The first guideline dated back to 1941 with the Bulletin n. 439 of the Bureau of Mine (Washington, U.S.A.) titled: Some essential safety factors in tunneling.

Since late '90 and early '00 consistent and solid H&S regulations were adopted worldwide, such as: BS 6164: 2011. Code of Practice for Health and Safety in Tunnelling in the Construction Industry (UK) or NIR (IT) (the guidelines are several and each volume, identified by a number, is related to a specific topic, i.e. n. 41 operation close to the face).

The NIRs (Norme Interregionali), are the Interregional Rules of the Regions Emilia Romagna and Tuscany in Italy, developed by the Occupational Health and Safety department of the two regions, aim to achieve the highest safety levels in full face tunnel work. Each of them is a monothematic technical treatise containing design, construction, organizational, technological and operational solutions.

The regulations have been developed on the basis of problems and events recorded during the construction of High Speed Railway Line (section Bologna to Florence, total underground excavation 104,3 km) and Variante di Valico (total underground excavation length 62,2 km), which is the upgrading section of the A1 Milan-Naples motorway between Sasso Marconi (Bologna) and Barberino di Mugello (Florence).

Since 1998 until now 45 documents have been drawn up. NIRs have been exported to other regions by clients and contractors and have been adopted by various regions in Italy. Various professional figures, designers, service companies, manufacturers of equipment, machines and communication and control systems have contributed ideas to the development of solutions which were subsequently formalised in the Interregional Notes.

The scope of the NIRs is to increase safety at the construction sites of the Large Underground Works and they cover all the issues related to a project development from the design stage to the construction stage focusing the attention on Safety issue.

4.3   The Italian experience

The NIRs have been defined on the basis of Italian experience, as mentioned previously in this paragraph. The standards consider only full-face excavation method; this design choice is supported by previous experience in which this solution proved to be successful compared to the "partial excavation".

The approach is based mainly on two concepts (summarised by Figure 4:

- the significance of the deformation response of the ground during excavation, which the tunnel engineer has to be able to fully analyse and then control;
- the use of the advance-core of the tunnel as a stabilization tool to control deformation response and to get a prompt stability of the excavation.

Control of deformation response can be achieved:

Ahead of the face - regulating the rigidity of the advance core by means of pre-confinement of the cavity using different techniques depending on the type of ground, the in-situ stress and the presence of water. There are two different kind of interventions:

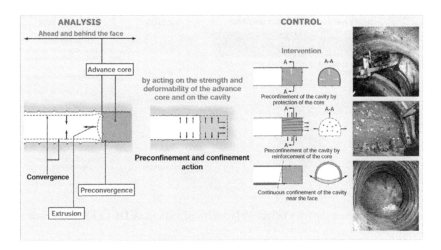

Figure 4.   The main concepts of A.DE.CO-RS Approach.

– protective conservation is the one set in advance, working around the perimeter of the cavity to form a protective shell around the core able to reduce deformation on the core-face system (Figure 5.a);
– reinforcement conservation is the one set directly into the core of advance, to improve its natural strength and deformation parameters (Figure 5.b).

Down from the face - regulating how the advance core extrudes by means of confinement of the cavity closing the ring and making the preliminary lining rigid close to the face (Figure 6).

The A.DE.CO-RS approach and the NIRs application fitting perfectly to hierarchical method applicable for risks reduction during tunnelling construction (Figure 7). The excavation phase is the most dangerous work phase and the main benefit of ADECO adopted by the NIRs is the high industrialization level. Each operation is well defined and executed by small

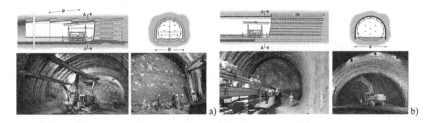

Figure 5.   Example of conservation technique: a) Sub-horizontal jet-grouting around the cavity & in the core; b) Reinforcement of the core using glass-fibre elements.

Figure 6.   Example of conservation technique: a) Invert; b) advance core GFRP reinforcement and invert.

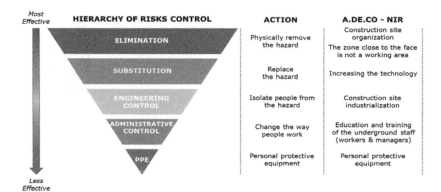

Figure 7. Relation between the risks reduction hierarchical method, A.DE.CO-RS approach and NIR.

number of experienced workers, using large equipment characterized by technological value which reduce the risk of human fault to a low level. The safety of workers during the excavation phase requires a unified approach and rational organization of the site in order to minimize the risks associated with the works and those related to the interference, the operating space and the execution times.

5   FULL FACE EXCAVATION & SAFETY ISSUES

With reference to the "full excavation face" method, the excavation proceeds through the use of an excavator equipped with a "hammer" or by drill and blast. Subsequently, after removing portions of unstable rock (scaling) a preliminary layer of concrete is sprayed on the excavated face and cavity portion. After this operation it is possible to install the primary lining constituted by steel arch and reinforced shotcrete.

Main advantages of the full-face excavation are related to the industrialization of the work phases: one of the great innovations is the simplification and cleanliness of the construction operations into the tunnel. There are always few workers moving into the tunnel (not more than six persons in each stage of construction), with few and powerful machines and equipment to be used), with a clear sequence of operations. There is never a superposition of different kind of interventions or operations. There is a direct relationship between the number of workers and of the machinery and safety: the less they are, the greater is the safety.

The relevant aspect with respect to the safety is for sure the layer dimensions of the working area. The full-face excavation method has a lot of benefits in term of safety but the concrete pre-lining spraied on the face and on the excavated portion of the cavity has a fundamental function preventing rock fall and local instability of the face and of the cavity. The huge dimension of the face (varying from 130 sm to 230sm) requires this procedure despite to the fact the area close to the face is forbidden to man works and operations. The NIR 37 titled "Guidelines for the safety of the excavation phase in tunnels bored with Conventional method" cover specifically this topic, underling the importance of this specific work phase.

5.1   *Work Phases*

With reference to the ADECO-RS Approach, a typical cycle-work for tunnel construction is reported in the following figures. The main phases are:

Ground reinforcement (Figure 8)

– Pre-confinement of the core-face, by means of fiber-glass reinforcements or grouting activities, or pre-support system, or drainages pipes ahead of the face (if necessary, according to face categories)

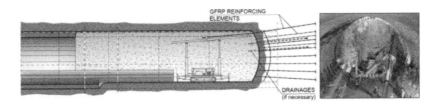

Figure 8.   Ground reinforcement work phase.

Excavation & first lining installation (Figure 9)

- Full-face excavation (top-down), with excavation step depending on the geomechanical conditions, ranging from 1.0 m up to 3.0 m. The excavation will be made by drill&blast or by mechanical system, with concave shape.
- Placing of fibre-reinforced shotcrete on the face and on the cavity (for each excavation step as a preliminary lining for safety provision of the excavated section).
- Confinement of the cavity, placing steel bolts or steel ribs.
- Completing the first lining, placing fiber-reinforced shotcrete (or reinforced with wire-mesh).

Invert casting & final lining (Figure 10)

- Excavation and casting of side walls and invert (at a defined distance from the face).
- Waterproofing installation.
- Placing vault reinforcement (if required) and casting the final lining (at a defined distance from the face).

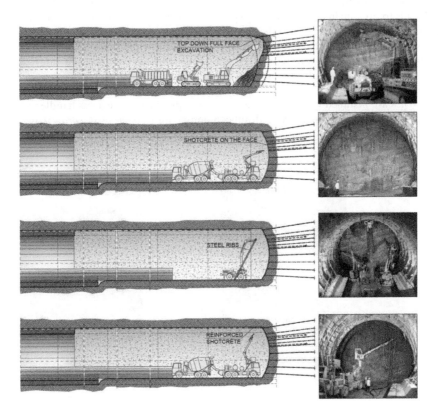

Figure 9.   Excavation & first lining installation.

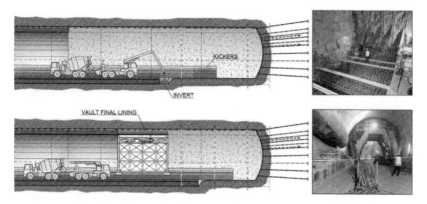

Figure 10. Invert casting & final lining.

The analysis of the monitoring data (mainly geological face mapping and extrusion-convergence measurements) will allow to confirm the predicted section type and regulate the intensity of the interventions and of the executive phases (such as, excavation step, distance from the face for invert casting and final lining casting).

5.2 *The evidence of the risk reduction*

The analysis of some specific work phases (Figure 11), clearly explain the risk reduction reached by the application of the concepts previously explained (cf. chapter 4). The full-face excavation method leads to operate in a wide working space using large and powerful equipment.

Face or cavity reinforcement installation is performed by a tunnel drilling machinery, the geometry of this equipment, characterised by two cradles that house the system of sliding blocks and telescopic arms, lets the worker operating far from the face in a safety area. Furthermore, each of the single drilling arms is fully independent, driven by two separated power and remote-control unit.

Steel arches are installed by using an excavator equipped the three arms handlers.

Scaling is always performed by hydraulic hammer and *concrete is sprayed* by shotcrete manipulator.

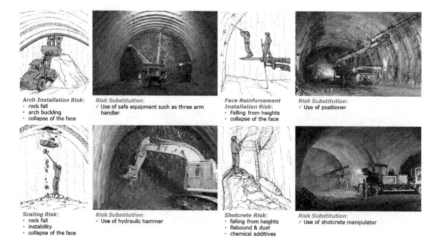

Figure 11. Example of risk reduction trough the application of a full-face excavation method.

6 CONCLUSION

The management of the risks associated with tunneling work follows a systematic process, which involves identifying hazards, assessing and controlling risks. Estimates of uncertainties and risks of the construction process are essential information for decision-making in infrastructure projects.

The risk is inherent all engineering applications and tunnel construction, is also prone to several hazards originating from different sources. In the early design and tender and contract negotiation phases, certain risks may be transferred, either contractually or through insurance, others may be retained and some risks can be eliminated and/or mitigated. In the construction phase, possibilities of risk transfer are minimal and the most advantageous strategy for both owner and contractor is to reduce the severity of as many risks as possible through the planning and implementation of risk eliminating and/or risk mitigating initiatives.

Safety Issue has been considered an important topic since a long time and around the world. The Occupational Health and Safety department of the Regions Emilia Romagna and Tuscany in Italy, developed the Interregional Rules based on Italian Experience referred to more than 150 km of tunnel bored using full-face excavation method. Each NIR is a monothematic technical treatise containing design, construction, organizational, technological and operational solutions developed achieving the highest safety levels. The Rules consider only full-face excavation method, this design choice is supported by previous experience in which this solution proved to be successful compared to the "partial excavation".

Main advantages of the full-face excavation are related to the industrialization of the work phases. There are always few workers moving into the tunnel, with a limited number but powerful machines and equipment to be used, with a clear sequence of operations. There is never a superposition of different kind of interventions or operations. There is a direct relationship between the number of workers and of the machinery and safety: the less they are, the greater is the safety.

The full-face excavation method has a lot of benefits in term of safety but the concrete pre-lining sprayed on the face and on the excavated portion of the cavity has a fundamental function preventing rock fall and local instability of the face and of the cavity. The huge dimension of the face (varying from 130 m^2 to 230 m^2) requires this procedure as the first safety provision despite to the fact the area close to the face is forbidden to man works and operations, because the best way to protect the worker is risk elimination.

Construction engineering and safety go hand in hand and everybody must come back home safely at the end of their shift.

REFERENCE

BS 6164: 2011. Code of Practice for Health and Safety in Tunnelling in the Construction Industry.

EN 1997-1: Eurocode 7. Geotechnical design - Part 1: General rules.

Eskesen S.D., Tengborg P., Kampmann J., Veicherts T. H. 2004. Guidelines for tunnelling risk management: International Tunnelling Association, Working Group No. 2, *Tunnelling and Underground Space Technology*, 19, (3): 217-237.

ITA-AITES WG5, 2011. *Safe working in tunnelling. For tunnel workers and first line supervision.*

ITA-AITES WG19, 2009. *General Report on Conventional Tunnelling Method.*

Lunardi, P. 2008. *Design and construction of tunnels: Analysis of Controlled Deformation in Rock and Soils (ADECO-RS)*. Berlin: Springer.

Špačková, O. 2012. Risk management of tunnel construction projects, Thesis, *Ph.D. Program: Civil Engineering Czech Technical University in Prague*, Prague: Faculty of Civil Engineering.

Tunnels and Underground Cities: Engineering and Innovation meet Archaeology,
Architecture and Art, Volume 9: Safety in underground
construction – Peila, Viggiani & Celestino (Eds)
© 2020 Taylor & Francis Group, London, ISBN 978-0-367-46874-3

Excavation with traditional methods through geological formations containing asbestos

N. Meistro, F. Poma, U. Russo, F. Ruggiero & C. D'Auria
COCIV, Consorzio Collegamenti Integrati Veloci, Genova, Italy

ABSTRACT: During the excavation of the Cravasco Adit of Terzo Valico dei Giovi Project, some geological asbestos formations were intercepted.As a consequence the COCIV Consortium has designed and realized a new and complex system,technical and procedural, able to allow the continuation of the activities and at the same timeguarantee the safety of workers, of the neighborhood and of the external environment. The measures apllied, a numerous air monitoring performed within and around the jobsite, inclose combination with the complex and rigid applied protocols, added to an active dialogue held with the Territorial Supervisory Authorities, have determined the complete success of the operations, identifying solutions that were afterwards replicated on the other jobsites of the Project.

1 PROJECT DESCRIPTION

The railway Milan–Genoa, part of the High Speed/High Capacity Italian system (Figure 1), is one of the 30 European priority projects approved by the European Union on April 29th 2014 (No. 24"Railway axis between Lyon/Genoa – Basel – Duisburg – Rotter-dam/Antwerp) as a new European project, so-called "Bridge between two Seas" Genoa – Rotterdam. The new line will improve the connection from the port of Genoa with the hin-terland of the Po Valley and northern Europe, with a significant increase in transport ca-pacity, particularly cargo, to meet growing traffic demand.

The "Terzo Valico" project is 53 Km long and is challenging due to the presence of about 36 km of underground works in the complex chain of Appennini located between Piedmont and Liguria. In accordance with the most recent safety standards, the under-ground layout is formed by twosingle-track tunnels side by side with by-pass every 500 m, safer than one double-track tunnel inthe remote event of an accident.

The layout crosses the provinces of Genoa and Alessandria, through the territory of 12 municipalities.

To the South, the new railway will be connected with the Genoa railway junction and the harbor basins of Voltri and the Historic Port by the "Voltri Interconnection" and the "Fegino Interconnection". To the North, in the Novi Ligure plain, the project connects ex-isting Genoa-Turin rail line (for the traffic flows in the direction of Turin and Novara – Sempione) and Tortona – Piacenza – Milan rail line (for the traffic flows in the direction of Milan-Gotthard).

The project crosses Ligure Apennines with Valico tunnel, which is 27 km long, and ex-its outside in the municipality of Arquata Scrivia continuing towards the plain of Novi Ligure under passing, with the 7 km long Serravalle Tunnel, the territory of Serravalle Scrivia (Figure 2). The underground part includes Campasso tunnel, approximately 700 m long and the two "Voltri interconnection" twin tunnels, with a total length of approximate-ly 6 km. Valico tunnel includes four intermediate adits, both for constructive and safety reasons (Polce-vera, Cravasco, Castagnola and Vallemme). After tunnel of Serravalle the main line runs out-door in cut and cover tunnel, up to the junction to the existing line in Tortona (route to

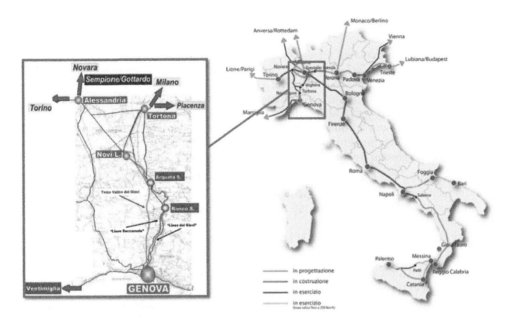

Figure 1.   High-speed Italian system.

Milan); while a diverging branch line establishes the underground connection to and from Turin on the existing Genoa-Turin line.

From a construction point of view, the most significant works of the Terzo Valico are repre sented by the following tunnels:

- Campasso tunnel 716 m in length (single-tube double tracks)
- Voltri interconnection even tunnels 2021 m in length (single-tube single track)
- Voltri interconnection odd tunnels 3926 m in length (single-tube single track)
- Valico tunnel 27250 m in length (double tube single track)
- Serravalle tunnel 7094 m in length (double tube single track)
- Adits to the Valico tunnel 7200 m in length
- Cut and cover 2684 m in length
- Novi interconnection even tunnels 1206 m in length (single-tube single track)
- Novi interconnection odd tunnels 958 m in lenght (single-tube single track)

The project standards are: maximum speed on the main line of 250 km/h, a maximum gradi-ent 12, 5 ‰, track wheelbase 4, 0 – 4, 5 m, 3 kV DC power supply and a Type 2 ERTMS signalling system.

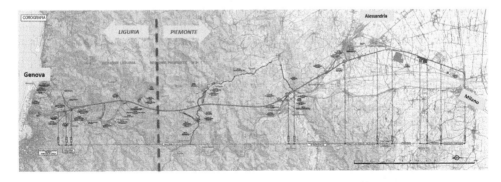

Figure 2.   Terzo Valico project.

2 CRAVASCO JOB SITE

Cravasco job site is articulated on a service area of reduced extensions, about 6,000 square meters, at an altitude of 293.50 m.s.l.m., and is connected to the provincial road S.P.n°6 through a service track with a slope of about 11%.

Machinery and equipment such as vehicle maintenance workshop, fuel distribution site, weigh-bridge, offices, oxygen and acetylene storage tanks, generators, tunnel muck temporary disposalarea, manual and automated big bags packaging system, tunnel water treatment and filtration system, ventilation groups are all included within Cravasco job site area.

The main underground works included in Cravasco work lots are the following:

1. Cravasco Adit – The Cravasco Adit, has a length of 1,260 m, a slope of 12.5% and anexcavation section of the order of 95 square meters.
2. Cravasco Cavern -Total length of 190 m and excavation sections of the order of 200 square meters.
3. Valico Tunnel – Total length of 9,319 m and excavation sections of the order of 90 square meters.

The Cravasco Adit, Cravasco Cavern and Valico Tunnel have been and still performed by traditional system through the Sestri-Voltaggio Line and a large number of different lithologies: from dolomites to chalk stones, from argilloschists to meta-basalts and serpentinites.

3 THE CROSSING OF ROCK FORMATIONS CONTAINING ASBESTOS

During the construction of Cravasco Adit, near pk 0+706 approx, when penetrating argilloschists,tunnelling work struck geologic formations that can be traced back to ophiolite sequences usually referred to as "Green stones" that contain asbestos fibres within a natural matrix, which led the Company to redesign a new and complex plant, technical and procedural system to enable excavations being restarted, minimising the risk of exposure for workers and the release of asbestos fibres into other work environments and the external environment.

After identifying the "source" and any potential "vectors" for the release and spread of fibres, the Company established and implemented technical, plant and procedural solutions for the continuation of activities in line with the following principles:

1. Prioritising at "source risk reduction" activities over passive protection measures for workers, preferring "collective" to "individual" protection measures.
2. Confine and minimize the job site areas affected by asbestos fibers contamination, separating and isolating the excavation area from the rest of the underground tunnel.
3. Reducing workers asbestos exposure to the minimum during all the activities.
4. No dispersion of asbestos fibers to the outside using a suction-type ventilation system equipped with special filters at the outlet, channeling all the waters of the site into a single purification plant and preventing the sludges from contaminating any other place in the tunnel and the job site.
5. Identification of work processes and operational methodologies that, in addition to enabling activities in safety, could guarantee sustainable productivity comparable to normal productivity in the absence of asbestos.
6. Making identified solutions and procedures easy to replicate also in other Terzo Valico sites potentially affected by excavations in asbestos-carrying rock formations.

Despite the initial and deeply-rooted mistrust of the local communities regarding the ability of the Consortium to protect the resident population in the vicinity of the site and also thanks to the constructive relationships with Local Supervisory Bodies for the Protection of Workers and the Environment, the excavation through the amiantiferous formations took place with the maximumprotection of the workers employed in the project and with a nearly "zero impact" on the environment.

The long-term impact of the initiative is simply to have acquired, at the company level, an excellent skill to deal with similar issues that may occur in the future. The numerous air monitoring performed within and around the job site, have shown the adequacy and effectiveness of what has been realized, as the exposure values of asbestos fibers to which workers were exposed, were below the values recommended by the ACGIH (2 ff/l after protection and for Italian Law 100 ff/l before protection) and the airborne fibers in the nearbyareas (close to zero), were always far below the WHO recommended exposure values for the population(1 ff/l). Also, the asbestos experience has induced a radical change of mentality in the workers, which can still be felt today, despite the return to normal excavation conditions.

There are few documented cases in the world of experiences similar to the one in question. From now on, the Company will be able to mobilize in a very short time all the resources necessary to face excavations in rocks containing asbestos of any size and nature.

4 THE MAIN ADOPTED TECHNICAL SOLUTIONS

Identified solutions entailed, as a first step, subdividing the tunnel into 3 physically separate areas, each associated with a specific risk with respect to the potential exposure to airborne asbestos fibres during excavation and progress activities.

A zone -"Contamination Area" is the tunnel section next to the excavation face, spanning max. 80 meters, where all processes linked to the excavation of rocks containing asbestos within amineral matrix are performed, such as excavation and the construction of reverse arches and side walls. All equipment required for the initial containment and abatement of asbestos fibres releasedduring excavation activities was concentrated in the A Zone.

B Zone -"Decontamination Area" is the tunnel section next to contamination Zone A, with variable length of 90 lm to 140 lm, where decontamination activities for all personnel, tools andequipment used in Zone A during contaminating excavation activities are planned and carried out.

C Zone -"Contamination-free Area" means the rest of the tunnel from the exit of the B decontamination Zone and the tunnel's entrance. This area is dedicated to all work activities not directly linked to the excavation of asbestos-contaminated materials and, therefore, not subject to the riskof exposure to asbestos fibres (such as waterproofing and final shotcreting of the tunnel).

The physical separation of the three areas was achieved by creating two "Physical Compartments" between the A/B and B/C Zones, built using dedicated demountable and quickly removable metal structures, which ensured quick and easy securing and sealing around the contours of the cable.

Both compartments were fitted with automated gates to enable the entry and exit of staff and vehicles along the tunnel and equipped, at their top, with flat spray nozzles that could produce an effective water blade for the entire time the gate is open, so as to guarantee the continuity of physical separation between the areas.

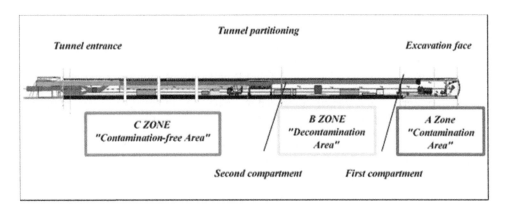

Figure 3. Tunnel partitioning.

Figure 4. Automated gates.

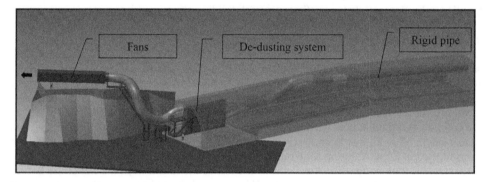

Figure 5. Ventilation system scheme.

At the same time as the tunnel was partitioned, the original pressure ventilation system was replaced with an extraction ventilation system, so as to capture the maximum number of airborne asbestos fibres released in the A Zone during excavation directly at source and quickly "decontaminate" the environment in the self-same A Zone after ending excavation activities and securingthe face by laying the relevant pre-coating.

The ventilation system was designed in such a way as to guarantee constant airflow in the tunnel. The air, running though the entire length of the tunnel, from the entrance to the extraction point close to the excavation face, may effectively contribute to preventing the spreading of released fibres to the rest of the tunnel and the external environment.

The air extracted by the ventilation system, conveyed by rigid metal pipes, runs through theentire length of the tunnel to be subsequently released into the atmosphere after undergoing afiltration treatment that captures any asbestos-carrying fibres.

Starting from the entrance of the tunnel and progressing to the excavation face, the new extraction ventilation system consists, specifically, in two fans fitted on an external metal support, with a power of 200 kW each, controlled by an inverter and equipped with suitable mufflers at airintake/outlet points, capable of extracting up to 60 m³/s of air from the tunnel.

Upstream of the extraction station and located in the first 50 m of the tunnel due to lack of space on the site's apron, three de-dusting units were designed and built in order to intercept and capture all airborne asbestos fibres present in the circulation air of the ventilation duct before it is finally released into the external environment.

Each de-dusting unit can filter an air flow of approx. 20 m³/s and is equipped with class K4 dryfilters with 99.96% efficiency (equivalent to HEPA H13). In each unit, the filtered material is conveyed, thanks to a compressed-air shaking system (vibrations), to an isolated container at the bottom of the de-duster, where it is picked up by an auger system that takes it to a mixer unit (Mixer Tank), which mixes dust with water. The resulting liquid, contaminated with asbestos fibres, is then sent to the waste water treatment plant.

In order to preserve the efficiency of dry filters, chosen because they are the only ones capableof intercepting powdery fibres of the same size as asbestos ones, a multi-stage water separator (Demister) was fitted upstream of the de-duster units. It consists of two filtration stages that can reduce the presence of water particles contained in the air extracted from the A Zone. This newsystem collects the separated water, which is potentially contaminated with asbestos fibres, to thewater treatment plant, using dedicated pipes.

Moving on towards the excavation face, the previous flexible hose, typical of a pressure ventilation system, was replaced along the entire tunnel section with a new Ø2200 mm rigid pipe until the extraction intake was brought close (max. distance of approx. 30 m) to the excavation face,piercing the two physical compartments that separate the tunnel areas described before.

A dust and fume extractor hood was designed and installed at the end of the pipe, near theextraction intake, initially using a PVC membrane supported by stainless steel rods for ease of execution. Its aim is that of increasing and facilitating conveying dust and fumes produced during rock breaking at the excavation face into the extraction pipe, thus limiting the concentration and spreading of asbestos fibres in the A Zone.

In order to balance the ventilation system and facilitate the washing and handling of the air on the excavation face, two "pressure boosters" were fitted near the physical compartment between the A and B Zones, connected to dedicated flexible hoses with Ø1000 mm, next to the excavation face (approx. 10 m), so as to move the air in the work area and improve its removal towards the extraction hood.

In the "Contamination A Zone", favourable effects of the new extraction ventilation system were coupled with the concurrent introduction of a high-efficiency water reduction system

Figure 6.   Ventilation system equipments.

aimed to increasing the capture and reduction of airborne asbestos fibres produced and released into theenvironment during excavation.

The standard spray system installed on the demolition hammer at the excavation face was further enhanced with 24 high-pressure spray nozzles, powered by a pressurisation unit consisting of pump with a flow of 50 l/min at a 120 bar pressure, in order to create a veritable air cone aroundthe demolition hammer's drill.

A few meters from the excavation front and next to the tunnel's ceiling, a spray arch fitted with approx. 40 nozzles tilted towards the excavation face was installed, to cover the entire section, with the main aim of keeping the "open" excavation face constantly wet after the removal of thepre-coating, which is the source of dust and, therefore, asbestos fibres.

Also in the A Zone, at the excavation face and near the two tunnel walls, compatibly with clearances and the manoeuvres of the vehicles employed, two directional "water fog cannons" were installed, each equipped with a crown of 60 high-pressure spray nozzles connected to a pressurisation unit consisting of an adjustable volumetric pump with capacity from 5 l/min to 35 l/min. The Fog Cannons pointed towards the muck pile progressively built up at the excavation face inthe breaking stages, to reduce the amount of fibres by increasing the wetting of the excavated rock.

The "B Zone – Decontamination Area", which, as already mentioned, was designed for theperformance of all decontamination operations for staff, vehicles and equipment used in the AZone during excavation activities and exiting towards the C Zone, entailed the installation of two separate decontamination units, one for construction equipment and the other one for operational staff transiting from the A Zone to C Zone.

Vehicle decontamination was designed and executed by making all vehicles coming from the A Zone go through a veritable, dedicated "wash tunnel", which, through a number of fan nozzles installed on the side walls, below the track and on the arches above, enable the accurate washingof the wheels, the chassis and all the external surfaces of the vehicle.

For the decontamination of all staff working on the excavation front, as well as of all professionals who have had to access the A Zone due to operational requirements, including those working for the Engineer and the Health Authorities, a dedicated "decontamination unit" was designedand built, taking as a starting point similar units used in asbestos decontamination projects in the building sector, and sizing it in a suitable way, taking into account the available space, to allow 10 workers (a full excavation crew and a further three individuals) to access the A Zone at the same time.

The area outside the tunnel, i.e. the site apron, was also considerably impacted by changes, additions and reconfigurations of the original layout in order to allow the most appropriate management and treatment of process waters from the tunnel that were particularly rich in asbestos fibres,as well as the delicate management of asbestos-contaminated excavation earths and rocks, to be disposed of as special hazardous waste pursuant to current legal provisions.

The original chemical/physical treatment system with quartzite filtration was adapted with the inclusion of a further stage consisting in a final ultra-filtration unit before waste waters are released into the stream. Ultra-filtration is a pressure filtration process that can split water-

Figure 7.   Wetting during excavation activities.

Washing vehicle tunnel

Workers decontamnation unit

Figure 8. Decontamination units in "B Zone".

insoluble particles and is used to separate solids suspended in the water itself, usually consisting in colloids, lime, bacteria and viruses. Solid separation is performed by using a "hollow-fibre membrane" with 0.02 μm pores capable of filtering asbestos fibres.

Another major change concerned the treatment plant's filter press, the final stage of the clarifloccuation of process waters from the site, which favours the dehydration of sludge through mechanical solid-liquid separation processes.

The known presence of large quantities of asbestos fibres in sludges made it necessary to fully contain this part of the system by building a negative-pressure room capable of preventing anyairborne fibres from leaking into the external environment.

The sludge produced by the filter press, therefore, was collected in a removable container andmanaged as waste, subject to prior classification, in line with applicable legislation.

The filter press room was also fitted with a staff decontamination unit and any equipment required for washing the external surfaces of the removable container.

The need to create an area for the unloading and safe management of asbestos-contaminatedearths and rocks produced by tunnel excavation had a major impact on the redesign of the site's original layout.

In line with current regulations, excavation earths and rocks that, after specific characterisation,show an asbestos content greater than 1 g per each kg of material, are classified as "special hazardous waste" to be sent to dedicated waste disposal sites after relevant management, treatmentand bagging.

It should be noted that, as an additional, partial change to the original industrial site's layout, itwas necessary to redesign and adapt all site industrial water circulation systems, both for the apronand the tunnel. This was due to the fact that dedicated pipes needed to be introduced to convey to the water treatment plant (equipped with an ultra-filtration unit as described above) all waters produced by: decontamination activities, wetting processes in the tunnel and the site apron, the filter press, as well as by the shed for the processing of excavated material, which were all potentially extremely rich in asbestos fibres.

During all excavation, loading and transport operations for rocky material, both ground crews deployed at the excavation face and machine operators (for excavators, loaders and dumpers) were protected from exposure to asbestos fibres by suitable PPE provided. Personal protectionequipment was suitably identified by applying the precautionary principle, depending on the risklevel assigned to work areas and so as to ensure the best level of comfort for workers engaged in various processes.

All staff members who accessed A and B Zones, apart from having to wear the aforementioned full double Tyvek suits, were issued with rubber boots and gloves with long oversleeve, both easily washable, as well as high-visibility jackets and hard hats.

With respect to respiratory tract PPE, ground staff operating mainly in A and B Zones were issued with TMP3 full-face APRs equipped with P3 EN143/02 filters, characterised by an FPN2000 Nominal Protection Factor and a FPO-400 Operational Protection Factor, i.e. capableof "reducing" the concentration of airborne fibres in the usage environment by 400 times. Staff members employed as machine operators only, on the other hand, were issued with half-faceAPRs equipped with a P3 EN143 filter characterised by an FPN50 Nominal Protection Factor andan FPO-30 Operational Protection Factor since operators, insulated in the cab, are already protected by the overpressure system that entails supplying the cab with air taken from the outsideenvironment and filtered through suitable HEPA H13 filters.

Work procedures and plant equipment planned and implemented for risk management purposes were then strictly applied to the usual, conventional excavation cycle that entails the use of an excavator equipped with a demolition hammer for processing the excavation face.

The excavated material produced and accumulated at the excavation front was then reloadedonto dumpers using a second excavator, placing the material into the hopper as delicately as possible and always making sure that the material produced at the excavation face and in the loading area was always kept suitably wet with the wetting systems provided. The dumpers for the transportation of excavated material were equipped with a dedicated tarpaulin to close the container and prevent the release of any fibres into the atmosphere and fitted with a system of seals betweenthe rear side of the dumper and the container itself, thus becoming "waterproof", so as to prevent the spillage of liquids (containing fibres, in part) along the route.

Once the vehicle was loaded and the ground crew had closed the tarpaulin, the driver who wasin the process of leaving the A Zone towards the portal of the tunnel, could open the gate with aremote control from his/her cab and, at the same time, activate a water curtain located on the partition that separates the A Zone from the B Zone. After leaving the A Zone, the dumper's driver, still insulated inside the vehicle's cab, proceeded towards the wash tunnel for the vehicle's full decontamination. After washing operations, the vehicle moved towards the

Figure 9.   Personal Protection Equipments.

Figure 10.   Big Bags packaging plant.

site apron, going through the second partition separating the B Zone from the C Zone, also equipped with an automated door and a water curtain remotely controllable from the cab, continuing without stopping until it reached discharge hopper of the automated shed, leaving the asbestos-contaminated excavation material to be processed in the subsequent stages of big bag packaging, decontamination, temporary storage on the apron and final delivery to authorised waste disposal sites.

In order to verify the suitability and the effectiveness of the aforementioned procedures and system solutions implemented, during the entire time it took to excavate through the asbestos geological formation, the airborne asbestos was subject to intensive monitoring in all work areas (A, B, C Zones and apron), in the cabs of the vehicles used during the excavation and excavatedmaterial removal/transport stages, as well as in clean and dirty changing rooms of the staff decontamination unit.

5 CONCLUSIONS

The "Cravasco Site", the first experience of tunnelling in rock masses containing asbestos within a natural matrix as part of the works for the Terzo Valico dei Giovi, was successfully concluded with the achievement of important milestones and provided key information for the development and improved management of future tunnel sites with similar features.

First and foremost, the main objective was achieved: the "minimisation" of the risk of exposure to asbestos fibre for all workers involved, thanks to intensive awareness-raising, information and training activities at all levels on the delicate issue of asbestos processing and to the timely and meticulous implementation of site-wide system and procedural solutions.

Another major objective was that of achieving "zero impact" on the external environment – the so-called "living environment" – throughout the excavation of asbestos-containing rock.

From a site organisation and production viewpoint, the most important results obtained were undoubtedly the identification of plant and procedural solutions that could be easily "reactivated" at the same excavation site and "replicated" at other sites of the Terzo Valico dei Giovi project potentially affected by the excavation of rocks containing asbestos.

Despite operational issues arising from newly introduced plant equipment and the adoption ofnon-standard work procedures, we also believe that the site managed to reach a more than "sustainable" level of productivity once fully operational, achieving a daily excavation progress of 2ml (excavation section of approx. 85 m2) and a monthly progress of up to 35–40 ml.

Finally, it should be noted how the major results obtained were also reached thanks to theprofessional, constructive and proactive input of Control Authorities which, as part of the relevantroles and responsibilities, effectively contributed to the achievement of a successful outcome.

Tunnels and Underground Cities: Engineering and Innovation meet Archaeology,
Architecture and Art, Volume 9: Safety in underground
construction – Peila, Viggiani & Celestino (Eds)
© 2020 Taylor & Francis Group, London, ISBN 978-0-367-46874-3

Ventilation of tunnels and cross-passages of the Ceneri Base Tunnel and similar projects

S. Nyfeler & P. Reinke
HBI Haerter AG, Bern, Switzerland

ABSTRACT: The Ceneri Base Tunnel (CBT) is part of the new Swiss Trans-Alpine Railway Link. Together with the Gotthard Base Tunnel and after 2020, the CBT will connect northern and southern Switzerland and Europe for high-speed passenger and freight trains. The twin-tube CBT is 15 km long. The single-track tubes are 40 m apart and connected every 325 m by cross-passages.

The tunnel has a state-of-the-art tunnel ventilation system (TVS) and an independent mechanical cross-passage ventilation system (CPVS). The TVS and CPVS shall provide acceptable conditions during maintenance works as well as safe areas during the self-rescue phase and smoke control for support of rescue and fire-fighting forces during an emergency.

Taking the CBT as reference, other contemporary TVS and CPVS concepts of twin-tube, high-speed rail tunnels are presented. The particular features are highlighted and the implications on civil design are described (need for shafts, space for fan plants, etc.).

1 INTRODUCTION

1.1 *Key features of Ceneri Base Tunnel*

The Ceneri Base Tunnel (CBT) is part of the new Swiss Trans-Alpine Railway Link (STS Swiss Tunnelling Society, 2016). The CBT is 15.4 km long and connects the Magadino region with the northern portal in Camorino and the Lugano region with the southern portal at Vezia. At the northern portal, branching caverns are built, which allow the connections to the existing railway line towards Bellinzona and Locarno. The underground branching caverns at "Sarè" about 2.5 km away from the southern portal allow a future continuation to the south.

The CBT consists of two parallel, single-track railway tubes with a slope from north to south of 6.3 ‰. The centre-distance of the two tubes is about 40 m. About every 325 m, the tubes are connected by cross-passages, which allow the self-rescue of occupants to the non-incident tube during an emergency and the accommodation of fixed equipment. The tunnel has no crossovers, emergency stops or multifunction stations. The design speed of CBT is 250 km/h.

1.2 *Tunnel ventilation system of Ceneri Base Tunnel*

1.2.1 *Tunnel ventilation system objectives*
The main tasks of the tunnel ventilation system of the CBT are to support the safety of occupants, the functionality of equipment and the health protection of the maintenance staff. This shall be ensured for the normal, maintenance and emergency operation mode of the CBT.

In normal operation, the following tunnel ventilation system objectives apply:

- Temperature in the single-track tunnel tubes to be limited to max. 40 °C
- Temperature in the cross-passages to be limited to max. 35 °C
- Relative humidity in the tunnels and cross-passages to stay below 70 %

In maintenance operation, the following ventilation objectives apply to the work areas:

- Velocity of air to be limited to max. 5 m/s
- Pressure changes to be limited to 10 kPa during stay in tunnel and 1.5 kPa within 4 s
- Temperature in cross-passages to be limited to max. 40 °C
- Air-quality regarding occupational health requirements to be ensured (e.g. regarding dust, pollution resulting from emissions of diesel engines)

In emergency operation with a fire and train stop in the tunnel, the following ventilation objectives apply:

- Smoke-free area to be assured, i.e. prevent smoke entering the non-incident tube during 90 min through cross-passages or by re-recirculation at portals
- Smoke concentration to be maintained as low as possible in cross-passages
- Temperature to be limited to max. 40 °C in all cross-passages by adequate ventilation to ensure trouble-free operation of the railway equipment
- Flow velocity of air towards the fire location of at least 2 m/s to be achieved in the empty cross-section of the incident tunnel tube
- Flow direction of air/smoke in incident tube to be selectable

1.2.2 *Tunnel ventilation system concept*

For the tunnel ventilation system, jet fans are installed inside the running tunnels (see Figure 1). The jet fans are arranged in pairs. The arrangement of the jet fans is determined by the function, the geometry and the operation needs. In order to create an overpressure in the non-incident tube versus the incident tube along the entire tunnel length, most of the jet fans are arranged between the first/last cross-passages and the portals. The train direction during normal operation leads to an asymmetric arrangement of fans in the two tubes.

1.2.3 *Tunnel ventilation system functions during different operation modes*

During normal operation of the CBT, the tunnel ventilation fans are not activated. The ventilation objectives are achieved by the train-induced air flow and thermal draft.

During the maintenance mode, an acceptable air-quality must be ensured by mechanical ventilation. Figure 2 shows the ventilation strategy for maintenance works in one tube, while in the other tube train operation can be continued. All jet fans create thrust in one direction (e.g. north to south) in order to generate the highest possible fresh air supply to work sites.

In the event of an incident, only the information about the incident tube (east or west) as well as the position of the trains in the tunnel system is known in a reliable manner. In the first phase of the event, the incident train cannot be confirmed with certainty in every case.

The main objective of the TVS is to create safe and stable egress conditions for the self-rescue as quickly as possible. This is ensured with the overpressure ventilation, i.e. by ventilation

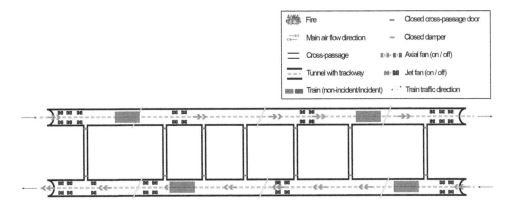

Figure 1. Ventilation concept of Ceneri-Base Tunnel (CBT).

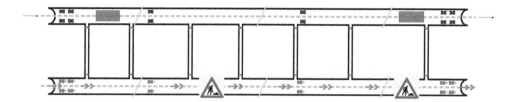

Figure 2. Ventilation concept during maintenance operation.

measures in the non-affected tube (see Figure 3). Jet fans near the opposite portals of the non-incident tube create thrust towards the tunnel centre by counter-directional operation.

Ventilation in the incident tube can have a negative impact on the self-rescue conditions for the following reasons (see Johnson et al. 2016 and Winkler & Carvel 2015):

– Active jet fans close to fire may destroy the thermal smoke stratification at the tunnel ceiling and lead to smoke in the entire tunnel cross-section.
– The TVS may actively carry smoke to evacuating passengers.
– Activation of jet fans may change prevailing flow directions and carry smoke rapidly to previously smoke-free areas leading to an additional risk for evacuees.
– Locally, jet fans may create overpressure in the incident tube.

For these reasons, fans in the incident tube are not activated in the initial emergency phase at CBT. Only the overpressure ventilation will be maintained in the non-incident tube. Later, when secure information is available and the self-rescue is terminated, the ventilation measures in the incident-tube are changed according to the information and requirements of the emergency services. Figure 4 shows for the later phase of an incident a ventilation strategy with the overpressure ventilation in the non-incident tube and longitudinal ventilation in the initial direction of train travel in the incident tube.

1.3 Ventilation of cross-passages of CBT

1.3.1 Cross-passage ventilation system objectives
The primary tasks of the CPVS are (see as well section 1.2.1):

– Continuous air renewal in cross-passages to limit high pollutant concentration and humidity

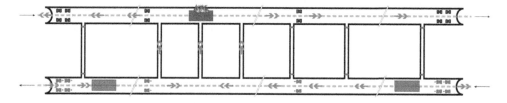

Figure 3. Overpressure ventilation in non-incident tube during self-rescue phase of emergency operation.

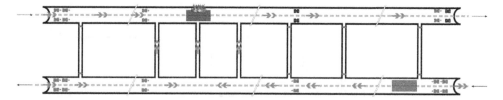

Figure 4. Ventilation concept in case of emergency after self-rescue phase with overpressure ventilation in the non-affected tunnel tube and longitudinal ventilation in the incident tube.

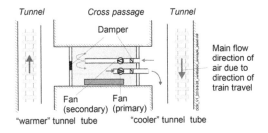

Figure 5. CPVS concept of CBT during normal operation.

– Cooling of the cross-passages by air-exchange with the "cooler" tunnel tube
– Support of tunnel ventilation system in case of an emergency

1.3.2 *Cross-passage ventilation system functions*

The CPVS in normal operation is sketched in Figure 5. The air is extracted from the "cooler" tunnel tube and flows through a supply duct to the end-wall of the cross-passages on the side of the "warm" tunnel tube. From there, the air flows back in the cross-passages and is carried out by an exhaust fan into the "cold" tunnel tube. The damper at the "warm" tunnel tube is closed. Each cross-passage is provided with temperature sensors. When the temperature limit is reached, the CPVS of the affected cross-passages is automatically switched on.

During emergency operation, the CPVS considers the location of the fire. If the incident tube is on the side of supply and exhaust of air during normal operation, the fans supplying air into the cross-passages are switched off and the damper in the end-wall at the non-incident tube are opened. If the incident tube is on the opposite side of air supply and exhaust during normal operation, the fans are switched to run normally. The dampers in the end-walls at the incident tube are kept closed.

2 OTHER VENTILATION CONCEPTS OF TWIN-TUBE, HIGH-SPEED RAIL TUNNELS

2.1 *Introduction*

Even though similar overall requirements for tunnel ventilation prevail, the concepts of modern, twin-tube high-speed rail tunnels vary considerably. The ventilation concepts are determined by the international, national and project-specific requirements.

2.2 *Objectives*

The primary objectives and purposes of the tunnel ventilation system (TVS) are to ensure:

– Acceptable levels of tunnel air temperature
– Acceptable air quality
– Smoke-free areas or smoke control and other fire-life-safety requirements

The secondary objectives of the TVS are to support:

– A limited negative impact on the environment (release of dust, aerodynamic, rail and ventilation borne noise at shafts and portals near TVS components, TVS energy demand).
– Minimized overall life-cycle costs of transportation system.

2.3 *Ventilation concepts for twin-tube tunnels*

Figure 6 presents concepts with ventilation shaft/adits (central ventilation):

a) Central ventilation by mid-tunnel ventilation shaft (passive, without mechanical ventilation)
b) Central ventilation by mid-tunnel ventilation shaft (active)

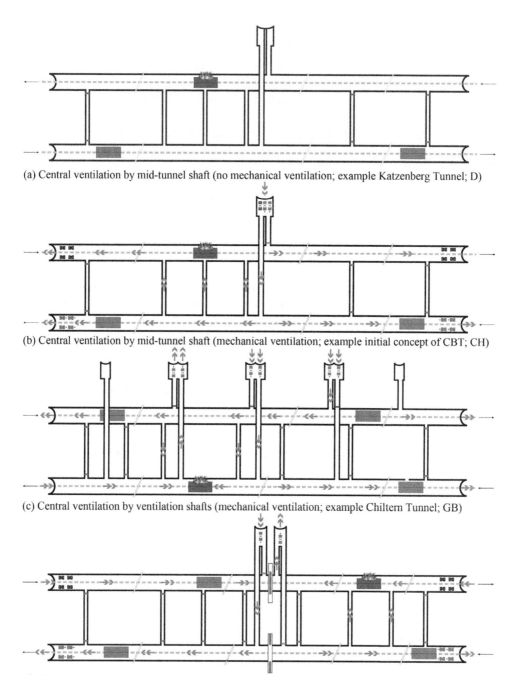

(a) Central ventilation by mid-tunnel shaft (no mechanical ventilation; example Katzenberg Tunnel; D)

(b) Central ventilation by mid-tunnel shaft (mechanical ventilation; example initial concept of CBT; CH)

(c) Central ventilation by ventilation shafts (mechanical ventilation; example Chiltern Tunnel; GB)

(d) Central ventilation by mid-tunnel shaft and support by railway tunnel doors (active mechanical ventilation; example for long tunnel with uniform inclination; USA)

Figure 6.   Tunnel ventilation system concepts with central ventilation shafts.

c) Central ventilation by several ventilation shafts along the tunnel (active)
d) Central ventilation by mid-tunnel ventilation shaft with railway tunnel doors (active)

The ventilation concept using a passive shaft in the middle of the tunnel is shown in Figure 6 (a) taking the Katzenberg Tunnel (D) as an example. The TVS for the Katzenberg Tunnel (10

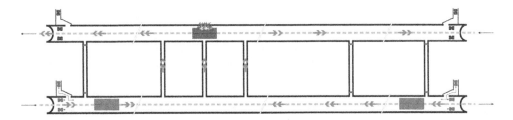

Figure 7. Longitudinal ventilation concept with Saccardo nozzles.

km) does not employ mechanical ventilation but a passive central shaft. In case of emergency, the smoke rises in the shaft by thermal draft. The shaft is neither used for egress nor for access.

The central ventilation by a ventilation shaft in the middle of the tunnel is shown in Figure 6 (b) using the initial ventilation concept of the Ceneri Base Tunnel (CH) as an example. The initial CBT ventilation concept features a central shaft with reversible axial fans and jet fans at the portals. This design provides a total of 4 ventilation zones and has no installations inside the tunnel (except jet fans near portals), simplifying the TVS maintenance. In case of an emergency, an overpressure is generated in the non-incident tube in the first phase. In the second phase of intervention by rescue forces, the fans are operated in exhaust mode from the incident tube (longitudinal ventilation in the emergency tube).

Another central ventilation concept employing several ventilation shafts is shown using the example of the future Chiltern tunnel in GB Figure 6 (c). The ventilation concept of the Chiltern tunnel provides a supply and exhaust of smoke/air by several ventilation shafts. The entire tunnel is divided into different ventilation zones. Depending on the incident tube and the ventilation zone in which the train on fire has stopped, air is supplied at one shaft and smoke/air exhausted at the adjacent shaft.

Another central ventilation concept employing tunnel doors is shown in Figure 6 (d) taking another project from the USA as an example. The main advantage is its increased efficiency because air can be guided through the desired tunnels legs adjacent to the shaft. Ventilation zones can be formed in the tunnel tubes. Also in this ventilation concept, an overpressure is generated in the non-incident tube.

The following concepts for longitudinal ventilation in the tunnel without ventilation shafts are presented (distributed ventilation):

– longitudinal ventilation with Saccardo nozzles in the portal area in Figure 7
– longitudinal ventilation with jet fans distributed along the tunnel in Figure 3

The concept of longitudinal ventilation with Saccardo nozzles in the portal area is shown in Figure 7 taking the example of the 28 km long Guadarrama Tunnel (E). The Guadarrama Tunnel uses Saccardo nozzles and jet fans near the portals. A national park above the tunnel does not allow using ventilation shafts. Saccardo nozzles are used to create an overpressure in the non-incident tube. During the self-rescue phase, there is no active ventilation in the incident tube.

The ventilation concept of longitudinal ventilation with jet fans is presented and shown for the example of the CBT in Section 1.2.3.

Table 1. Advantage and disadvantages of different tunnel ventilation concepts.

TVS – Concept	Advantages	Disadvantages
Central ventilation by mid-tunnel shaft (passive, without mechanical ventilation) Figure 6 (a)	• low operating costs • no equipment in tunnel • acceptable performance for most of operation time	• no overpressure in non-incident tube during self-rescue

(Continued)

Table 1.   (*Continued*)

		• uncontrolled flow in incident tube relying on natural flow only including the risk of flow reversal • high construction costs for shaft • no controlled ventilation during maintenance • limited means for environmental control during normal operation
Central ventilation by mid-tunnel shaft Figure 6 (b)	• simple mechanical ventilation concept • assurance of safe area in parallel tube under all conditions • no system components in tunnel except jet fans at portals • high availability • smoke extraction possible later • controlled ventilation during maintenance possible	• high construction costs for shaft and fan plant • high operation costs • no simultaneously controlled longitudinal ventilation in incident tube and overpressure in non-incident tube
Central ventilation by several ventilation shafts Figure 6 (c)	• assurance of safe area in parallel tube under all conditions • no system components in tunnel except jet fans at portals if any • simultaneously controlled longitudinal ventilation in incident tube, overpressure in non-incident tube and ventilation zones • high availability • controlled ventilation during maintenance possible	• most complex mechanical ventilation concept • high investment and operation costs for civil works and TVS equipment • complex control in case of an incident • high cost/benefit ratio
Central ventilation by mid-tunnel shaft and railway tunnel doors Figure 6 (d)	• simple forced ventilation concept • assurance of safe area in parallel tube under all conditions • no system components in tunnel except jet fans at portals and doors • high availability • smoke extraction possible any time • controlled ventilation during maintenance possible	• high construction costs for shaft and fan plant and doors • additional equipment in tunnel to be maintained • additional constraint for train operation
Longitudinal ventilation by Saccardo nozzles Figure 7	• simple forced ventilation concept • assurance of safe area in parallel tube under all conditions • no system components in tunnel except jet fans at portals and doors • controlled ventilation during maintenance possible	• high construction costs for fan plants at each portal • thrust of one fan plant only in one direction • no simultaneous longitudinal flow in incident tube and overpressure in non-incident tube • high pressure differences of tubes • no ventilation zones
Longitudinal ventilation by jet fans Figure 3	• simple forced ventilation concept • assurance of safe area in parallel tube under all conditions • independent ventilation in tubes • maintenance ventilation possible • high availability	• high costs for cabling • limited access for maintenance of the fans in railway tunnel • no smoke extraction • no ventilation zones

2.4   *Comparison of tunnel ventilation concepts*

Table 1 compares the advantages and disadvantages of different ventilation concepts. The relevance and evaluation of advantages and disadvantages depends on each individual project.

3 OTHER CROSS-PASSAGE VENTILATION CONCEPTS

3.1 *Requirements and functions during normal operation*

During normal operation, cross-passages are generally not accessible. Their functions are mainly as follows:

- Housing of technical equipment
- Protection of technical equipment requiring a controlled environment in terms of temperature, air quality (in particular dust concentration) and pressure fluctuations
- Cable transit
- Support of tunnel ventilation (e.g. prevent internal or portal re-circulation of air)

3.2 *Requirements and functions during maintenance operation*

Maintenance can be carried out while one or both tunnel tubes are closed to or, in traffic particular cases, with trains circulating in both tunnel tubes. The requirements for cross-passages are as follows:

- Appropriate air-quality for worksites
- Protection of staff during train passages
- Further requirements as for normal operation

The requirements on the quality of the working environment concern primarily temperature, humidity, pollutant concentrations, flow velocities of air and pressure fluctuations.

3.3 *Requirements and functions during emergency operation*

The main requirements for cross-passages are related to self-rescue and protection of occupants as well as the support of emergency forces:

- Egress passage provision for occupants during self-rescue
- Shelter/waiting area for occupants
- Access provision, logistic support and protection for emergency services
- Protection of occupants in the parallel tunnel tube until full evacuation to the exterior is completed
- Protection of safety-relevant equipment in cross-passages, which must operate during emergency

Full protection and functionality of all cross connections is required at least for the time required for full evacuation to the exterior. Requirements arising from intervention are related to the safety concept adopted for the specific tunnel (intervention strategy, equipment of rescue services).

3.4 *Ventilation concepts*

The following ventilation concepts for cross-passages are presented (see Figure 8):

a) Passive ventilation driven by train-induced pressure differences
b) Ventilation by an independent ventilation system (air-exchange with both tunnel tubes)
c) Ventilation by an independent ventilation system (air-exchange with one tunnel tube)
d) Ventilation by overpressure in a tunnel tube

The cross-passage ventilation concept of passive ventilation due to train-induced pressure fluctuations is shown in Figure 8 (a). The train-induced pressure fluctuations create airflow through the openings in the end-walls of the cross-passages. The effect of ventilation depends on the train operation in the tunnel tubes. The adequate air-exchange, especially during breaks of train operating, might not be guaranteed. In the event of an incident, the ventilation dampers can be closed (if present) and the cross-passages can be sealed off. Otherwise, the tunnel ventilation provides an overpressure in the cross-passages and prevents the entry of smoke into the cross-passages.

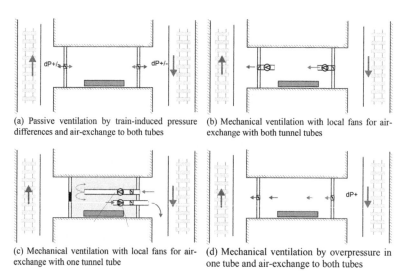

(a) Passive ventilation by train-induced pressure differences and air-exchange to both tubes

(b) Mechanical ventilation with local fans for air-exchange with both tunnel tubes

(c) Mechanical ventilation with local fans for air-exchange with one tunnel tube

(d) Mechanical ventilation by overpressure in one tube and air-exchange to both tubes

Figure 8. Ventilation concepts for cross-passages.

The cross-passage ventilation with air-exchange to both tunnel tubes is shown in Figure 8 (b). The air is extracted from the cooler tunnel tube and flows through a supply air duct into the cross-passages. From there, the air is carried by an exhaust fan into the "warm" tunnel tube.

The CPVS concept with air-exchange with one tunnel tubes is shown in Figure 8 (c). The air is extracted from the cooler tunnel tube and flows through a supply air duct to the end-wall on the side of the "warm" tunnel tube. From there, the air flows back in the cross-passages and is taken out by an exhaust fan and duct into the "cold" tunnel tube. The ventilation damper in the supply and exhaust air are fully opened, the ventilation damper to the "warm" tunnel tube is closed.

The CPVS concept with overpressure ventilation by the TVS is shown in Figure 8 (d). If the temperature in the cross-passages becomes critical, an overpressure is generated in a tunnel tube, driving air into the cross-passages. The effectiveness is influenced by the train traffic.

Table 2. Advantages and disadvantages of ventilation concepts for cross-passages.

CVPS	Advantages	Disadvantages
Passive ventilation by train-induced pressure differences Figure 8 (a)	• simple concept • low investment and operating costs • simple civil requirements for cross-passages	• limited availability due to uncontrolled air flow by train traffic • relatively high dust accumulation • no particular support of TVS
Ventilation by an independent system Figure 8 (b)	• tunnel environmental control • air-exchange to the cooler tube • tube separation/aerodynamic support in case of an incident • air-conditioning during maintenance work	• complex ventilation system (fans, ducts, etc. necessary) • demanding maintenance • high aerodynamic loads on fans • aerodynamic coupling of tubes
Ventilation by an independent system Figure 8 (c)	• tunnel environmental control • air-exchange to the cooler tube • aerodynamic tube de-coupling • support in case of an incident	• complex ventilation system (fans, dampers, ducts, etc.) • demanding maintenance
Ventilation by overpressure in a tube Figure 8 (d)	• tunnel environmental control possible • low investment and operating costs • simple civil requirements for cross-passages	• concept and effect dependent on train operation • comparatively high accumulation of dust in cross-passages • aerodynamic coupling of tubes

3.5 Comparison of cross-passage ventilation concepts

Table 2 lists the advantages and disadvantages of the respective CPVS concepts.

4 CONCLUSIONS

Ventilation concepts of modern, twin-tube, high-speed rail tunnels are typically determined by the requirements for smoke control upon fire and the requirements for air-quality control during maintenance works in the tunnel. Comparing contemporary rail projects of different countries, the following trends for tunnels of medium length are noted:

- The non-incident tube provides a safe area during a fire by preventing the entry of smoke from the incident tube. In most new tunnels, this is achieved by creating an overpressure in the non-incident tube by forced ventilation.
- The key target of the tunnel ventilation system is to support the self-rescue of train occupants. Considering the particular conditions in rail tunnels (i.e. high density of people, long trains and egress as well as access path, thermal stratification of smoke at tunnel ceiling), it appears most reasonable to avoid any major longitudinal flow in the tunnel but to keep flow conditions calm during the self-rescue phase of a fire incident. Therefore and in many cases, the mechanical ventilation is activated in the incident tunnel only, after the self-rescue phase has finished. The longitudinal ventilation of air/smoke is switched on during the phase of intervention by emergency forces.
- Amongst others, the emergency ventilation concepts are influenced by train operation (e.g. type, number and headway of trains) and the preferences of the tunnel operator (e.g. strategy of reducing non-rail equipment in tunnel to minimum). In addition, project-specific requirements affect the tunnel ventilation system layout (e.g. topography along tunnel). Those aspects influence the choice of implementing a tunnel ventilation system concept with distributed jet fans in the tunnel or with central fan plants at shafts/adits or at portals.
- A complex tunnel ventilation system results if the longitudinal ventilation is required already during the self-rescue phase, if both flow directions in the tunnel shall be feasible and if distinct ventilation zones for an incident train shall be provided.
- Increasingly stringent occupational health criteria require controlled air-exchange during maintenance works in tunnels in order to reduce concentrations of pollutants or dust. While the emergency mode of operation is often size determining for the tunnel ventilation, the maintenance mode is the most frequently activated tunnel ventilation system operation mode.
- Depending on the environmental conditions, ventilation of cross-passages is achieved sufficiently by train-induced air-exchange only or by forced mechanical ventilation.
- Cross-passage ventilation systems need to be designed to support the tunnel ventilation system and to assure the functionality of equipment installed in cross-passages during all operation modes.

5 ACKNOWLEDEMENTS

The authors wish to thank the management of AlpTransit Gotthard AG, the project owner of CBT, for the permission to publish this paper and for being involved in the TVS design of CBT.

REFERENCES

Johnson, P., Woodburn, P. & Henderson L. 2016. Evacuation of Modern Rail Tunnels – Emergency Ventilation and Large Fires or Not?, 7th Int. Symp. on Tunnel Safety and Security, Montreal, Canada

Winkler, M. & Carvel, R. 2015. The effect of longitudinal ventilation on tenability during egress from passenger trains in tunnels during fire emergencies, 16th International Symposium on the Aerodynamics, Ventilation & Fire in Tunnels, Seattle

STS Swiss Tunnelling Society/FGU Fachgruppe für Untertagebau, 2016. Tunnelling the Gotthard – The success story of the Gotthard Base Tunnel, ISBN 978-3-033-05803-3

Tunnels and Underground Cities: Engineering and Innovation meet Archaeology,
Architecture and Art, Volume 9: Safety in underground
construction – Peila, Viggiani & Celestino (Eds)

Novel safety concept at the Austrian construction lots of Brenner Base Tunnel

E. Reichel & K. Bergmeister
Brenner Base Tunnel BBT-SE, Innsbruck, Austria

ABSTRACT: Planning and implementation of underground safety installations and tactical specifications in conjunction with emergency services for optimum processing of incidents. Technical and organisational planning during construction of the project area on the Austrian side of the Brenner Base Tunnel. Communication becomes very important; therefore a novel procedure have been developed. Most important are a transparent decision making process with responsibles who takes clear decisions.

1 INTRODUCTION

With a length of 64 km, the world's longest underground railway link, the Brenner Base Tunnel, is being built below the Brenner Pass from Innsbruck, Tulfes in Austria to Franzensfeste in Italy. Three tunnels are being built in the standard cross-section, with two single-track tunnels intended for rail travel and a third, smaller tunnel serving as an exploratory tunnel which is intended to provide information about the mountain. During the operational phase, the exploratory tunnel will serve as a service, rescue and drainage tunnel. The tunnel system, which comprises a total length of 230 km, is being driven both conventionally and by means of tunnel boring machines. The BBT is divided into several independent construction lots, 80% of which have already been put out to contract and awarded. The overall coordination of project from construction through to commissioning is the responsibility of the European company BBT SE.

2 OCCUPATIONAL HEALTH AND SAFETY

The coordination of the health and safety of workers, from the above-ground construction site to the installation of the railway technology, will be managed and headed by BBT SE. For this purpose, the construction site coordinators and the supervisory body, the competent labour inspectorate, are also required on the Austrian territory in addition to BBT SE.

The occupational health and safety plans also cover major events affecting third parties such as fire, the environment and technical events.

The construction of underground facilities is characterised by high risks during execution. Workers are exposed to a wide range of hazards, as they are subjected to limited lighting and ventilation conditions, limited access and escape routes and an increased risk of fire. Planning and resource coordination (workers, materials and machines) are constant challenges. If safety-related problems arise during the project implementation phase, this usually results in an interruption to, or delay in, the execution.

3 SAFETY PLANNING SCENARIOS

Safety planning scenarios range from classic cave-ins to fire, injuries with on-site medical treatment, gas leaks, explosion hazards, water ingress, technical malfunctions, radiation and exceedance of limit values. In principle, fires are dealt with most intensively, as only a minimal amount of time can pass before the safety-related and impact-oriented measures begin. In the other scenarios, these measures can be adapted accordingly with far fewer critical time frames. Fire seems to be the most complex yet most likely form of major accident. A wide range of safety-related measures are therefore aimed at early fire detection and reduction of the impact of fire. Specifically, the following scenarios are considered in the safety planning: Blast accidents, chemical accidents, explosions, fire smoke, gas leaks, oxygen depletion, particulate matter, low pressure, snow, avalanche, power failure, traffic and water ingress.

4 PROTECTION OBJECTIVES

4.1 *Personal protection*

4.1.1 *Self-rescue*
In the event of an incident in a tunnel, the self-rescue phase is particularly crucial. Above all, in a fire, a dangerous climate forms very quickly due to the smoke and heat. The quality of breathable air declines rapidly and becomes toxic very quickly. Rescue or external support is only possible after a long period due to the response times in long tunnel systems.

The primary protection objective is clearly defined as self-rescue of personnel. The quality of the air is essential for self-rescue. This must be taken into account for the design of the emergency ventilation system. In order to consider the direct impairment on an individual's capacity to act, the narcotic and suffocating effects of fumes or oxygen deficiency due to a fire must be taken into account. In particular, compared to a conventional fire outdoors, the spread of fumes in a tunnel must be considered. In addition to the toxic effects of fire smoke, temperature also has a significant influence on self-rescue. This burden must be kept as low as possible during the period of self-rescue. Air flow and clothing also play an important role in addition to the air temperature, humidity and duration of exposure.

A wide range of technical installations can facilitate and support self-rescue. Lighting is fundamental here, as well as electronic personnel detection systems and communication equipment.

Provisions such as refuge containers, escape apparatus with a duration of use of at least 90 minutes when moving, lighting, quick access to alarm systems, ventilation design/control and control room.

If self-rescue does not result in all persons being evacuated from the tunnel system, an emergency rescue must be carried out.

4.1.2 *Emergency rescue*
In the event of an incident in a tunnel, it is of vital importance for the emergency rescue that the deployment concept and emergency personnel equipment take into account the conditions regarding the accessibility of the tunnels, thus enabling an optimal deployment. An important requirement for a successful emergency rescue is the shortest possible response time from the point the alarm is triggered to the emergency personnel reaching the tunnel portals and subsequently the area of the incident. Most of the BBT portals can be reached relatively quickly by the emergency services (fire brigade, paramedics, hazmat team, etc.).

However, in scenarios involving a fire or an explosion, the approaching fire brigades can only intervene if they reach the site of the incident within the first 20 minutes (Bergmeister 2013).

As the client, BBT SE has directly concluded contracts with the experienced local fire brigade and rescue services for the entire construction period. This is an efficient concept to ensure a uniform execution of emergency rescues for the entire construction phase, as different contractors will be responsible for the different construction phases of the different construction lots. The framework conditions for the contractors are provided by BBT SE. The contracts with the fire brigade have been concluded with the respective municipal fire departments along the Austrian route. One particular focal point of the contracts is the equipment required for use underground as well as education and training. A special education and training program for tunnel construction sites is of the utmost importance for inexperienced emergency teams. For this, one of BBT's objectives is to ensure that the rescue services "grow" together with the construction project from the start of the construction process enabling them to undergo constant development in terms of safety and support the project. Education and training, practical exercises, inspections and knowledge of the local area in terms of underground characteristics are integral parts of this which ensure that the response teams are well-trained, well-equipped and well-staffed.

The technical installations for emergency rescue operations have also been planned in collaboration with the emergency services. For example, a multi-service vehicle (MSV) for rescue operations - a self-propelled refuge container on wheels (see also Lussu 2019). This is a vehicle which can be regarded on the one hand as a temporary safe zone and is equipped with a self-contained air supply of at least 10 hours at full occupancy. The MSV is equipped with sensors, is self-propelled and has internal and external gas detectors (O_2, CO_2, CO, H_2S, NO_2, EX), external temperature detection, radar and a laser distance measuring device, communication equipment and location systems.

The design of the refuge containers (duration of the oxygen supply) is based on the course of the rescue measures. The refuge containers are described below.

In order to provide the emergency services personnel, in particular the decision makers (incident commanders), with the necessary safety for their decisions, operating limits have been defined together with the emergency service organisations. When and under what conditions is an underground operation justifiable with regard to the safety of the emergency services personnel? For this, the following limits have been established:

a) minimum visibility of 100 m upon entering the tunnel system
b) maximum average temperature in the tunnel of 60°C
c) minimum oxygen content of 17%
d) at least one functioning communication channel

A rescue mission by the fire brigade will only commence after a consultation with the client's and contractor's representatives. In principle, the tunnel system will not be entered to put out a fire. A burning vehicle or object can be left to burn out until the fire extinguishes itself. The risk of putting out a fire is too high for the emergency services in this case. Therefore, the specified lengths of stay in the refuge containers have been implemented with the appropriate times.

This ensures the safest possible deployment and means that the rescue concept is consistently traceable.

4.1.3 Protection of property

The main focus of the protection objective is the rescue of persons and the self-protection of the emergency services; Protecting the environment and property is considered secondary. In particular, in the event of a fire, a deployment to protect property is not the primary focus.

For the protection of property and infrastructure, such as tunnel installations, BBT SE is pursuing the following objectives:

- Avoidance of interruptions to operation
- Avoidance of costly and complex reconstruction work
- Avoidance of the loss of machines and materials

5 ORGANISATION IN THE EVENT OF A DEPLOYMENT

The established hierarchy plays a key role in the organisational handling of missions. The competencies and responsibilities must be defined in emergency plans, which are clarified before the start of the construction project. Subsequently, crisis teams must be formed and a process organisation created. The classic areas here are personnel management, the evaluation of the situation, the rescue options, the logistical requirements, decision making process and external reporting, i.e. press service.

The organisation and deployment of the emergency service personnel generally has the following definitive characteristics:

- Arrangement of the fire brigade incident commander with the client and contractor on the potential risks
- Assessment of the fire brigade incident commander on the situation and possibilities
- Order to the emergency services after considering the deployment limits

For underground operations, the interaction between the fire brigade, police and rescue services is important and must be constantly practised and the tactics must be adapted to the structural situation. The involvement of the construction companies and the monitoring centre control room is also important in this regard. The communication of the emergency services is to be defined in the alarm and deployment plan.

6 TECHNOLOGY

6.1 Communication/Radio/Alarms

In the event of a deployment, a functioning, efficient means of communication is central to the success of the deployment. Different and, above all, independent communication systems are installed in the tunnel system for this purpose. These include:

a) the classic, wired tunnel telephone,
b) a mobile network,
c) an analogue radio system for the fire brigade,
d) a Wi-Fi network, as well as
e) the digital radio network for the emergency services.

In particular, communication between injured or trapped persons and the rescue services is crucial and for this reason, the systems must be designed to be redundant and fail-safe.

6.2 Underground location system

Anyone entering or leaving the tunnel is usually registered. For complex underground transport routes, such as in the BBT, precise positioning within the tunnel system is also very important. The drift, segment or refuge container which the persons happen to be in is information which the emergency services can use when planning their tactics. When combined with the communication installations, these systems help to streamline a deployment operation. The location system used throughout the entire the Brenner Base Tunnel system is based on an RFID system with active personal and vehicle units. In the event of an emergency, this location system is able to quickly locate all persons, thereby ensuring a safe and thorough evacuation or supporting a targeted search for missing persons.

The personal detection system is not only used for locating employees underground but also for directing emergency services teams. Not only can they be monitored from the outside, they can also receive their own position and distance to the site of the incident at any time, since the lengths (up to 18 km) can lead to a loss of orientation. These systems counteract this problem.

6.3 Ventilation

Special attention was paid to the topic of ventilation in the event of an emergency as early as in the call for tenders. The following core assumptions must be fulfilled with regard to the ventilation concept:

- the ventilation system must be set up in such a way that even during the construction phase, when one tunnel has already been damaged by an accident another tunnel must be identified as a safe area in order to minimise escape routes and emergency routes for the rescue service. The connective system together with the safe tunnel must be kept free of smoke through suitable measures (e.g. ventilation bulkheads),
- the ventilation system must blow air toward the working face (air circulation concepts can be implemented upon proof of equivalence and after approval by the occupational health and safety authorities),
- the ventilation system must be created with a supporting sucking ventilation system to reduce fumes. The air quality must be continuously documented using automatic recording devices. The ventilation planning is geared to ensure that the driving systems are supplied with enough fresh air during operation and that the exhaust air from the driving systems does not exceed the permissible limits in the transport area, tunnel tubes, connective structures and intermediate access tunnels. In the event of an emergency (fire), the ventilation concept must be planed and designed in such a way that guarantees that the driving areas, tubes which are unaffected by the incident and the escape and rescue routes are kept free of smoke.

6.3.1 *Ventilation in the event of a fire*

Fires in the air circulation system are more complex compared to in a blowing ventilation system. Since there is a supply and exhaust pipe, two cases must be distinguished in the air circulation system.

In the event of a fire alarm, the following measures are implemented immediately:

- Complete deactivation of the ventilation system. This is performed from the control room.
- Through the intervention from the control room, no one else can operate the ventilation controls (established hierarchy)
- All smoke section gates are closed to prevent natural ventilation.
- In the areas which are unaffected by the fire, a "yellow alarm" is triggered. → Receive information from the control room about where the shortest, safest escape route is and leave the tunnel system.
- In the area of the incident, a "red alarm" is triggered → immediate escape outside or to a temporary safe zone (refuge container)
- In addition, information on the damage data is provided to foremen and employees in the tunnel system via SMS and email on the mobile phones.
- After clearing the tunnel system, the foremen must send a reply to the control room stating whether all employees have left the tunnel. A cross-check with the location system is performed.
- The transfer of the information to the first arriving rescue workers on the location of the incident and how many people are still in the tunnel system is prepared and performed

An incident command is set up and the rescue operation begins.

6.4 *Refuge chambers*

6.4.1 *Protection objective*

The refuge chambers (e.g. in Figure 1) serve as a temporary safe zone for self-rescue. A protective or rescue container is pulled along with the calotte driving system at a length of 500 m. For mechanical driving systems, the rescue container is located on the trailer.

Rescue containers offer protection against explosive gases and, above all, smoke in the event of a fire. They should be used if it is not possible to escape to a permanent safe zone (above or below ground) with an oxygen self-rescuer.

Figure 1.   Refuge chamber.

6.4.2 *Types of protection*
The rescue container is designed independently of the ambient atmosphere. It must be possible to stay inside the container without personal protective equipment. This can be achieved with either pure air flush technology or regeneration technology.

The respective minimum requirements for both possible technologies are listed below in Figure 2.

6.4.3 *Number of persons to be protected*
The size of the container is tailored to all persons situated in the area of the containers location (client representatives + visitors as well as shift workers). A minimum space requirement of 0.5 m^2 per person must be provided.

6.4.4 *Protection – continued*
Where technically feasible, the refuge chamber will be connected to a compressed air line (external air supply). In particular, this applies to the TBM. The air must be cleaned by means of a filter system beforehand so that the air is breathable.

6.4.5 *Protection – self-contained*
Refuge containers with **REGENERATION TECHNOLOGY** are also used: The air in the container is freed of carbon dioxide (CO_2) and the consumed oxygen is regenerated. For this, the air in the container must be fed through soda lime (CO_2 absorber and CO absorber) using a fan in a recirculation process to ensure sufficient air movement in the container. The amount of oxygen supplied can be regulated (depending on the number of persons) to counteract over or under-dosage. Medical oxygen is used.

Containers with **AIR FLUSH TECHNLOLOGY** are also used:

The air supply is guaranteed for the entire operating time via self-contained compressed air cylinders. It is operated using an adjustable air dosing system – primary pressure (storage pressure) and adjustable flushing pressure (flow). There must be a minimum supply of 40 L/min/person

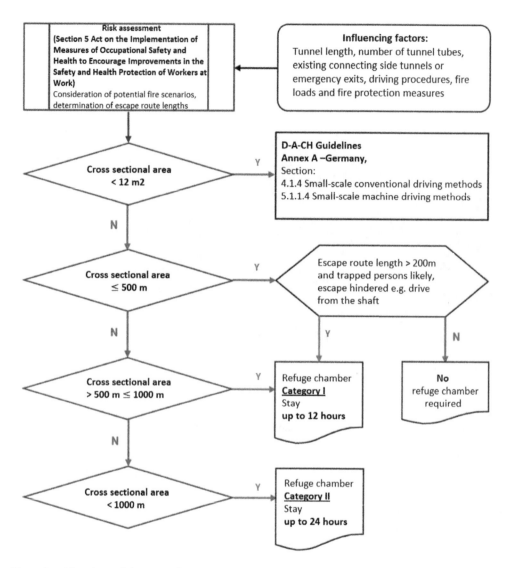

Figure 2. Flowchart of the constraints for the refuge container.

Pollutant reduction:
Upon entering the refuge chamber, pollutants can also be brought inside. A purge air supply, which uses compressed air cylinders, must be provided which is capable of exchanging the air in the container 3.5 times within 30 minutes. For the time when the atmosphere is not breathable, oxygen self-rescuers should be used.

6.4.6 *Positioning of the refuge chamber*
An immediate effect of the heat of the fire is not taken into account, as the refuge chamber is to be positioned at a sufficient distance from fire loads. Areas around the rescue container are to be kept free from fire loads. The locations of the rescue containers are marked with permanent green light. To find the rescue containers when there is smoke in the tunnel, off-grid strobes and acoustic broadband devices are used (which must be disengageable). The rescue containers must be as close as possible to the working face (risk of entrapment by fire) and as far away as possible from fire loads. Depending on the design, a minimum distance of 400 m

and a maximum distance of 800 m must be maintained from the working face for blasting operations (runners for easy relocation). The location of the rescue containers must be indicated in the escape and rescue route plan.

6.4.7 *Safety center*

The construction site safety center must be staffed at all times (even during shut-down periods/holidays/exiting). It is configured as the central control and monitoring center for the tunnel construction site. All monitored tunnel technology is displayed with actual status evaluations (personnel detection, fire protection systems, gas measurement displays, e-installation system, cooling system, ventilation system) and the ventilation, alarms and smoke-tight doors in the logistics connecting side tunnels can also be remotely controlled from this location. The monitoring of the tunnel technology and the emergency telephone as well as the alerting systems based on alarm plans and action lists are all tasks of the safety center. In addition to the safety center, the operations center is housed in the construction office. Here, in the event of an emergency, the representatives of the emergency services, the client and the contractor meet for briefings. The key for the operations center is located in the security center and is unlocked by the security center in the event of an alarm.

7 CRISIS MANAGEMENT – PRESS

An emergency folder contains the written records of all the elements of the emergency management system for specific construction projects in a simple, clear and unambiguous manner. The aim is to present a uniform system for all BBT construction sites. The emergency folder is available at every construction site and is updated in the event of changes in responsibilities, construction phase or other requirements which concern the emergency management system. The incident command comprises the decision makers and consists of representatives of BBT SE (client), the construction company (operational incident commander), emergency services and public authorities. Depending on the size of the incident, an incident task force should be formed which serves as support for the incident command. It is coordinated by the head of the committee or by the operational incident commander.

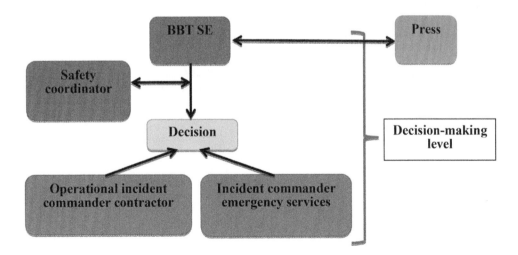

Figure 3. Decision flow chart.

Decision making and responsibilities:

Depending on the size of the incident, an incident task force should be formed to support the incident command (decision-making level).

The strategy is defined in the rules of procedure for the incident task force. The operational incident commander of the client or BBT SE representative gives the order to summon the task force. The order is given to the aforementioned persons according to the specialist area of the task force. Depending on the location, additional persons will be appointed accordingly. The summoning of the task force is ordered in the control room. The head of the task force is chosen based on the scope of the mission. They decide on the size and scope of the task force (as large as necessary, as small as possible) and assign the different specialist areas. The contractor may also be the head of the task force depending on the size of the incident and their competence (scope).

The staffing of the different specialist areas depends on the type and size of the incident. Persons employed by the contractor, client and, if necessary, other experts can be used to handle the different tasks in the specialist areas. The naming of persons and the detailed description of the tasks of the incident command and the task force are listed under the tasks of the incident command. The press and media work runs exclusively through BBT SE's press spokesperson or the board in agreement and coordination with the incident command. Depending on the size of the incident, there may be a press spokesperson on site with a liaison officer.

8 CONCLUSION

In order to ensure the safety of the 230 km long tunnel system of the Brenner Base Tunnel, BBT SE introduced a new safety concept during the construction phase. Under the auspices of BBT SE, a new safety concept will be developed in coordination with the construction site coordinators, contractors, labour inspectors, local emergency services and operational centers. The key aspects are the direct management of contracts with the emergency services, quick response times not only by the emergency services but also the client and contractors, as well as the coordination of rescue installations and concepts together with the emergency services in order to ensure an optimal course of action.

REFERENCES

Bergmeister K. 2013, Sicherheit und Brandschutz im Tunnelbau [Safety and fire protection in tunnel construction], Betonkalender 2013, *Verlag Ernst und Sohn*

Lussu, A. & Kaiser, C. & Grüllich, S. & Fontana, A. 2019 Innovative TBM transport logistics in the constructive lot H33 - the Brenner Base Tunnel. *ITA-AITES World Tunnel Congress 2019*, Naples, Italy.

DAUB Guidelines 2018, Empfehlungen für den Einsatz von Fluchtkammern auf Untertagebaustellen' Deutscher Ausschuss für unterirdisches Bauen e. V. *German Tunnelling Committee (ITA-AITES)*

Tunnels and Underground Cities: Engineering and Innovation meet Archaeology,
Architecture and Art, Volume 9: Safety in underground
construction – Peila, Viggiani & Celestino (Eds)
© 2020 Taylor & Francis Group, London, ISBN 978-0-367-46874-3

Damage assessment of final linings in metro tunnels due to fire action

K.M. Sakkas, N. Vagiokas, A. Kallianiotis & G. Loupa
Laboratory for Enviromental Engineering, Democritus University of Thrace, Xanthi, Greece

ABSTRACT: A crucial element of a tunnel lining design in metro tunnels is the time and the way till the final lining will begin to spall due to the high temperatures, which may be developed during a fire. Although the structural elements of the tunnel lining are considered inflamable, in a real fire situation the concrete lining can be spalled largely, sometimes perhaps entirely, with very serious consequences on cost and safety of people. The spalling rate of concrete as well as the spalling depth are two main values that cannot be measured since they are affected by a series of factors including mechanical properties, permeability of concrete and age of concrete. A lot of research has taken place in order to calculate the spalling of the concrete using numerical models, which however is not validated sufficiently by the test results. So the target of the specific research is to investigate experimentally the rate and the depth of spalling for several types of concrete under high temperatures. On this purpose specimen of all types of concrete that can be found in metro tunnels were manufactured and exposed at high temperature in a special furnace which has the ability to simulate all kind of fire scenarios and at the same time monitor the spalling depth and time. The temperature of the exposure was selected after modelling of a 25 MW fire in a metro tunnel. The modelling simulated the distribution of the temperature along the tunnel in case of a 25MW fire and representative values of the temperature for 10, 20, 50 and 100m away from the fire ignition were selected.

1 INTRODUCTION

1.1 *General*

A number of serious metro tunnel fire incidents have been reported worldwide that have led to injuries and life losses, heavy damage in the concrete lining, excess material damage, and significant time periods of tunnel restoration during which the tunnels were unavailable for traffic. Fires in tunnels can seriously damage their concrete lining rendering it to collapse. The damage is caused particularly by the spontaneous release of great amounts of heat and aggressive fire gases, resulting to spalling of concrete. Spalling is described as the breaking of layers or pieces of concrete from the surface of a structural element when it is exposed to the high and rapidly rising temperatures experienced in fires. The most severe spalling phenomenon is the so called explosive spalling, be described as a violent showering of hot pieces of concrete, which beyond to the detrimental effect on concrete lining can additionally cause very serious problems to the firefighting service personnel rendering their work substantially more difficult and dangerous (Jönsson & Herrera, 2010). The spalling phenomena are expected at several temperatures depending on the strength of the concrete. The American Society for Testing and Materials (2001) mentions that explosive spalling can be expected at temperatures between 300°C and 450°C, while it is generally accepted that concrete exposed at temperatures higher than 380°C is considered as damaged and should be removed and repaired (Khoury, 2000). This temperature is near to the calcium hydroxide dehydration temperature (400°C) which is a process that causes a significant reduction in the mechanical strength of the concrete (Fletcher, 2007). The latter is considered to be significantly reduced at exposure temperatures higher than 300°C (Sakkas, 2013). The spalling rate of concrete as well as the spalling depth and time are two main values for the numerical analysis, which are very difficult to assume or calculate. The reason is that spalling is affected by a series of factors including apart from the compressive and tensile strength, also the following: a) permeability of concrete, b) age of concrete, c) accurate mix of concrete, d) size of aggregates, e) pore pressure, f) existing cracking on the concrete, g) reinforcement of the concrete and h) size and shape of the exposed specimen (Boström and Jansson,2004). A lot of research has taken place in order to calculate the spalling of the concrete using numerical models, which however is not validated sufficiently by the test results. First of all, the spalling theory is not yet clear. Others support that spalling is attributed to the pre-existing cracking, others to the moisture content and others to the thermal stresses (Shuttel-worth,2002). In the last years, scientists have concluded that spalling is a combination of thermal stresses and pore pressure. According to the opinion of various international experts, the current numerical modelling capabilities for predicting the possibility of concrete spalling under a tempera-ture scenario at the intrados of the tunnel lining, is very limited. Recently, some thermo-poro-mechanical constitutive models have been appeared in the literature, but all these require extensive validation procedures before applied in practice. Also, no constitutive model has been published in the literature which is able to account all the factors leading to concrete spalling (Akhtaruzzuman & Sullivan,1970). A numerical procedure that can be followed in order to estimate the spalling vulnerability of a specific concrete has been suggested by Kodur (Kodur, 2008). According to this, the model is based on pore pressure calculations in concrete, and spalling is considered to occur by comparing the computed pore-pressures with the temperature dependent tensile strength of con-crete. A similar (and perhaps a little more advanced) procedure has been used by the Hatch Mot McDonald engineers by using FLAC3D (Franssen et al, 2007). Both Kodur and HMD engineers validated their models with the experiments of Phan (2008). However even these models worked great in specific types of concrete but in other types seemed to overestimate the spalling vulnerabil-ity of concrete. Also, the models that have been developed for the calculation of the spalling show great weakness at concrete with high moisture content and low permeability because the moisture migration due to pressure gradients leads to a completely saturated state of the concrete pores and as a result the required thermodynamic laws do not apply any more. Also, according to other studies the calculation of the spalling is a very complicated issue, since is not very easy to simulate the behavior of concrete under fire conditions. On the other hand an essential issue on the design of the evacuation of a metro tunnel is the spalling time. It is one of the main values which define the available time of the passengers to evacuate safely the tunnel. As a result and in order to meas-ure the spalling values and design the evacuation of a metro tunnel it is proposed to test the con-crete at several temperatures for the spalling depth and time.

2 EXPERIMENTAL

2.1 Types of concrete

In total, three different types of concrete were investigated, which were found to be the most usu-ally used in metro tunnels. More specific these types of concrete were C20/25, C30/37, C35/45 with latter two to be the most usual. The syntheses of these three types of concrete are described below:

i) Category C20/25

After a test mix, it was found that the composition for the concrete required consists of 40% gravel, 10% fine grain gravel, 50% sand by weight of dry materials with 320 kg/m^3 cement CEM II/AM (WL) 42.5R and additive superplasticizer at a rate of 0.63 kg per 100 kg of cement. The W/C ratio was equal to 0.60. The compressive strength of the 15 cm cubic edge sample made from this mixture was found to be 42.8 MPa at 28 days of age.

ii) Category C30/37

After a test mix, it was found that the required concrete composition consisted of 40.4% gravel, 10.4% fine grain gravel, 49.2% sand by weight of dry materials with 350 kg/m^3 cement CEM II/AM (WL) 42.5R and an additional superplasticizer at a rate of 0.6 kg per 100 kg of cement. The W/C ratio was equal to 0,51. The compressive strength of the 15 cm cubic edge sample made from this mixture was found to be 46.7 MPa at 28 days of age.

Figure 1.   Set up of fire test.

iii) Category C35/45

After a test mix, it was found that the composition for the concrete required consists of 40.4% gravel, 10.4% fine grain gravel, 49.2% sand by weight of dry materials with 380 kg/m³ of CEM II/AM cement) 42.5R and an additional superplasticizer at a rate of 0.6 kg per 100 kg of cement. The W/C ratio was equal to 0.48. The compressive strength of the 15 cm cubic edge sample made from this mixture was found to be 53.5 MPa at 28 days of age

2.2   Temperature of investigation

The temperature and time of exposure was selected by modelling a 25 MW (figure 2) fire in a single and double tube metro tunnel. The tunnel length is 762 m, which is the maximum limit of tunnel length without cross passage obligation. The evolution of the fire is based on the FIT project – fire curve which is the worst case. The power of fire was selected to be ignited in the middle of the metro tunnel. The modelling of the fire simulation the distribution of the fire along the tunnel in case of fire and representative values of the temperature for 10, 20 50 and 100m away from the fire was selected.

2.3   Equipment of the tests

The tests were performed in the laboratory by using a test furnace (figure 1) which was designed according to European Federation of National Associations representing producers and applicators of specialist building products for Concrete (EFNARC) guidelines. The furnace has the ability to simulate the temperature-time curves employed in several international standards. For this test a 30 cm × 1 cm × 5 cm specimen was prepared. The test was performed 28 days after the production of the specimens.

3   RESULTS

3.1   Fire tests

Initially the most severe fire test was conducted for each concrete type. The most severe scenario referred to 10m distance from the fire ignition in a single tube tunnel. The temperature increased at 600 °C after 22 min and remains at that temperature for 30 min. In the C20/25 and C30/37 concrete types no spalling occurred. This was attributed to the fact that it was very low time for the concrete to be fully thermally treated. This can be easily understood from figure 2 where the temperature of the back surface of the specimen was measured during the test. As it can be easily observed the temperature did not exceed the temperature of 50 °C, meaning that the concrete created a gradient of 55°C/cm, which is attributed to the low thermal conductivity value of the concrete. As a result the concrete was not treated uniformly at the whole specimen and this avoided the spalling phenomena. In order to find the temperature that the specific type of concretes udergoes spalling phenomena, specimens from C20/25 and C30/37 were subjected to higher temperatures. More specific three more tests were conducted, at temperatures of 700 °C, 800°C and 900°C. From the testing of the specific concrete types it was realized that the C30/37 spalled at the temperature of 800 °C while the C20/25 concrete type spalled at the temperature of 900 °C. The type of spalling was similar to the C35/45 with a sudden explosion at 25 min after the beginning of the test for the C30/37 and 18min for the C20/25. This is something that it was expected mainly according to Khoury (Khoury, 2000) who mentions that one of the most crucial parameters for the occurrence of spalling is the high compressive strength. Concrete with high compressive strength is more dense and as a result the entrapped water does not find the suitable void to be evaporated and explodes making the explosive spalling.

In the case of the C35/45 spalling phenomena occurred 22 min after the beginning of the test, while the temperature was at 600 °C (figure 3). The spalling that occurred was typically explosive spalling with a piece of 10 cm to be removed from the surface of the concrete specimen at a depth of 1 cm. The same type of concrete was then subjected to another fire curve referring to a distance from the fire ignition (20m – max temp. 500 °C). As it can be observed in figure 3b the concrete did not undergo spalling phenomena. This result is very crucial for the design of the evacuation. Based on these results the engineer should also take into account the spalling of the concrete by proposing two solutions: a) to evacuate in less than 20 minutes to ensure that all the users will evacuate the tunnel b) Oblige users close to the fireplace to walk through the train and reach the point where the temperature is lower (20 m) and therefore there is no risk of spalling the concrete with the corresponding wound users.

In general, in order to avoid spalling phenomena at the concrete various methods of passive fire protection have been developed which however have not yet been applied to metro tunnels. At this point it should also be mentioned that apart from the concrete steel rebars necessitate protection because at temperature over 600 °C loose part of their load bearing capacity.

Therefore, steel and concrete are both fire sensitive construction elements requiring passive protection against fire in order to be capable of withstanding of fire for an appropriate period

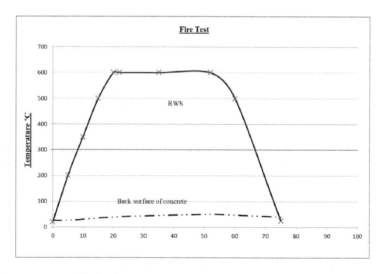

Figure 2. Temperature profile for fire test.

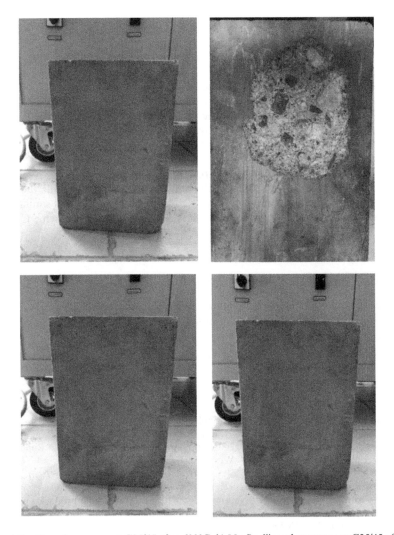

Figure 3. a) Spalling phenomena at C35/45 after 600°C, b) No Spalling phenomena at C35/45 after 500°C.

of time without loss of stability (Fletcher, I. 2007). Passive fire protection methods are generally divided in two categories: external (insulation) and internal (concrete design). The former are more advantageous being applied in new as well as in existing tunnels and consist of the cladding of the concrete by a fire resistant material which creates a protective external insulation envelope. These methods are a) cementious fire protection material applied either sprayed or as a board which, due to its low thermal conductivity, keeps the temperature of the concrete below the requirements and b) adding during the production of the concrete, polypropene fibers which during the thermal exposure melt thus creating passage to the evaporated water to be removed from the concrete without acting explosively.

3.2 Conclusions

From this study the following conclusions were excluded:

a) Concrete with the smallest strength did not show signs of spalling, which is also comes in accordance with the references. For this reason various methods and materials have been developed to prevent spalling, but this has not yet been applied to metro tunnels. This is a) coating the concrete with a fire protection material which, due to its low thermal conductivity, keeps the temperature of the concrete low and b) adding during the production of the concrete, polypropene fibers which during the thermal exposure melt thus creating passage to the evaporated water to be removed from the concrete without acting explosively.

b) In the case of concrete C35/45 it appeared that the design of the evacuation should also take into account the spalling of the concrete by proposing two solutions: a) to evacuate in less than 20 minutes to ensure that all the users will evacuate the tunnel b) Oblige users close to the fireplace to walk through the train and reach the point where the temperature is lower (20 m) and therefore there is no risk of spalling the concrete with the corresponding wound users.

c) Concrete types C20/25 and C30/37 appeared spalling phenomena under higher temperatures. More specific the C30/37 spalled at the temperature of 800 °C while the C20/25 concrete type spalled at the temperature of 900 °C. The type of spalling was similar to the C35/45 with a sudden explosion at 25 min after the beginning of the test for the C30/37 and 18 for the C20/25

ACKNOWLEDGEMENT

This research was funded by the project encoded EDVM 34 and title "Support of researchers with emphasis on young researchers" of the priority axis "Development of Human Resources, Education and Lifelong Learning" in Priority Axes 6, 8 and 9, co-funded by the European Social Fund (ECT) and the National Strategic Reference Framework.

REFERENCES

Akhtaruzzuman, A.A. & Sullivan, P.J.E. 1970. Explosive spalling of concrete exposed to high temperature. *Concrete Structure and Research Report*, Imperial College, London, UK.

American Society for Testing and Materials, 2001. Standard Methods of Fire Test of Building Construction and Materials. *Test Method E119–01*. West Conshohocken, PA, 2001.

Boström, L. & Jansson, R. 2004. Spalling of Concrete for Tunnels. *Proceedings from the 10th international Fire sience & Engineering Conference (Interflam '04), Edinburgh, Scotland.*

EFNARC. 2009. European Federation of National Associations Representing producers and applicators of specialist building products for Concrete. *Specification and Guidelines for Testing of Passive Fire Protection for Concrete Tunnels Linings.*

Fletcher, I., Welch, S., Torero, J.L., Carvel, R.O. & Usman, A. 2007. Behaviour of concrete structures in fires. *Journal of Thermal Science*: 37–52.

Franssen, J.M., Hanus, F. & Dotreppe, J.C. 2007. Numerical evaluation of the fire behaviour of a concrete tunnel integrating the effects of spalling. *Proceedings of the Fire Design of Concrete Structures from Materials Modelling to Structural Performance Workshop, 8–9/11, Coimbra, Portugal.*

Jönsson, J. & Herrera, F. 2010. HGV traffic – Consequences in case of a tunnel fire. *Fourth International Symposium on Tunnel Safety and Security, Frankfurt am Main, Germany, 17–19/3.*

Khoury, A. 2000. Effect of fire on concrete and concrete structures. *Progress in Structural Engineering and Materials:* 429–447.

Kodur, V.K.R. & Dwaikat, M. 2008. A Numerical model for predicting the fire resistance of reinforced concrete beams. *Cement and Concrete Composites 30(5):*431–443.

Phan, L.T. 2008. Pore pressure and explosive spalling in concrete. *Materials and Structures, 41(10):* 1623–1632.

PIARC guidelines, 2004. ITA Guidelines for Structural Fire Resistance for Road Tunnels. *Purkiss JA, Fire Safety Engineering: Design of Structures, Butterworth-Heinemann.*

Sakkas, K., Nomikos, P., Sofianos, A. & Panias, D. 2013. Inorganic polymeric materials for passive fire protection of underground constructions. *Fire and Materials, 37:* 140–150.

Shuttelworth, P. 2002. Fire Protection of Concrete Tunnel Linings. *Tunnel Management International, Vol 3, Issue:* 239–244.

Tunnels and Underground Cities: Engineering and Innovation meet Archaeology,
Architecture and Art, Volume 9: Safety in underground
construction – Peila, Viggiani & Celestino (Eds)
© 2020 Taylor & Francis Group, London, ISBN 978-0-367-46874-3

The common operative rules for health and safety during the construction of base tunnel of Montcenis between France and Italy

A. Sorlini
TELT – Tunnel Euralpin Lyon Turin SA, Torino, Italy

J. L. Borrel & H. Guirimand
DIRECCTE, Chambery, France

B. Galla, M.G. Pregnolato, P. Picco & G. Porcellana
ASL-TO3 – SPRESAL, Rivoli, Italy

G. Cianfrani & C. Saletta
ISPETTORATO NAZIONALE DEL LAVORO, Torino, Italy

ABSTRACT: The railway project between Lyon and Turin was developed to unite and standardise design choices, between the countries in light of the strategic importance of this work for the EU, and adopting the best practices of both countries.

In the early 2000s, when French exploratory tunnels started, this demand for uniformity created an opportunity to draw up a shared bi-national document between France and Italy, which established the "Common Operating Rules" for protecting the health and safety of workers. In 2016, the Rules were updated after some meetings between the labour inspectorates and the project sponsor. In 2017 it was validated by both countries.

1 THE TURIN-LYON HIGH-CAPACITY RAIL PROJECT

The new line between Turin (Italy) and Lyon (France) will be the central and key part of the East-West "Mediterranean Corridor", one of three major rail lines passing south of the Alps planned by the European Union (see figure 1).

The bi-national border crossing between Italy and France includes a 57.5 km, 8.70 m diameter single-way twin-tube base tunnel, the longest in the world, which crosses the Alps at an elevation of approximately 600 m, roughly 45 km in France and 12.5 km in Italy (see figure 2).

Design of the base tunnel includes four exploratory adits and tunnels

Three French exploratory adits were completed between 2007 and 2010: Saint-Martin-La-Porte (2.4 km), La Praz (2.5 km) and Modane (4.0 km).

The Italian exploratory tunnel of La Maddalena (7.1 km) was completed in February 2017 and the French exploratory tunnel of Saint-Martin-La Porte (9 km) has been under construction since 2014 and has excavated more than 5 km.

The Mont-Cenis base tunnel is a complete geological section of the Western Alps, which are crossed from west to east with an overburden of over 2,000 metres.

The excavations will face complex challenges due to the geotechnical and hydrogeological conditions, the nature of the rocks and the conditions of access to the workplaces, in sections dozens of kilometres away from the entrances and under extreme microclimatic conditions.

The thousands of construction workers spread across six main work sites, the need for accommodation in the Alpine valleys and the impact on the local population and the environment are other major issues that will characterise the entire project.

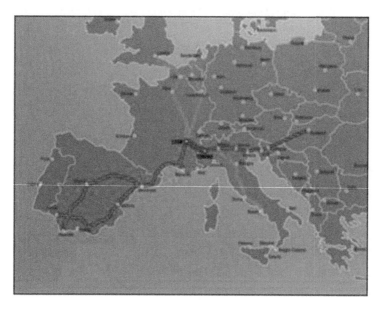

Figure 1.   Mediterranean HCR corridor.

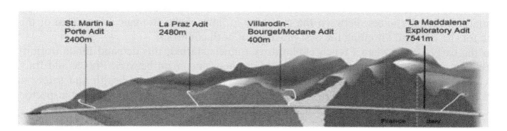

Figure 2.   Base tunnel through Alps and adits.

2   THE CREATION OF A BI-NATIONAL COMMITTEE AND THE ORIGIN OF THE DOCUMENT

The construction sites for the future Turin-Lyon railway link represent a crucial objective for the inspection services of the Ministries of Labour, and of the Piedmont Region for Italy (through the Azienda Sanitaria Locale - ASL) of both countries. They have been mobilised for several years to guarantee an effective response to the social and economic concerns that accompany the execution of such an important project.

Cooperation between the departments has proved necessary for the "border crossing section", located between Susa-Bussoleno in Italy and Saint Jean de Maurienne in France, whose specific characteristics fall well outside ordinary problems. Furthermore, the transnational character of this section has raised a number of questions in terms of both the law applicable to relations and working conditions, and the methods of intervention of those inspection services. Added to this is the likelihood of the contracting companies being of different nationalities, if not from outside the European Union.

For these reasons, the "Direzioni Territoriali del Lavoro" (Territorial Labour Inspectorates) of Turin and Alessandria, the "Servizio Prevenzione Sicurezza Ambienti di Lavoro" (Workplace Safety and Prevention Services) ASL TO3 of the Piedmont Region, the "Direction Régionale des Entreprises, de la Concurrence, de la Consommation, du Travail et de

REGLES COMMUNES OPERATIONNELLES DE SECURITE POUR LA CONSTRUCTION DU MÉGA-TUNNEL DE LA LIGNE FERROVIAIRE LYON-TURIN

SECTION TRANSFRONTALIERE

REGOLE OPERATIVE COMUNI PER LA COSTRUZIONE IN SICUREZZA DEL MEGATUNNEL SULLA LINEA FERROVIARIA TORINO – LIONE

TRATTA TRANSFRONTALIERA

Figure 3. The cover of the final version of the "Common Operating Rules" of April 2017.

l'Emploi" of the Rhône-Alpes region (DIRECCTE), as well as the "Direzione generale per l'attività ispettiva" (General Directorate for inspective activities (DGAI), from January 2017 "National Labour Inspectorate") in Rome and the "Direction générale du travail" (General Labour Directorate) (DGT) in Paris have collaborated in responding to such requirements, launching an innovative initiative in which the client (initially LTF, then TELT) and some of the companies present during the French exploratory works in the early 2000s are heavily involved.

The work carried out between 2002 and 2010 has been recognised by the respective governments, by including the resulting document in the bilateral agreements between France and Italy and requiring compliance with it as part of article 10.2 of the Treaty of January 2012.

The article defines the applicable rules in terms of working and employment conditions of workers and provides for cooperation actions between the inspection bodies of both states. This expressly envisages that the public operator sponsoring the works must attach the final document to the contract and that failure to comply with these rules shall be accompanied by financial penalties.

The Common Operating Rules define the common rules applicable in terms of occupational health and safety during the works and attest to the willingness of both countries to implement uniform and coordinated procedures relating to the prevention and control of professional risks, as well as the working conditions of workers.

The Rules must be applied for the works carried out both in Italy and France. They were drawn up by comparing European, Italian and French regulations and "good practices" observed on the excavation sites, with reference to the shafts already carried out.

The work of drafting the document was divided into two phases:

- From 2002 to 2012: following a series of meetings between the inspection services, held alternately in France and Italy (the first of which in Modane in Savoy, in 2002), the first versions were drafted. This work was referred to in the Treaty of January 2012 and was the subject of a bilateral meeting between the French and Italian Ministries of Labour, held in Rome on 11 October 2012.
- From 2016 to 2017: through further meetings between the services and, in the final stage, the participation of TELT, the document was sent to the Technical Safety Commission (CTS) of the Intergovernmental Conference. Following the favourable opinion of the CTS, the last bilingual version was approved by the respective ministries in spring 2017, in time for it to be used in the contracts for implementation of the works.

The first phase of the meetings also involved two French organisations, the OPPBTP (Organisme Professionnel de Prévention du Bâtiment et des Travaux Publics) and the CARSAT (Caisse d'Assurance Retraite et de la SAnté au Travail) with whom the first contacts were made with the equivalent CPT (Comitato Paritetico Territoriale) of Turin.

Some technical visits were also important, especially the one at the Lotschberg tunnel construction site in June 2004.

3   OBJECTIVES AND MANAGEMENT TOOLS

The main objective of the work was the definition of uniform rules common to both countries, with the aim of choosing the most conservative rules of the two legal frameworks. The objective was therefore to "raise the bar", aiming to improve safety and the health of workers.

It was not always possible to define a single solution and, in some cases, the national rules prevailed, albeit often relating to the same European directive.

A second objective, to protect and verify the application of the rules, was the formation of a common inspection commission, which can act jointly throughout the development of the work.

The regulatory harmonisation of a project involving more than ten years of construction was based on the design stage of the project and on known and disseminated technologies. However, the same document defines updating methods, which are inevitable at a time when technological development is taking place very rapidly, and when anticipating what tools will be available in the immediate future is very difficult.

In addition to this, realising that living conditions off site are also important, it touched upon issues linked to accommodations and services for the workers, thousands of whom will have to live along the Maurienne valley and in Val di Susa for over 10 years.

The operating management of relations between the inspection services of France and Italy was entrusted to a "Coordination Structure" that will see the participation of the services of each country, the French and Italian security coordinators, and the client. This structure will meet periodically to analyse the situation and decide on common bi-national coordination actions.

The client is also instituting a "Health Coordination Structure" composed of the physician from each company for the various lots, which will meet at half-yearly intervals and with the "Coordination Structure" referred to in the previous point.

At the level of each operative lot, it was decided to also extend the Hygiene and Safety Committee between Companies, a French institutional body, to the Italian sites. This committee is instituted by the client and chaired by the safety coordinator. The works management, the representatives of all the contractors and the inspection and insurance bodies (DIRECCTE, CARSAT, etc.) will participate in it.

The committee meets approximately every two months and analyses the situation of the site, proposing or validating solutions to specific problems. It also meets in the event of a serious accident in order to understand the causes and determine prevention or protection measures.

Figure 4a.   Meeting in Bardonecchia, December, 2003.

Figure 4b. Visit to Modane adit, September, 2006.

4 SOME EXAMPLES

4.1 *Border management*

In most cross-border projects, the regulation of the national jurisdiction is an issue to be faced in the design phase in order to clearly define the territorial responsibilities of each state, especially in terms of administration of the applicable law and the legal jurisdiction.

Rules must be included in the agreements between the states so that they have regulatory value: technical rules alone are not sufficient to manage these complex situations.

Taking inspiration from other recent cases, considering that the planning of the tunnel excavation will result in the tunnels carried out by the French side entering Italy for several kilometres, it was decided that the nationality of a site should follow so-called "portal law". If the site advances and the entrance is on the French side, the entire site will be French. The same applies on the Italian side.

This eventuality has already occurred with the excavation of the Maddalena exploratory shaft, which crossed the border by around 100 m.

Once the tunnels have been drilled, the border returns to its geographic position and each part of the site will fall back into its own territory.

4.2 *Mandatory documents*

A series of documents and registers must therefore be kept for all the lots, such as:

- The daily coordination register, represented by the logbook of the coordinator on duty.
- The safety register, which relates to the machines and equipment.
- The notification of underground works, prior to starting the excavations or works in the tunnel.

4.3 *Worker training*

The parallels between French and Italian standards have enabled the already existing high standards to be maintained. However, it was agreed to call for the collaboration of the CPT (Comitato Paritetico Territoriale) in Italy and OPPBTP (Organisme Professionnels Prévention Bâtiment Travaux Public) in France for definition of the programmes.

Special emphasis was placed on the training of immigrant workers, which should take place after checking their understanding of the language of reference (French or Italian).

4.4 *Mechanised excavations with TBM*

In the part regarding recommendations and technical instructions during excavations and supports, part of which was taken from the Italian standard DPR 320/56 with a few updates, it is worth citing the TBM requirements, such as, for example:

- Full compliance of the machines with directive 2006/42/EC.
- Instruction Manual in the language of the country of operation (French or Italian).
- Availability of the contractor vis a vis work inspections.
- Arrangements for passing through sections with a risk of asbestos.
- Maximum attention to the design of footpaths and escape routes, ergonomics and comfort of the workstations, the most advanced technological choices available and the proper management of the assembly and dismantling phases.
- Standard provision for the employees: refectory/lounge, survival chamber, self-rescue equipment to reach the survival chamber, toilets with running water.

4.5 *Ventilation*

The ventilation system must follow the French CNAM recommendations, which envisages the use of pressure and suction ventilation, reducing situations where pressure alone is used to a minimum. This is a substantially new system for Italy compared to the practice of using pressure ventilation alone.

4.6 *Microclimate*

The provisions of the Italian standard have been adopted, since in France there is no reference to the maximum permitted temperature for work environments. For underground works in France this is new. The ventilation and air conditioning systems must therefore match these parameters.

4.7 *Rescue and emergency*

The rules call for coordination with the public rescue services starting from the planning stage according to the model of the French procedure. This preventive organisation must also be reflected in the contents of the Safety and Coordination Plans of the individual contracts.

Some requirements for the organisation and availability of resources or equipment refer to:

- Systems for evacuation of injured persons from the tunnels.
- Landing areas for the helicopter rescue service.
- Control and positioning systems for staff working underground.
- Installation of survival chambers at a maximum distance of 1,000 m from each other.
- Provision of self-rescue equipment for all the employees.
- Fume abatement systems such as water curtains.

Of course, many other issues have been addressed, some of them left to the application of the respective national standards, such as, for example, explosive atmospheres, the presence of toxic or explosive gases, the presence of asbestos, free silica, and radioactive minerals.

5 CURRENT SITUATION

The "Common Operating Rules" had already been implemented in the final issue of the Project published in 2013, albeit adopting a version of the document that was not yet official.

Following the work carried out in 2016 which led to formal validation by the respective Ministries of Labour in Spring 2017, the final version of the new edition of the Project, which was in the process of being reworked both in France and in Italy, was implemented.

The document is also included in the tender dossier as an annex to the contract, as required by art. 10.2 of the Treaty of January 2012 between France and Italy.

As of the date of this article, the bi-national Coordination Structure is in the organisation phase and will start at the beginning of 2019.

6 CONCLUSIONS AND ACKNOWLEDGEMENTS

The work begun in 2002 and concluded in 2017 has seen the participation of many officials from the French and Italian inspection services. In particular, seven French and ten Italian inspectors alternated in participating in the first drafting of the document, the latter in part of the prevention services of Azienda Sanitaria Locale 3 of the Piedmont region.

Three French and eight Italian inspectors participated in the final stage of reviewing and issuing the document, as did the directors of the French Ministry of Labour and the Italian National Labour Inspectorate.

TELT, the company tasked by the French and Italian Governments with the direction of the works, had the opportunity to collaborate in the final review phase with its own proposals and assessments, some of which were implemented in the final version.

The authors wish to thank all those who have participated in various capacities in the writing and publication of the "Common Operating Rules", which represent a rare example of cross-border collaboration and coordination in a project of common European interest like the Mont Cenis base tunnel.

REFERENCES

Ministère du Travail, Direction General du Travail & Ispettorato Nazionale del Lavoro, 2017. Règles communes opérationnelles de sécurité pour la construction du megatunnel de la ligne ferroviaire Lyon Turin, section transfrontalier; Regole comuni operative per la sicurezza per la costruzione del mega-tunnel della linea ferroviaria Lyon Torino, sezione trasnsfrontaliera.

Tunnels and Underground Cities: Engineering and Innovation meet Archaeology,
Architecture and Art, Volume 9: Safety in underground
construction – Peila, Viggiani & Celestino (Eds)
© 2020 Taylor & Francis Group, London, ISBN 978-0-367-46874-3

A novel watcher system for securing works at tunnel face

T. Tani & Y. Koga
Taisei Corporation, Tokyo, Japan

T. Hayasaka & N. Honma
Nikko Denki Tsushin, Tokyo, Japan

ABSTRACT: In mountain tunneling, construction works are in the proximity of tunnel faces where falling rocks and collapse of ground are often feared. It is however difficult to improve immediately such work environments, e.g. by applying full remote mechanized construction. In reality, a full-time watch guard stands in place and supervises safety at the tunnel face, when charging/connecting explosives and installing tunnel supports. When the guard judges that the continuation of tunneling work is in danger, he gives prompt warning to the workers, so they can immediately leave from the work area. A newly-developed face monitoring system utilizes an image recognition technology and a high-speed camera set beside the watch guard. It promptly catches fall of pebbles and flaked pieces of sprayed concrete as symptoms of possible tunnel face collapse possibly leading to disaster. It gives warning within a mere 0.1 seconds after the detection of movements of falling objects and gets the construction workers to leave immediately from the dangerous spot. The system has now been implemented in two projects. It is contributing to reducing workload from the tunnel face watch guard and to securing safety of the construction workers.

1 INTRODUCTION

In mountain tunneling, construction workers carry out various operations near the tunnel face, depending on the stage of construction. In particular, during preparations for blasting to excavate hard rock and during installation of steel arch supports (bent H-section steel beams) to support the surrounding rock after excavation, it is necessary for the construction workers to be close to the tunnel face where the risk of injury is high. Actual operations at the tunnel face are affected by rock failure or fragility of the ground, which can cause falling rocks or fragments of sprayed concrete to fall. Of the 47 persons that were killed or injured in the 44 falling debris accidents that occurred from the year 2000 to the year 2010, 6% were killed, and 42% were out of work for one month or longer, according to a technical document (National Institute of Occupational Safety and Health, Japan 2012).

Although development of technologies to carry out these operations automatically or by remote control of construction machinery is advancing, at present the technical level has not reached the stage where these technologies can be applied to all sites. In operations at the tunnel face, measures are therefore taken such as applying sprayed concrete to the natural ground that is exposed during excavation, or measures such as installing a rock fall protective mat above the construction workers in order to ensure safety of the workers carrying out operations close to the tunnel face. In addition, a full-time tunnel face watch guard is deployed to watch out for abnormal occurrences at the tunnel face and immediately warn the workers to evacuate, in order to avoid a serious accident (see Figures 1, 2).

Although the initiatives as described above are being taken, the authors have developed a tunnel face watcher system to reliably issue commands to the workers at the tunnel face

Figure 1. Workers and safety measures at tunnel face.

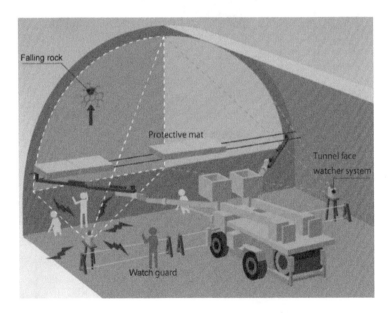

Figure 2. Workers and safety measures at tunnel face.

during times of danger, to create an even safer working environment. In addition to the tunnel face watch guard, this system includes a falling rock detection and alarm device that detects falling debris from the tunnel face by mechanical eye, and sets off an alarm within 0.1 seconds of occurrence by both light and sound. The system also includes a projection device that displays safety information on the tunnel face, such as the locations of occurrence of the falling rock and weak ground at the tunnel face. This paper describes the functions and characteristics of the equipment developed and the image recognition technology in the falling rock detection and alarm device and issues that have been recognized through operations on site, together with prospects for development in future.

2 FALLING ROCK DETECTION AND ALARM DEVICE

2.1 *Equipment configuration and specification of the device*

Figure 3 shows a configuration of the equipment in the falling rock detection and alarm device. During operation, a camera with lighting is installed at a location (5 to 10 m) away from the tunnel face so as not to obstruct the operations. Falling objects such as falling debris are detected from images taken by the camera, to form a warning system. Figure 4 shows the device installed within a tunnel and a front view of the device.

This device monitors small falling rocks of about 1 cm size, as a forewarning of larger rock falls or collapse of the tunnel face. A frame difference method is used for detection of falling rocks. In this method images taken at high speed (30 to 50 times per second) by a high-speed camera are used to generate difference images between two consecutive frames (images), to

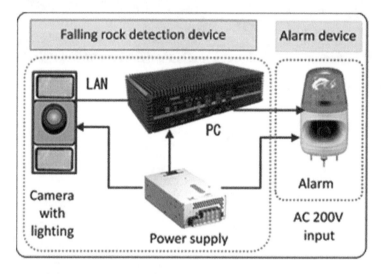

Figure 3.   System configuration diagram of the face monitoring system.

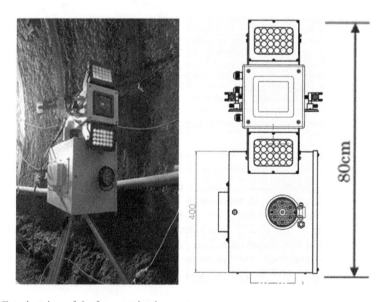

Figure 4.   Exterior view of the face monitoring system.

detect objects moving in the vertical direction as falling rocks. Note that the image processing and image detection technology used in this device will be described in detail in Section 3.

A Light Emitting Diode (LED) lighting is used to provide illumination over the area being monitored for falling rocks. Near infrared light is irradiated so that it does not interfere with laser pointers used by the workers or (visible) lights from the construction machinery, etc. A high-performance personal computer (PC) is included in the system for sensing and detecting falling rocks, by analyzing the images taken by the camera with an optical filter that allows only near infrared light to pass through the lens installed. This PC has necessary information on the conditions for taking images to detect falling rocks. This includes, for example, camera angle of view, resolution, frame rate, and distance from the tunnel face to the camera. The PC is also equipped with information on the objects to be detected. This includes, for example, size of the falling rocks to be detected (minimum and maximum) and information on falling speed. The specification of the PC used is shown in Table 1.

When a falling object (falling rock or peeled sprayed concrete, *etc.*) is recognized, the alarm device immediately informs the workers of the danger by flashing LED warning lights and the sound of a siren.

2.2 *Issues in development*

During development, the following three issues were extracted, as a result of a study of similar equipment operation.

(1) The equipment has to be able to withstand the severe environment of a tunnel construction site, such as dust, high humidity, and vibration. Also the equipment has to be consice and light so that it can be easily handled.

Table 1. Specification of the PC used.

External dimensions	W 264.2 × D 156.2 × H 66.5 (mm)
Power supply voltage	DC input 9–36 V
Mass	3 kg
CPU	Intel Core i7-6700TE (2.4 GHz)
Memory	PC4-17000 (DDR4-2133) SO-DIMM 8 GB
SDD	2.5 inch MLC 120 GB
Usage environment	Temperature 0 to 50°C, humidity 10 to 80% (40°C)

Figure 5. Sample stones in a variety of sizes for the falling rock detection tests.

Figure 6.   Rock fall tests in two different sensitivity cameras under various illumination at the tunnel face.

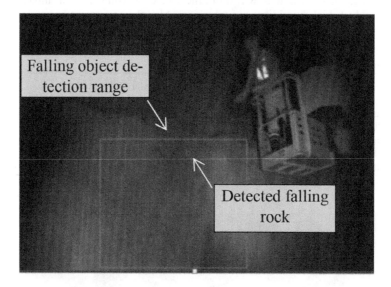

Figure 7.   Exterior view of the face monitoring system.

(2) The equipment has to be capable of reliably detecting small falling rocks down to about 1 cm size, regardless of the color of the stone or the background color.

(3) The equipment must not respond to movements other than falling rocks or to noise, and must not generate false alarms.

As for issue (1), each item of equipment was designed to have a protective enclosure (housing) with a rating of IP55 or higher in accordance with the IP standard for waterproof protective structures. Also, it was decided that the overall equipment including the PCs used for equipment control and image measurement and the LED projectors would have a structure completely free from fans. Details of equipment size reduction are described in Section 2.3.

Regarding issue (2), before the design of the device, using tunnel faces on rest days, rocks of various sizes (Refer to Figure 5) and colors were artificially dropped while varying the level of illumination and the image capture speed of cameras in two different sensitivities and resolutions, as shown in Figure 6. This was to determine the necessary and ideal conditions for detection, which could then be reflected in the design. Note that the same tests were carried out when the sprayed concrete on the tunnel face forming the background for the detection images was both wet and dry. Figure 7 shows the monitor screen for verification when detecting a falling rock during testing. In the figure the rectangular frame is the area for detection of falling rocks, while the small red square indicates that a falling rock has been detected by the image processing. It can be seen that when small rocks are dropped from above, they are immediately detected within the frame. At present a popular 2 million pixel high-sensitivity camera is used in verification operations on site. It has been confirmed that falling rocks of size down to 1 cm can be detected under various imaging conditions.

For issue (3), there are concerns that if the device reacts to movements other than falling rocks and triggers an alarm, the workers will become used to the alarm and may not take appropriate evacuation measures when there is actually a danger. In connection with this issue it was decided that image recognition and judgment would have to be carried out using advanced image processing. These technologies are described in detail in Section 3.

3 IMPROVEMENT IN DETECTION ACCURACY BY IMAGE RECOGNITION

In this section, the result of development in image recognition technologies to prevent false alarms from being issued is described. This is the most important development key of the three main development issues.

3.1 *Method of preventing false detection*

3.1.1 *Overview*
Figure 8 shows a configuration of a falling rock detection algorithm. For detection of falling rocks, the images captured by the camera are input to an image acquisition unit as data in real-time. The falling rock detection is carried out based on this image data. First a falling rock detection area calculation unit determines falling rock detection exclusion areas based on difference images between frames in the past N frames (in practice N=3 to about 10). Then

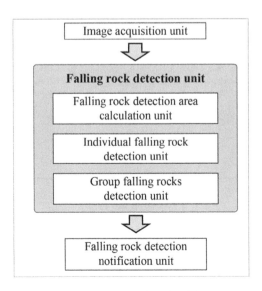

Figure 8.   Schematic diagram of the falling rock detection algorithm.

two falling rock detection units (individual and group falling rock detection units) detect falling rocks in the falling rock detection area, excluding those in these exclusion areas. If a falling rock is detected, data is transmitted to a falling rock detection notification unit indicating that a rock fall has occurred, and the alarm is issued.

By this method even if a worker or construction machinery is captured in the images, by setting that part as a falling rock detection exclusion area in advance, false alarms are eliminated, and only falling rocks can be accurately detected and the alarm can be correctly issued.

3.1.2 *Recognition of moving bodies other than falling rocks*

The image recognition algorithm for falling rock detection includes a falling rock detection area calculation unit that recognizes moving bodies other than falling rocks such as workers or heavy machinery near the tunnel face, and identifies a falling rock detection area, excluding areas of these moving bodies; and two types of falling rock detection units that recognize moving bodies appearing in the falling rock detection area as falling rocks. In this section, the method of recognition of moving bodies and the processing method of excluding false detection are described.

The falling rock detection area calculation unit carries out processing on difference images between consecutive frames in a time series, extracts moving body areas from each of the difference images in the past N frames, and carries out a process to determine a falling rock detection area that is the area excluding the collective range of each moving body area. The method of calculation of the area for detection of falling rock candidates is explained using Figure 9. Difference images (middle row in Figure 9) are generated for two consecutive frames in the time series of the images for the last N frames (top row in Figure 9), and for each difference image a moving body area is extracted by carrying out a binarization process using a threshold value. Figure 9 shows a person raising one arm, and mainly the raised arm area is extracted as the moving body area. Note that in this case the trunk, head, the other arm, *etc.*, apart from the raised arm have moved slightly, so these parts have also been extracted as the moving body area.

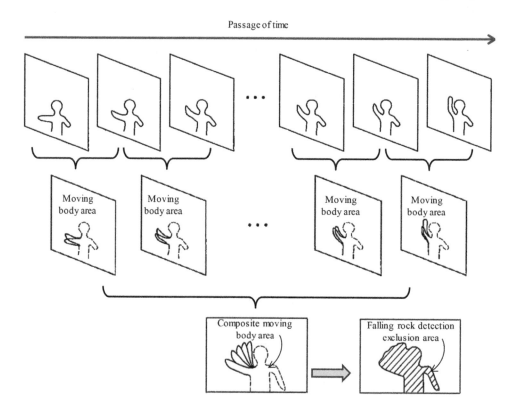

Figure 9. Recognition method for moving objects such as workers and machines.

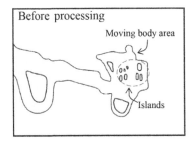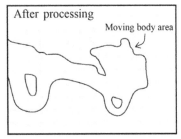

Figure 10.   Expansion processing for moving area.

Next, the falling rock detection area calculation unit obtains a composite moving body area (bottom row in Figure 9) by combining these moving body areas, and obtains the falling rock detection exclusion area as the composite moving body area by performing a labeling process allocating the same number to connected pixels as one group and an expansion process that is described later, so the falling rock detection area becomes the area apart from the falling rock detection exclusion area. In Figure 9, the area of the arm, the trunk, the head, the other arm, etc., that have moved in the past N frames mainly becomes the composite moving body area. In addition, the expansion process is carried out to enlarge the composite of each moving body area in predefined regions, interpolating the gaps between the moving bodies and the parts that were not detected as a moving body because by chance they had not moved in the past frames. This way islands that should belong to the falling rock detection exclusion area but are recognized as part of the falling rock detection area (refer to Figure 10) are reduced, and the occurrence of false detection of falling rocks is minimized.

Also, parts of the images that are bright and have a certain area are excluded from the falling rock detection area. This is to include reflective vests or hardhats worn by the workers in the falling rock detection exclusion area.

3.2   Improvement in falling rock detection accuracy

The falling rock detection unit includes an individual falling rock detection unit that identifies specific single moving bodies that appear in the falling rock detection area as falling rock candidates, tracks them through multiple frames, and judges whether or not they are actually falling rocks. The individual falling rock detection unit is described with reference to Figure 11. The top row in Figure 11 shows images taken in each of the three frames, and a falling rock appears in each of them. Here time passes from left to right. Also, the Y axis in Figure 11 shows the distance in the vertical downward direction as positive. As shown in Figure 11, the falling rock moves downward (natural fall) with time. The bottom row of Figure 11 shows the difference images for each frame, and in each difference images a group of falling rock candidates is included. The falling rock candidate on the upper side corresponds to the falling rock before it moves, while the falling rock candidate on the lower side corresponds to the falling rock after it has moved.

When two falling rock candidates have been detected in a group, the individual falling rock detection unit obtains the centroid calculated from the centroid positions the two falling rock candidates., It is a mathematical centroid, when the upper and lower falling rock candidates are considered in two-dimensional images. Centroid is calculated based on the shape, area, etc., of the falling rock candidate. Next, a judgment is made whether or not the position of the lower falling rock candidate in the difference image shown on the right side of the bottom row in Figure 11 comes within the natural fall area calculated from the previous difference image. The natural fall area indicates a movement range of the falling rock between frames for natural fall. For example it comes within a fixed angle and a certain fall distance relative to the previous position of the falling rock candidate. For this reason the natural fall area generally comes below the centroid of the falling rock candidate on the previous frame. For example, if the position of the falling rock candidate is within the natural fall area over multiple consecutive frames, the falling rock candidate is judged to be truly a falling rock.

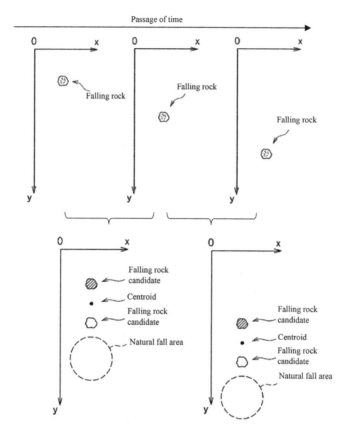

Figure 11.   Detection method of individual falling rock.

Figure 12.   A result of moving object detection.

In addition, the falling rock detection unit includes a group falling rock detection unit that detects all moving bodies appearing in the falling rock detection area as falling rock candidates, tracks all the centroids of the falling rock candidates over multiple frames, and judges whether or not they f are truly falling rocks. Consequently, it is possible to detect true falling rocks even when a multiple rock falls occur at the same time, and correctly to prevent over-looking falling rocks.

3.3 *Verification*

Verification was carried out into whether or not the moving bodies were correctly judged from the various image processing and the image recognition scheme described above and whether falling rocks only are properly notified, based on images acquired on site near the tunnel face to date and those on falling rock tests. Among 50 hours of video used in the verification, only representative verification images captured are shown here.

3.3.1 *Moving bodies*

Figure 12 shows a recognized image of a worker on a work platform (man cage) at the tip of a drill jumbo boom and the boom during movement. It can be seen from this figure that the boom, the worker, and randomly falling seepage water are not judged to be falling rocks. It was confirmed that image recognition could be carried out without issuing an alarm. Note that in the center of the figure, the area enclosed by the red line indicates the falling rock detection exclusion area resulting from the moving body recognition, whereas that enclosed by the green line outside the red line indicates the falling rock detection exclusion area after performing the expansion process. On the right-hand side of the photo, the position of water seepage is outlined in green lines. In this area, however, recognition of individual falling objects was carried out as described in Section 3.2, and this process was capable of judging that the seepage water falling continuously as multiple water drops should be included in the falling rock detection exclusion areas.

3.3.2 *Time required to judge falling rocks*

Using the same equipment configuration, laboratory tests were carried out to confirm the time required to detect a falling rock. This was carried out by reading the position of the falling rock on the recognition image using a ruler that appeared in the same video at the time it was detected, and calculating the time until the detection of the falling rock.

The position at which detection occurred as read from the image was 5 cm. According to the equation of free fall, it is equivalent to 0.1 seconds from the start of the fall to detection.

4 SUMMARY AND OUTLOOK FOR THE FUTURE

The falling rock monitoring system of the tunnel face watcher system has been completed as a technology for ensuring safety at the tunnel face, and its performance has been confirmed using test images obtained on site and laboratory tests. In particular, a method has been devised to ensure that falling rocks only are reliably detected for issuing an alarm, by using image recognition. In addition, at present the equipment size was reduced down to one third compared with the prototype. This concise equipment has been used on site since July 2018, contributing to ensuring safety of operations at the tunnel face.

For the future we are further slimming down the equipment to achieve major size reductions. In addition, it is considered necessary for specifications to be adapted to the conditions at the various sites. Verification tests will be carried out at various mountain tunneling sites with different cross-section size and geological conditions.

In January 2018 the Ministry of Health, Labour and Welfare (MHLW) has amended the "Guidelines for Falling Debris Accident Prevention Measures at Tunnel Faces in Mountain Tunneling Construction" (MHLW, Japan 2018), and there is a perception that safety initiatives in the industry as a whole have accelerated. It is intended that devices for ensuring safety of construction and workers be actively and continuously developed in future in monitoring not only for rock falls at tunnel faces, but for slopes on working roads in mountainous areas.

REFERENCES

Ministry of Health, Labour and Welfare, Japan, 2018. Guidelines for Falling Debris Accident Prevention Measures at Tunnel Faces in Mountain Tunneling Construction (Translation from Japanese title).

National Institute of Occupational Safety and Health, Japan, 2012. Analysis of Survey of Labor Accidents due to Falling Debris at Tunnel Faces and Proposal of Prevention Measures (Translation from Japanese title), JNIOSH-TD-No. 2, 5–6.

Tunnels and Underground Cities: Engineering and Innovation meet Archaeology,
Architecture and Art, Volume 9: Safety in underground
construction – Peila, Viggiani & Celestino (Eds)
© 2020 Taylor & Francis Group, London, ISBN 978-0-367-46874-3

How to refurbish a tunnel: Shop open or closed?

B.A. van den Horn, A.M.W. Duijvestijn & C.S. Boschloo-van der Horst
Arcadis Netherlands B.V., Amersfoort, The Netherlands

ABSTRACT: Renovation or refurbishment of an in-service road tunnel can be as complex as the commissioning of a new road tunnel. Mostly, there are two tubes with an escape tunnel in between and space for the cabling of the tunnel safety equipment. The tunnel installations, providing the safety for the whole system, work as one system operating both tubes. Therefore, working on tunnel installations in one tube while operating in the second tube might interfere safety systems there. For instance, refurbishment works in one tube could lead to loss of lighting in the operated tube, which is dangerous. So, a complete closure of the tunnel would be preferred during refurbishment. However, sometimes the tunnel is so crucial in the traffic network, that the tunnel must be at least partly available. In such cases a tube-after-tube-approach is an option. After commissioning the first refurbished tube, the next tube is upgraded. This was the case for the Maastunnel in Rotterdam. In this project, the objective (and challenge) was to gradually increase the safety level from the existing level prior to the upgrade works to the final safety level in the refurbished situation. In the case of the combined Piet Heintunnel (one rail tube and two road tubes) in Amsterdam the rail connection will be operated during refurbishment of the road tunnel, starting in 2019. In this paper the pros and cons and the different lessons learned regarding simultaneous refurbishment and operation of tunnels are shared.

1 INTRODUCTION

Many road tunnels in Europe need refurbishment of the civil structure or technical installations. Additionally, the legislation and requirements for tunnel safety have changed considerably in the last decade. Therefore, refurbishment and upgrade projects have been carried out in many tunnels or will be carried out soon. In these projects the desired safety level will often be based on the current legislation and requirements where possible. The constraints of the existing situation and choices from the past, poses challenges to the refurbishment and upgrade projects.

In the Netherlands several refurbishment and upgrade projects of road tunnels have been carried out in the last years, are ongoing or planned for the near future. An important topic to be dealt with during refurbishment projects is the reduced availability of the road leading through the tunnel.

A complete closure of the tunnel would be preferred during refurbishment. However, most of the road tunnels are crucial in the traffic network. In general, the acceptance of non-availability of road infrastructure by commuters is very limited. For instance, the increasing pressure on the road network in major cities like Amsterdam and Rotterdam would lead to disruption in traffic flow after one of the city tunnels would be taken out of service. For this reason, many road tunnels are only partly closed for traffic during refurbishment works to meet the minimal availability criteria.

Partial tunnel operation during refurbishment works may lead to additional risks for the tunnel users. Therefore, tunnel managers are responsible for a dedicated operational safety plan to manage these additional risks and to guarantee a safe tunnel operation during refurbishment. The safety plan describes the agreements between the tunnel manager, the project delivery manager and the contractor on temporary measures for safe operation during the refurbishment works.

2 SAFE TUNNEL OPERATION DURING REFURBISHMENT PROJECTS

2.1 *Minimum safety: compliance with the Dutch Tunnel Act*

All Dutch in-service road tunnels with a length of more than 250 meter need to comply with the Dutch Tunnel Act before 1 May 2019. The Dutch Tunnel Act imposes requirements on the tunnel management organization, direct requirements on the safety features and implicit performance requirements on the use of the tunnel. The latter ones follow from the Quantitative Risk Analysis (QRA) as prescribed by Article 6 of the Dutch Tunnel Act, which states that the societal risk shall not exceed the test line $0,1/N^2$ per kilometer tunnel tube annually, where N is the number of fatalities for $N > 10$.

Unfortunately, the Dutch Tunnel Act does not provide rules for a phased renovation, during which one tunnel tube is being operated during refurbishment. In such a case, each construction phase would be a change in tunnel operation, which requires compliance with the Dutch Tunnel Act, viz. a societal risk profile that meets the test criterion after the changes in the tunnel system. This would mean that formally at the start of each construction phase a commissioning permit must be granted. However, during the construction phases it might not be possible to meet the test line, which was one of the main reasons to start the refurbishment. This would lead to a situation where the administrative authority would not grant a commissioning permit. Therefore, a tailor-made safety decision process was designed that allows the safety enhancement of Dutch in-service tunnel connections. The tunnel manager and administrative authority must agree that in each construction phase the societal risk profiles are compared with the profile of the previous phase and shall not be higher than the previous phase, see Figure 1.

Also, the safety processes traffic flow, accident control, egress and emergency response are included in a qualitative safety assessment. This approach has been successfully applied for the Maastunnel in Rotterdam.

2.2 *General safety framework for simultaneous tunnel operation and refurbishment works*

Figure 1 (right) shows a general framework based on the safety trias for the safe operation of a road or rail tunnel in general and during refurbishment works in particular. The trias is determined by three pillars:

1. the *organisation* of the tunnel management emergency response and traffic control,
2. the tunnel *infrastructure* consisting of the tunnel geometry and the technical safety equipment,
3. the *use* of the tunnel depending on the admission regime. To pursue a safety optimum beyond the minimum level, these pillars with their interfaces are considered simultaneously.

To meet the agreed safety level, a proper balance must be found between the pillars, as shown in Table 1. For example, the choice of a fire detection system (infrastructure) can support the tunnel operation manager (organization) for a fast in-house first response to control the accident. Conversely, extra supervision can be applied if a fire detection system is not installed. Another example is a cargo truck fire that can lead to heat releases of more than 200

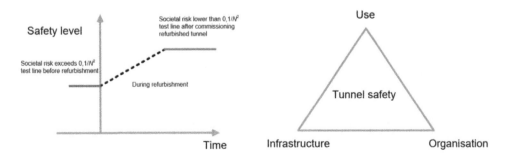

Figure 1.   Safety level before, during and after the refurbishment works (left) and the safety trias (right).

Table 1. Examples of cases with possible choices.

Example	Use	Infrastructure	Organization
1	No restrictions	Fire detection	Basic supervision in control room
2	No restrictions	No Fire detection	Extra supervision in control room
3	No admission of trucks	Reduced ventilation capacity	Basic supervision in control room
4	One of two lanes closed between 23:00h – 06:00h	• No ventilation in supporting tube • Temporary alarm system during the refurbishment project	• Supporting tube for emergency response is a working area • Dedicated on-site fire fighter team between 06:00h – 23:00h • Extra operational safety procedures for tunnel manager and contractor

MW in the tunnel. A ban of heavy cargo trucks may allow a reduced ventilation capacity. In the case of a so-called tube-after-tube approach one tube is being operated while the other tube is not available for emergency response, a dedicated on-site fire fighter team at the tunnel with fast intervention may result in the mitigation of the effect of a tunnel fire. Lane closures and bus prohibition means less potential victims in the tube during a fire and may therefore, also lead to a reduction of safety equipment or less safety procedures.

2.3   Safety organization during refurbishment

2.3.1   Responsibilities and reporting lines during refurbishment
Figure 2 illustrates the principal roles and reporting lines for the management of road tunnels during a refurbishment project. The description of these roles is not exhaustive, only the key responsibilities are outlined from a refurbishment point of view.

The administrative authority grants the permission to commission the tunnel after refurbishment.

The tunnel manager is responsible for all the facilities for the *operation tube* and supervises the tunnel maintenance manager and the tunnel operation manager. He will also directly or

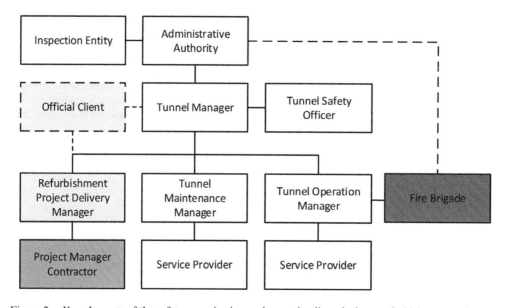

Figure 2.   Key elements of the safety organisation and reporting lines during a refurbishment project.

indirectly - via an official client - supervise the project delivery manager. The tunnel manager reports on accidents during the refurbishment works to the administrative authority.

The project delivery manager is appointed by the official client (or the tunnel manager) to undertake the refurbishment project development, design, and construction, commissioning and handing over to the tunnel manager.

The tunnel operation manager is in place at road tunnels that are staffed and provide an in-house traffic control and a first in-house response to accidents. The tunnel operation manager will support the tunnel manager in the creation, implementation and review of operating and emergency procedures during the refurbishment project. In addition, the tunnel operations manager coordinates appropriate and comprehensive training for the tunnel operators, first responders and breakdown crews. The role of tunnel operation manager can be carried out by a managing agent company (Duijvestijn et al., 2019) or an individual.

The tunnel maintenance manager supervises the installations, maintenance, repair or modification of the road, equipment or systems for the road tunnel. He is responsible that an effective plan is in place for appropriate maintenance and comprehensive training for those using or maintaining the tunnel system. The tunnel maintenance manager can also be carried out by a managing agent company.

The tunnel safety officer is functionally independent and responsible for the coordination of procedures and the liaison with the emergency services. He or she takes part in the preparation of operational schemes, emergency procedures, emergency exercises, emergency operations and the definition of safety schemes. The safety officer should be involved in the specification of the structure and equipment of refurbishment of in-service tunnels.

The inspection entity is functionally independent from the tunnel manager. It will carry out evaluations and tests. The inspection entity can be appointed by the administrative authority or carried out by the administrative authority itself. The inspection entity undertakes tunnel safety inspections during the refurbishment works.

Service providers are those organizations that are appointed or contracted by the tunnel manager to facilitate the design, build, maintenance and operation of the tunnel. The contractor of a refurbishment project can be regarded as a service provider for which the project manager contractor is responsible. The project manager of the contractor is responsible for the work tube. In case of a tunnel accident with evacuation the tunnel manager will take over the responsibility for the whole tunnel system.

The fire brigade is responsible for safety on the construction area in the work tube as well as the safety in the operation tube. The fire brigade is also the consultant of the administrative authority.

2.3.2 *Additional safety procedures during refurbishments*

During normal operation, the escape routes may be used as a working area. For instance, in case of a fire in the operating tube, the work tube may serve as an escape route and safe zone (smoke and dust free) for evacuees. The escape route may extend over the building site to the adjacent ground level on the outside. Hereto, it is of paramount importance that the safe zone is a real safe zone, in the sense that people will not harm or injure themselves because of building activities.

Dedicated safety plans must be made by the tunnel manager and the contractor with the procedures how to ensure that a safe escape route for road users and working staff is provided in the work tube in case of an emergency. The following safety conditions are pursued for the road tunnel evacuees:

1. *Obstacle free escape route*: In case of an evacuation in the operating tube, the working staff of the contractor will leave the work areas with all equipment obstructing the escape route immediately or within a few minutes, depending on the project specific needs.
2. *Unrecognizable escape doors*. The escape doors shall not be opened by the contractor from the working areas except when needed to evacuate from a (fire) accident in the working areas. An escape door that is being worked on is not suitable for road users to escape and therefore it is rendered unrecognizable by the tunnel management staff in the operating tube. An unrecognizable escape door is also unusable, or not suitable for evacuation, in case of accidents in the working areas.

3. *Dust free and available escape doors.* The escape doors shall be reliable to open. Hereto, it must be prevented that dust and dirt from building activities hamper the operation of the escape doors. Therefore, the escape doors are made dust free by the contractor on a frequent basis.

3 EXAMPLES OF DUTCH ROAD TUNNEL UPGRADES

3.1 *Maastunnel*

3.1.1 *Concrete* repair, new equipment *and upgrading ventilation system*
The listed Maastunnel in Rotterdam is the oldest Dutch in-service underwater road tunnel (1942) and is still a vital part of the road network in the city. In July 2017 an extensive refurbishment and upgrade project has been started with the purpose to realize a safe road tunnel compliant with the Dutch Tunnel Act and Building Codes, to restore the concrete construction and apply state-of-the art technical installations and finally to do justice to the listed status by bringing back the 1942 time sense (Noordijk et al., 2018). This project will be finished in 2019.

One of the most extensive tasks was to install a new emergency ventilation system. Before the refurbishment of the Maastunnel the transverse ventilation system consisted of an ingeniously designed system of supply and exhaust ducts for the different sections and tubes which enter the tunnel through the two ventilation buildings between the river tunnel and land tunnel. This system was not capable to control tunnel fires larger than a passenger car fire. A drawing and picture of one of the ventilation buildings of the Maastunnel is shown in Figure 3.

For the Maastunnel, the supply and extraction ducts in the land tunnel sections could only locally be removed above the traffic tube to install one cluster of jet fans near each of the four portals. The old transverse ventilation system was not designed for emergency ventilation. So, the emergency and emission ventilation consists of the newly installed jet fans composing the longitudinal ventilation system, where in the original situation only transverse ventilation was present

3.1.2 *The challenge of simultaneous exploitation and refurbishment*
For the Maastunnel project it was decided to follow a tube-after-tube-approach for the refurbishment. The project was phased in a way that a traffic flow in south - north direction is always possible. The reason for this is the location of the Erasmus Medical Centre at the north bench of the river Maas.

Construction phase 1 of the Maastunnel project in which the west tube was refurbished is finished. Construction phase 2 (2018 – 2019) is ongoing: the east tube is being refurbished and

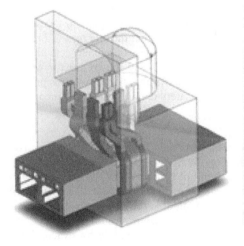

Figure 3.   Ventilation system of the Maastunnel before refurbishment with supply and exhaust.

the west tube is in operation. After the end of construction phase 2 the project will be delivered and the renewed Maastunnel will be commissioned.

Simultaneous exploitation and refurbishment pose a challenge on the traffic flow, incident control, egress and emergency response and during the works. To understand this challenge the evacuation concept will be outlined in the following sections.

Evacuation concept after completion of the project
The evacuation concept in the Maastunnel is based on creating a smoke free zone upstream from the fire accident location. The key points in the evacuation concept are:

1. *Clear marking of the green escape doors in the wall between the traffic tubes via led lighting.*
2. *Controlled enhanced unfoldable staircases to the escape doors via an elevated inspection path.*
3. *Emergency signs and clear illuminating indications of the escape doors.*
4. *Ventilation in the driving direction of the accident-tube to prevent smoke to lay-back to the people upstream of the accident.*
5. *Ventilation in the non-accident tube, which is cleared from working staff, in the same direction as in the accident tube via reversible ventilation to prevent smoke recirculation from the accident tube and to create overpressure in the non-accident tube.*

To prevent high numbers of potential fatalities during a tunnel fire, an adaptive Traffic Jam Management System will be installed: this system consists of a speed discrimination system that traces the traffic speed. For too low a traffic speed the systems controls the out and in flow of traffic to prevent traffic jams in the tunnel.

Figure 4 shows that the escape doors are elevated so the evacuees first must climb an unfoldable stairway onto the inspection path before using the doors. Then, the evacuees will reach the inspection path of the non-accident tube and return to the road surface via the stairs and then exit the tunnel. After the last evacuee has left the tunnel, the evacuation is complete.

Evacuation concept during refurbishment of the Maastunnel (2017–2019)
During the refurbishment work fires may occur in the operating tube or in the work tube. In both cases people must be evacuated from the accident tube via the non-accident tube to a safe zone as will be outlined in the following.

Fires in the operating tube During a fire in the operating tube egress and emergency response are crucial when it comes to life safety and reducing the consequences of a fire. The prerequisites for egress of road users have been translated by the contractor into safety measures to safely escape from the operating tube via the work tube. The most important adjustments to the

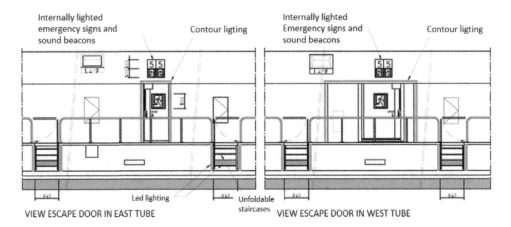

Figure 4. Escape route from the accident tube to the non-accident tube in the Maastunnel.

escape route are explained below. Figure 5 shows a project phase during which the demolition of the driving deck of the west tube was prepared.

The escape route from the east operating tube leads through a new climbing niche with a temporary staircase to the inspection path through the escape doors. And then from the inspection path in the work tube via a temporary staircase to the floor (see Figure 6).

All evacuees walk to a safe zone outside the tunnel via the southern tunnel portal over the driving floor and the mobile, temporary, reversible building ventilation to control the air quality of the workers is stopped by an instructed employee.

The building traffic route leads through the work tube, which is marked by a (white) alignment until the moment where new asphalt is applied. Within this marked route, a (green) alignment is placed to mark the escape route in the work tube to the south side. The width of this escape route is at least 1.2 m. When building traffic that travels from south to north comes to a stop, it should be in such a way that the escape route is not blocked.

During the phases where the driving deck is removed, the floor is repaired, and the new driving deck is placed in the river tunnel, the existing inspection path is widened from 0,7 m to 1,2 m and fitted with a robust handrail.

In the land tunnel the escape route leads from the operating tube via a new climbing niche with a temporary staircase to the inspection path through the escape doors and then from the inspection path in the work tube via a staircase to the floor (see Figure 6, left). At the transition from river to land tunnel, the evacuees are led via a staircase to the driving floor and further to the southern tunnel portal (see Figure 6, middle). In case of accidents, the air control ventilation in the work tube is stopped by an instructed employee. Again, the traffic that travels in and out from both sides of the work tube is stopped so that the escape route is not blocked.

All evacuees walk over the driving floor to the southern tunnel portal and the evacuees in de land tunnel north are led via a staircase (see Figure 6, right) to the widened inspection path of the river tunnel and move in the same way as described at the river tunnel section above.

Fires in the work tube The contractor has provided an evacuation system which aims to alert its employees in case of a fire accident in the work tube, after which they will leave the

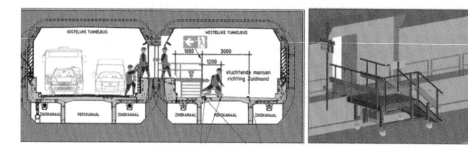

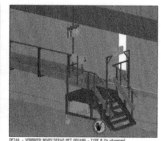

Figure 5.   Escape route during phase where demolition of the driving deck of the west tube was prepared.

Figure 6.   Temporary stairs in the escape route during phase driving deck is removed in the river tunnel.

work tube via the operating tube. The system consists of a so-called slow-whoop system (sound) and a visual system (red light signal). This system can be activated at each emergency door in the work tube (see Figure 6). Activation is done by pressing a button by the observer of the fire. A signal is going to the traffic control centre and the tunnel operator will to close the operating tube. The sound and the light signal are noticed by anyone, including the front-desk at the south entrance. The frontdesk will call the tunnel operator to confirm the alarm. After the traffic has left the operating tube it will function as a safe zone for the workers. The evacuation for workers from the work tube in the river tunnel during the period that the old driving deck has been destructed will be for via a ladder and movable handrail.

Emergency response
Since the work tube is a building area, it is not fit for the purpose as a supporting tube for the fire brigade to approach and fight the fire. Therefore, a dedicated in-house fire fighter team is positioned at a portacabin at the southern entrance of the operating tube with the purpose to compensate the absence of a supporting tube with a fast intervention aimed at "keeping a small fire small". The in-house fire fighter team is operational during the traffic peak hours. When the in-house fire fighter team is absent, the traffic is directed via only one traffic lane. The other free lane is red-crossed and reserved for the regular fire brigade. In this way each accident can be approached with normal rescue vehicles supported by ventilation to create smoke free approach conditions.

3.2 *Piet Heintunnel: a combined road and rail tunnel*

3.2.1 *Refurbishments*
The Piet Heintunnel (see Figure 7) is an underwater tunnel in Amsterdam consisting of three tubes which were built at the same time: two for road traffic (with an escape tube in between) commissioned in 1995 and one for rail traffic commissioned ten years later in 2005 (Duijvestijn et al., 2008). Reason for this was that the preparations in 1999 for building a light rail connection through the rail tube were delayed, because of major tunnel accidents in Europe in that period. Because of this, the requirements for tunnel safety increased steadily over the years. To establish a safety concept for the rail tunnel one had to cope with the as-built geometry of the existing tunnel structure. This resulted in a concept where in case of a rail tunnel fire, the tram passengers were required to use the traffic tube for evacuation.

Refurbishment in 2014
In 2014, twenty years after commissioning the road tunnel a first refurbishment project was carried out for on the Piet Heintunnel. The work consisted of replacing the top layer of the road surface in the closed part of the tunnel, including repairing the detection loops in and outside the tunnel. Besides, concrete repairs to the walls, ceiling and joints were done and some leakage problems were resolved. These refurbishment works took two months in the summer of 2014.

Because detour routes for the traffic were available, it was decided to close the Piet Heintunnel for normal road traffic during the project. Only the emergency services could make use of

Figure 7. Schematic impression of the Piet Heintunnel in Amsterdam during refurbishment.

the tunnel in the case of top priority circumstances where every second counts. The tram in the adjacent tram tube remained in operation.

The tunnel manager used a dedicated safety plan to warrant the conditions for a safe use of the road tunnel by the emergency services and a safe tram operation. The safety plan contained procedures to ensure:

- The non-working tube for the emergency services is forbidden to enter for the contractor.
- The escape doors in the rail tube shall not be blocked without possibility to remove the block in case of an emergency because they are part of the escape route for the tram passengers during a rail tube evacuation.
- The escape doors in the rail tube must be cleared during work within 1 minute.
- Working tools and material should not be drawn by the contractor within a radius of 15 meter from the escape doors for rail passengers in the adjacent road tube.
- The escape doors in the road tubes to the escape tube shall not be blocked because of possible intervention by the fire brigade in case of a fire in one of the road tubes.
- Emergency vehicles must always be able to drive through the working tube.
- Working tools and material should not be drawn outside the contractor's working-time.
- Safety critical installations for alarming working staff in the road tube adjacent to the rail tube must always be available.
- Safety critical installations to facilitate the evacuation of the rail tube and working staff in the road tube adjacent to the rail tube must always be available.
- Safety critical installations in the road tubes for a safe use of the emergency services must always be available.
- Emergency posts always need to be closed.

If these conditions for the safe operation of the Piet Hein tram tunnel were not fulfilled, the tram operation was immediately stopped. During the refurbishment project no significant accidents took place.

4 EVALUATION

In this section the pros and cons regarding simultaneous refurbishment and operation of tunnels based on the project experience are shared.

When all tubes are closed for traffic, there is full freedom for the contractor to use the tunnel as a working area and the lack of additional operational risks in the tunnel for the tunnel manager. The downside is that the detour needed for the traffic may give rise to additional safety risks in the surroundings of the tunnel and the connection is unavailable for emergency services.

When one tube is closed for normal road traffic and one in operation the advantage is that the detour needed for the traffic is much simpler depending on te direction of the traffic direction in the operation tube and that a crucial connection is available.

A disadvantage could be that the contractor must make preparations to create a safe zone for road users or tram passengers in the working area during an evacuation. This generates extra costs because the working plan might be more complex. Furthermore, a lot of extra safety procedures are needed for interfaces between the operation tube and the work tube, because the prerequisites for a safe escape to the work tube need to be managed. This means extra work: the tunnel management staff, the project organisation, the contractor and the emergency services must organise repeated multi-disciplinary training during the project. Besides weekly meetings are organised to discuss safety inspection forms with observations on the prerequisites for a safe escape for road users from the operation tube to the work tube. Figure 8 illustrates some examples of situations where blocking of the escape route was observed and subsequently actions for improvements were taken in case of the Maastunnel project.

It is important for the tunnel manager to specify the requirements for a safe tunnel operation during refurbishment at an early stage in the procurement process. Since modern contracts allow a lot of freedom to te contractor, it is important to pose the right questions when it comes to

Vehicle parked in front of escape stair   Blocked escape route by a stair

Figure 8.   Examples of observations during the safety inspections.

operational safety in the operation tube and health and safety for the working staff in the work tube. Health and safety are a mixed responsibility for the tunnel manager and the contractor.

During the Piet Heintunnel refurbishment in 2014 no accidents occurred. In construction phase 1 of the Maastunnel project no fire accidents took place in the operation tube, only false fire alarms in the work tube and one incident where the tunnel operators noticed dust on the CCTV camera's. These small incidents helped the tunnel management organisation to evaluate the correct handling of the procedures.

5   CONCLUSIONS

To meet the agreed safety level during and after a refurbishment project, a proper balance must be found between organization, infrastructure and use of the tunnel. It can be concluded that:

- Refurbishment with the shop closed is preferred from a safety point of view, when short and safe detour options are available. In the case of the Piet Heintunnel in 2014 it was decided to close the tunnel for normal traffic and only allow emergency response vehicles on a high priority mission.
- In service road tunnel refurbishments with the shop open will lead to additional project costs and it may take more time. These costs must be weighed with possible cost benefits for society. A tube-after-tube approach with a tailor-made safety decision process is needed to meet legislation. In the case of the Maastunnel, the tunnel manager and the administrative authority agreed that a stand still principle of the safety in each construction phase would be enough to proceed. This approach has worked so far.
- Extra safety procedures are required by a tube-after-tube approach and extra training and education for operation and emergency response are needed for all parties.
- During operation the tunnel manager and the contractor are responsible for the safety of road users and personnel, respectively. Therefore, multi-disciplinary training at the beginning of the project with frequent repetition have appeared to be crucial for a safe operation and health and safety for personnel.

REFERENCES

Noordijk, L. & Van den Horn, B.A. & Duijvestijn, A.M.W. 2018. The Art of Refurbishment of In-Service Tunnels. In Anders Lönnermark and Haukur Ingason (eds), *Proc. intern. Symp. on Security and Safety of Tunnels 2018, Borås, 14–16 March 2018*. RISE Report 2018: 13.
Duijvestijn, A.M.W. & Boschloo-van der Horst, C.S. & Thewes, J. 2008. Strategic Safety Planning for Rail in a Combined Rail and Road Tunnel. In Kanjlia, V.K. (eds), *Proc. World Tunnel Congress, Agra, 22–24 September 2008*. New Delhi: Central Board of Irrigation and Power.
Duijvestijn, A.M.W. & Leek, W.P.G. & Quirijns, G. 2019. Optimizing, Operate and Maintain Road Tunnels Using Reliability Centered Maintenance and Big Data. *Proc. World Tunnel Congress, Napoli, 22–24 September 2019*. Rotterdam: Balkema.

Tunnels and Underground Cities: Engineering and Innovation meet Archaeology,
Architecture and Art, Volume 9: Safety in underground
construction – Peila, Viggiani & Celestino (Eds)
© 2020 Taylor & Francis Group, London, ISBN 978-0-367-46874-3

Excavation through contaminated ground by gasoline with EPB-TBM

H.A. Yazici, M.B. Okkerman & C. Budak
Aga Enerji, İstanbul, Turkey

ABSTRACT: This paper concerns of the measures taken to pass the gasoline contaminated area in Bakirkoy –Kirazli Metro Line, Istanbul. The project consists of two lines having a length of 8165 m each and two EPB-TBMs. During side investigations, it was discovered that an area close to the tunnel portal was contaminated by gasoline leakage coming from a petrol station, which is situated nearby to tunnels. According to the gas station authorities, fuel oil contamination was also found under another gas station approximately 60m North East of the present gas station. Within 100 m radius of the working area, there are residential apartments, parks, 4 schools and a big shopping mall. For that reason, to keep the hydrocarbon threshold below the exposure and explosion limits, several measures were taken. First, the pressure chamber was filled with bentonite in order to dilute the concentration of the hydrocarbons within the chamber. After a while, this application was abandoned since it caused the clogging of the cutterhead which effected the performance of TBM. As a second step, a gas monitoring system was designed and several gas sensors were installed at critical points of the cutterhead. In case of the gas increases in the environment, reaching to a predetermined level, the air velocity forced to increase in order to dilute the gas content. It is believed that the experience gained will serve a basis to overcome similar problems.

1 GENERAL INFORMATION ABOUT THE PROJECT

Bakırkoy-KirazliMetro (BBKM) line, which is being constructed within the Istanbul Metropolitan Municipality, is planned as an extension to Basaksehir-Olimpiyat and Basaksehir-Metrokent metro lines and is going to be 8.8 km long. This line is one of the most important part of the planned Istanbul transportation network as the line will intersect or interact with the other completed or planned transportation projects that are undertaken throughout Eastern and Western boundaries of the Incirli region. The project will get through 7 individual stations and will finish at Bakirkoy İDO area. Construction works within the scope of the project are two metro line tunnels of 6.2 km, which is currently excavated with 2 TBM's. The last station of the BBKM line is being built by cut and cover method. The station excavation size is 28 × 185 meters at a depth between 20-23.50 meters.

2 HYDROCARBONE CONTAMINATION OVERVIEW

Hydrocarbon contamination mechanism is shown in Figure 1. Typically, hydrocarbons form a layer that floats on the water table in a light non-aqueous phase (LNAPL) [II]. Due to its higher concentration of contaminants and lack of dissolution, the LNAPL presents the larger threat to the construction workers. However, where the tunnel crown passes at distance of more than 5 meters from the phreatic surface level the groundwater acts as a buffer zone eventually protecting the tunnel from contaminants. Notwithstanding, dissolved contaminants and hydrocarbon vapors may also pose a probable threat to the workers' health and a possible explosion risk depending on the environmental conditions. In particular, petrol components such as benzene, which is soluble

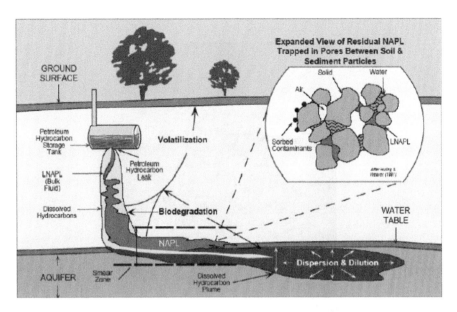

Figure 1.   Typical hydrocarbon contamination mechanism (Environmental Science in the 21st Century).

in water, may change back again into vapor when extracted with the screw conveyor and reach atmospheric pressure, becoming potentially dangerous to human health. Furthermore, the geo-logical characteristics of the ground to be excavated in the BBKM project consisting in possibly weathered limestone make the potential spread of the contamination larger since the ground has interconnected cavities and voids, which aid in the expansion of the LNAPL. Thus, every time one of the TBMs crosses the surface of the water table at an area where contaminants may be present there is a risk of hazard to human health and of explosion.

3   GEOLOGY AND GROUND CHARACTERISTICS AROUND CONTAMINATED AREA

Northern section of Istanbul province contains various Earlier Paleozoic and Neogene age rock formations. Side investigations shoved that, between the İDO station and the İncirli station mostly Bakırkoy formation and partly Gungoren formation are found. Bakırkoy formation is mostly defined by the limestone. Between the limestone layers, thin marL and clay bands are found.

Determination of the geology around the contaminated area which is concentrated around the first cross passage are determined by the drilling studies and during one of these studies on the İSK-3 drill hole, petroleum products were discovered as seen in Figure 2. According to İSK-2 drill hole, top of the drill has 3,5 m thick embankment. Beneath the embankment, clayey sand with limestone was discovered until 7 m depth. Beneath the clayey sand, wea-thered limestone with low plasticity was found just until depth of 13,5 m. And beneath that, clayey limestone was observed until the depth of 47 m. In İSK-3 borehole, 3.25 m of embark-ment was observed on top of the drilling core. Beneath that, silty sand with high plasticity was found until the depth of 6 m. Beneath the silty sand layer, weathered limestones with low structural strength was found until the depth of 19,5 m. Beneath the limestone layer, clayey limestones were discovered until 29 m. During the studies to determine the ground water level, at -0,46, petroleum products were encountered just above the water level.

3.1   Contaminated ground region and the tunnel line intersection

During side investigations in the area, which passes through the south of the working site, pet-roleum products were discovered in the core sample approximately 20m southeast of the site as seen in Figure 2.

Figure 2.   Gasoline found in the İSK-3 borehole.

According to statement of the gas station authority, fuel oil contamination had been discovered under another gas station approximately 60m North East of the gas station, decontamination system had been established and the studies regarding decontamination were in progress.

Within 100 m radius of the working site, as seen in Figures 3 and there are residential apartments, parks, at least 4 school and a big shopping mall. Additionally, there is Bakirkoy-

Figure 3.   Top view of the contaminated area with Line 1 between 229th and 333rd rings.

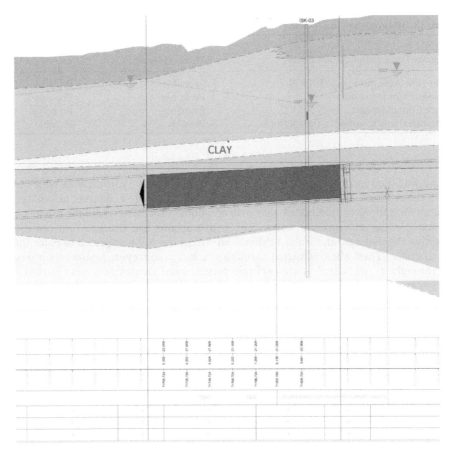

Figure 4.   Contaminated area intersection with the tunnel line between 7 + 706th and 7 + 826th km.

Bahcelievler-Kirazli Metro Project construction site which is located in the South of the gas station. Contaminated area intersection with the tunnel line between 7 + 706th and 7 + 826th km is seen in Figure 4.

In cases of contaminant, leakages like gasoline there are 3 ways for the contaminants to reach the receptors: through open air, through underground soil and underground water. According to National Institute of Occupational Safety and Health, limit exposure limit to gasoline for a human being is 300 ppm for 8 hours, short term exposure is 500 ppm.

Regarding the excavation process, contamination is related to the porosity, permeability of the soil and the ground water. If the ground is porous, permeable enough, underground water can carry the contaminant to larger area. And during the geotechnical studies, it has been observed that the soil around the metro line was permeable. Thus, it was safe to say that the leakage around the gas station has contaminated and spread through the surrounding area. For that reason, to keep the hydrocarbon threshold below the exposure limits, explosion limits, several precautions and systems were implemented.

4   PRECAUTIONS TAKEN FOR A SAFE EXCAVATION

During the excavation, contaminant could be found as in liquid form solved within the water, as non-aqueous phase on the water surface or as solved in the excavated muck; appropriate methods are needed to prevent or lower the risk of hydrocarbon hazard. Hydrocarbon phases and risk analysis is given in Table 1.

Table 1. Hydrocarbon phases and risk analyses.

Hydrocarbon Contamination Phase	Construction Workers Health	Explosion Risk
Light Non-Aqueous Phase (LNAPL)	Probable	Probable
Dissolved Contamination	Probable	Possible
Vapor Contamination	Probable	Probable
Contamination Absorbed in the soil/rock	Possible	Possible

4.1 Bentonite Unit

Bentonite was used in order to dilute the concentration of the hydrocarbons which are dissolved in the muck or in a not dissolved state within the excavation chamber in case the TBM encounters formations contaminated with hydrocarbons. Especially where the tunnel crown is closer than 5 m to underground water level, hydrocarbons like benzene in a not dissolved phase within the underground water or dissolved within the soil is expected to enter the excavation chamber. Thus when extracted through the screw conveyor, hydrocarbon gases can spread throughout the tunnel in atmospheric pressure and creates poisoning hazard for the workers or creates explosion hazard in the tunnel.

Bentonite mixture was passed through $1m^3$ mixer and sent to bentonite tanks with a pump of 10 kW and 50 m^3/h. In those tanks, bentonite was mixed with air and sent to TBM bentonite tank with 5.5 kW slurry pumps. Then bentonite was pumped in to the excavation chamber with the help of the foam lines.

This application was abandoned since montmorillonite within the bentonite swelled much quicker than anticipated because of the water income inside the excavation chamber from the soil and swelling of the bentonite mixed with clayey limestone formation inside excavation chamber effected advance performance of TBM and consistency.

4.2 Gas Sensors

Apart from the original sensors (Trolex) of the TBM as seen in Figure 5, additional gas sensors to detect hydrocarbons were integrated to TBM and periodical measurements were taken throughout the TBM and the tunnel.

TBM gas sensors are connected to operator cabin and additionally they are completely integrated to electronical control system of the TBM. In case these sensors detect gases -which are hazardous- above the programmed limit, they can stop TBM advance and the discharge gate of the screw conveyor from opening. In BBKM project sensors that are installed on the TBMs are;

- $1 \times O_2$
- $1 \times CO$

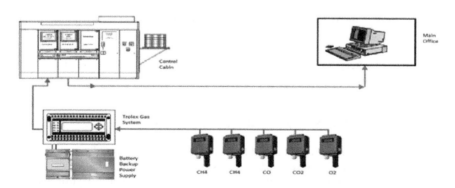

Figure 5. TBM gas sensor system (TROLEX).

- $1 \times CO_2$
- $2 \times CH_4$

Although there is no sensor specific for the benzene gas, since the chemical structure of the methane (CH_4) is similar to benzene gas, methane sensor can detect benzene (C_6H_6).

To be on the safe side, additional gas sensors which can detect hydrocarbon gases (TORA) were installed and integrated with the TBM as shown in Figures 6 and 7 and given in Table 2. According the hydrocarbon levels;

- If the gas concentration in the tunnel is between 10% - 20%, sensors give visual and audial warning. (Blue light)
- If the gas concentration is between 20%-40%, sensors that are integrated with the TBM PLC stops the advance and personnel are evacuated from the tunnel. And tunnel is ventilated until the hazardous gasses are diluted below the threshold. (Orange light)

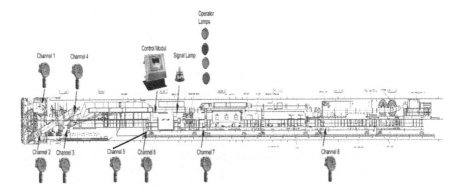

Figure 6.   Hydrocarbon sensors (TORA) on the TBM.

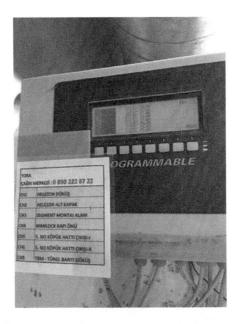

Figure 7.   TORA hydrocarbon sensor monitor.

Table 2. Hydrocarbon sensors that are integrated to TBM.

CH1	Under the screw conveyor discharge gate
CH2	Above the manlock door
CH3	Segment assembly area
CH4	Screw conveyor discharge gate
CH5	Segment feeder
CH6	Hydraulic motors
CH7	Next to chiller motor
CH8	TBM belt discharge

• Above 40%, main breaker immediately turns of the power and personnel is evacuated at once. (Red light)

Also, periodical gas measurements were carried out with portable gas detectors from various points of the TBM and throughout the tunnel by the health and safety personnel. Portable gas detectors are seen in Figure 8.

Protocols in case of hazardous gas detection is given in Table 3.

4.3 *Gas measurements taken instantaneously inside the excavation chamber*

Instantaneous gas measurements were taken inside the excavation chamber with a line setup from the foam line to the bulkhead of the TBM as seen in Figure 9. Foam line is connected to bucket in which a gas sensor is placed. When the valve on which the line is connected to on the bulkhead is opened, with the help of the EPB pressure that is created in the chamber, gas in the chamber runs through the flowmeter and reaches to the gas sensor. Flowmeter is adjusted to $25m^3/min$ and flow rate can be monitored from the operator cabin. To avoid blockages in the line, an extra water line is connected to the measurement line and in case of blockage, pressured water is pumped in to the line to clean it.

Figure 8. Portable gas detectors.

Table 3. Protocols in case of hazardous gas detection.

Observation	Explanation	Cause of Happening	Intervention
General Detection System **ALARM** 1st Level (%10)	In the case of a gas sensor detectsgas concentration above 10%	• Explosive gas built up inside the tunnel. • Activities inside the TBM • Sensor malfunction	Measurements are taken inside the TBM where the sensor gives 1st level alarm with portable sensors.
General Detection System **ALARM** 2nd Level (%20)	In the case of a gas sensor detects gas concentration above 20%	• Explosive gas built up inside the tunnel. • Activities inside the TBM • Sensor malfunction	Tunnel has to be evacuated and exproof ventilation must be made.
General Detection System **ALARM** 3rd Level (%40)	In the case of a gas sensor detects gas concentration above 40%	• Explosive gas built up inside the tunnel. • Activities inside the TBM • Sensor malfunction	TBM PLC stops the advance and stops all electrical and hydraulic systems except the emergency system. (Fire extinguishing, camera, etc...)
General Detection System Fault	In the case of a gas sensor fault (showing false alarms, incorrect values etc..)	• Cabling error between the control panel and the sensors • Longtime power shortages • Exhaust gas from the MSV engine.	Health and Safety superintendent is called.

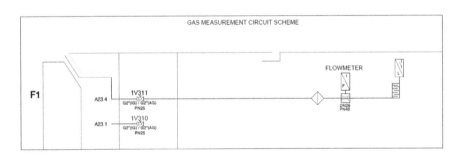

Figure 9. Gas measurement line inside the excavation chamber.

4.4 İDO cross passage dewatering area

To enter the excavation chamber and control the cutterhead, a dewatering area has been established at the exit of the cross-passage area with Ø100 and Ø65 bored piles as seen in Figure 10. Then an empty forage is bored on 85th bored pile and water has been evacuated through there. Since the danger of the petroleum products lower down to tunnel excavation elevation due to water evacuation, after the dewatering process, empty forage was washed with clean water and then evacuated again through piezometer hole.

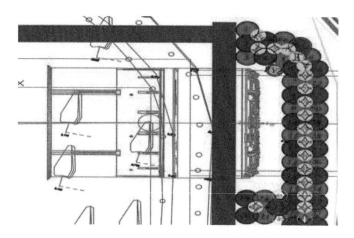

Figure 10.   Bored piles at the exit of the cross passage for Line 1.

4.5   *Other measures*

Since the contaminated area was close to the shaft, no additional ventilation systems were required. Exhaust ventilation wasn't chosen since the toxic gas could spread along the tunnel and the air circulation wouldn't be enough to dilute hazardous gas inside the tunnel and the TBM.

A key crew operated inside the TBM with full protection gears during the advances; 1 TBM engineer, 1 TBM operator, 1 segment erector operator, 1 belt conveyor operator, 1 grout injection operator was inside the TBM during the advances. During ring assembly belt crane operator and segment assembly personnel was going inside the tunnel.

Key crew had fireproof overalls, gas masks, fireproof boots, fireproof working gloves and working glasses with them. Aside the key crew no one was allowed inside the tunnel without protection gear or without the permission of the shift engineer.

5   CONCLUSIONS

Istanbul faces frequent earthquakes, which loosens the pipe connection of underground petroleum tanks. During underground works it is sometimes faced with petroleum contaminated ground which causes a tremendous risk for workers health. This paper is a typical example how the situation is tackled. In case of hazardous gas was detected the problem was solved with adequate ventilation.

REFERENCE

Stewart, Environmental Science in the 21st Century – An online Textbook.

Tunnels and Underground Cities: Engineering and Innovation meet Archaeology,
Architecture and Art, Volume 9: Safety in underground
construction – Peila, Viggiani & Celestino (Eds)
© 2020 Taylor & Francis Group, London, ISBN 978-0-367-46874-3

The design and evaluation of tunnel health monitoring system

M. Yu, M. Hu & L. Zhou
SHU-UTS SILC Business School, Shanghai University, Shanghai, China
SHU-SUCG Research Centre for Building Industrialization, Shanghai University, Shanghai, China

ABSTRACT: Nowadays, macrostructures such as bridges have relatively complete monitoring systems and evaluation systems, while the Tunnel Health Monitoring System (THMS) is still in its early stage of development. The design of THMS has not yet formed an effective paradigm, and the data offered by THMS has not been fully used by the existing evaluation system. Therefore, a complete THMS needs to be rebuilt from the perspective of real-time monitoring. Based on previous researches, this paper proposes an immobilized-mobile combined THMS(IM-THMS), which can gather data comprehensively and cost-effectively. The immobilized sensor network gathers data on the health of the tunnel environment, while the mobile inspection gathers data on surface defects. Given the high data refinement and high time-series of real-time monitoring and automatic inspection system, this paper rebuilds the tunnel health evaluation system. The proposed methods can provide more accurate guidance for the maintenance of the tunnel.

1 INTRODUCTION

With the continuous development of urban tunnel construction, the service life of the tunnel has increased, and the tunnel engineering has transitioned from the comprehensive construction period to the inspection and maintenance period. Different types and degrees of distresses will appear on the tunnel structure during operation period, and the distresses will affect the normal operation of the tunnel structure to a certain extent. Therefore, in order to realize the real-time monitoring of the operation of the tunnel structure, domestic and foreign scholars have proposed the concept of tunnel health monitoring, that is, real-time evaluation of the operational status of the structure through the structural response information obtained by sensors installed on site. At present, the research work on the tunnel structure health monitoring system has just started. The mature engineering cases and safety evaluation indicators at home and abroad are relatively few. The design of the whole health monitoring system has not yet formed a paradigm.

In recent years, the monitoring technology has developed rapidly, and the monitoring method has changed from traditional manual monitoring to sensor network monitoring and mobile monitoring. Relatively speaking, the research on tunnel health monitoring system is lagging, and the construction of tunnel health monitoring system is still more dependent. Sensor network monitoring without the integration of other monitoring technologies, the existing evaluation system has not fully utilized the information provided by the tunnel health monitoring system. Therefore, based on the existing research, this paper reconstructs a complete tunnel health monitoring system and tunnel structure health evaluation system from the perspective of tunnel real-time health monitoring. Immobilized-mobile combined THMS (IM-THMS) has the characteristics of comprehensive and extensive collection information and reasonable cost and its engineering application is convenient. The tunnel health assessment system combines the automatic monitoring data and the trend indicator information to construct the tunnel structure degradation damage model, which not only reflects the current state characteristics of the tunnel, but also reflects the evolution trend of the tunnel structure. This more accurately reflects the health of the tunnel, which is conducive to the development of preventive maintenance strategies.

This paper is organized as follows. The related work on the existing tunnel monitoring methods, tunnel health monitoring systems and tunnel evaluation systems are introduced in Section 2. The design of the tunnel health monitoring system is presented in Section 3, including the improvement of tunnel structure health evaluation system. A simple example is provided in section 4. Finally, conclusions are discussed in Section 5.

2 LITERATURE REVIEW

This part reviews the literature from three aspects, existing tunnel monitoring and detection methods, research and development of tunnel health monitoring system, and study of tunnel structure health evaluation system.

The traditional methods of tunnel monitoring and detection are mainly artificial, mainly relying on artificial naked eye detection and artificial instrument measurement. This method has the advantages of high accuracy, mature and reliable technology, but at the same time, it has the disadvantages of high subjectivity of detection results, high detection human cost and low safety of detection personnel. With the increasing number of tunnels in China, the environmental conditions of tunnels become more and more complex, and the workload of detection is increasing. Obviously, artificial detection technology has been unable to meet the basic requirements, and the time required to complete the entire detection has become longer and longer. In view of the disadvantages of current tunneling structure detection technology, new technologies have been introduced and developed at home and abroad for the detection and monitoring of tunnel structure health, such as sensor network technology, digital camera technology, 3D laser scanning technology, handheld disease recording technology and automatic measurement robot technology. Comparison of these detection techniques is shown in Table 1.

Tunnel Structural Health Monitoring (TSHM) is the application of SHM in Tunnel engineering. It involves civil engineering, testing technology, analysis technology and information technology, and is a comprehensive system engineering with interdisciplinary. In recent years,

Table 1. Comparison of tunnel detection techniques.

Technical Name	Practical Applications	Advantages	Disadvantages
Sensor Technology	Building sensor network full coverage detection Fiber Bragg Grating Sensor, MEMS New Sensor, etc.	Improve the accuracy of tunnel structure disease detection, and facilitate remote, dynamic, real-time detection.	Sensor installation cost is high and there is power supply problem.
Digital Photography	Digital photography calibration detection combined with image processing technology	The collected image information is rich and accurate.	Image processing workload and high personnel costs.
3D Laser Scanning Technology	3D laser scanner	A relatively comprehensive information on the tunnel environment and the ability to obtain tunnel deformation.	Post-processing workload is large, requiring manual handling of instruments.
Handheld Disease Recording Technology	Hand-held disease recorder	Digital, information, timeliness and accuracy, reducing intermediate processing	Manual inspection is subjective.
Automatic Measurement Robot Technology	Fully automatic measuring robot	High precision, high degree of automation and easy operation.	Expensive

the research and application of tunnel structure health monitoring have been developing rapidly. Some scholars have adjusted the structural health system of tunnels from the perspective of the development of monitoring technology to make it more in line with the current tunnel monitoring. Some of them start from the establishment of evaluation method and evaluation system and explore from the safety evaluation of tunnel operation period.

At present, the research on the evaluation of tunnel status mainly focuses on the establishment of the index system of the performance evaluation of tunnel structure service. The index system of the performance evaluation of tunnel structure service includes integrity evaluation index, structure safety evaluation index and structure durability evaluation index. The integrity index is an index system which evaluates the safety and function of the tunnel structure from the whole Angle along the tunnel longitudinal. The safety evaluation index is the index system which evaluates the safety of the tunnel structure from the longitudinal and transverse perspectives. Durability evaluation index is an index system to evaluate the durability status and remaining service life of tunnel structure components. In terms of the study of tunnel integrity evaluation indicators, most literatures adopt AHP to combine theoretical analysis with expert experience, qualitative judgment and quantitative analysis. Such as patent "one kind of shield tunnel service performance evaluation based on variable weight method, this method first initial weights are obtained by using expert investigation method, and then according to the index data of initial weights variable weight adjustment, finally using the method based on fuzzy comprehensive evaluation to evaluate structural unit, and then according to the unit structure evaluation level comprehensive judging tunnel overall service performance level. However, Li Xiao Jun et al. of Tongji university creatively put forward a comprehensive service performance evaluation index TSI. This method only considers measurable variables. The safety evaluation of tunnel structure is to establish the macroscopic damage model of tunnel structure, use effective identification method to identify the damage parameters of structure, understand the damage situation of tunnel structure, and judge the damage degree of tunnel structure, to evaluate the safety of tunnel structure. The durability evaluation of tunnel structure is to predict the tunnel life by establishing the reliability model of tunnel structure and using the accumulative damage model based on historical data, to provide reference for the establishment of maintenance strategy of tunnel system. For example, Yuan Yong et al. from Tongji university have devoted themselves to the study of reliability mechanism of tunnel structure and its risk assessment methods in recent years, put forward the concept of tunnel predictive maintenance, applied the reliability modeling based on data-driven state maintenance, and conducted the analysis of remaining life of tunnel through the cumulative damage model.

The tunnel evaluation system in this paper is mainly based on the research of Huang Hongwei et al., which makes some improvements, and puts forward the dynamic prediction index and USES the more objective entropy weight method as the method to establish the index weight. The establishment of the tunnel structure health evaluation index system can not only reflect the current health of the tunnel but also predict the trend of the tunnel state by using the data of automatic monitoring, and evaluate the tunnel structure health more comprehensively and accurately, which is helpful for managers to make operation and maintenance decisions.

3 DESIGN OF TUNNEL HEALTH MONITORING SYSTEM

3.1 *Overall system composition*

The tunnel health monitoring system proposed in this paper is a comprehensive system of data collection, management and analysis based on long-term tunnel monitoring, taking data analysis and model analysis as the main means, and taking verification of key design parameters and indicators, structural service performance evaluation and prediction, and rational allocation of maintenance resources as the main objectives. The most important core technologies of IM-THMS are monitoring technology and evaluation and analysis technology. The monitoring technology can collect the key parameters (such as deformation, structure stress, etc.) of the environment and structure of the tunnel in a long term, real-time and comprehensive way. The evaluation and analysis technology construct the relation model of tunnel structure and environment based on the monitoring system information, and calculates and analyzes the response and characteristics of tunnel under specific environment, to judge the safety and health status of tunnel structure according to the variation of measured parameters.

The basic components of IM-THMS proposed in this paper include monitoring and detection, diagnosis and evaluation. The monitoring and detection of the system is mainly based on the combination of immobilized sensor network monitoring and mobile inspection to monitor and detect the structural state of the tunnel. The combination of the two methods makes the collected data more comprehensive and more accurately reflect the healthy state of the tunnel structure. Combined with the theory and calculation, the collected tunnel health status data were analyzed and the damage identification was compared with the structural health standards. The systematic evaluation part uses the more objective entropy weight model to calculate the index weight in the evaluation index system established by AHP. The basic structure of tunnel structure health monitoring system proposed in this paper is shown in Figure 1.

3.2 *Contents of tunnel structure detection*

Monitoring technology in IM-THMS combines immobilized and mobile monitoring methods, immobilized sensor network combines real-time monitoring and regular detection to collect tunnel health environment information, and mobile inspection to collect tunnel surface disease

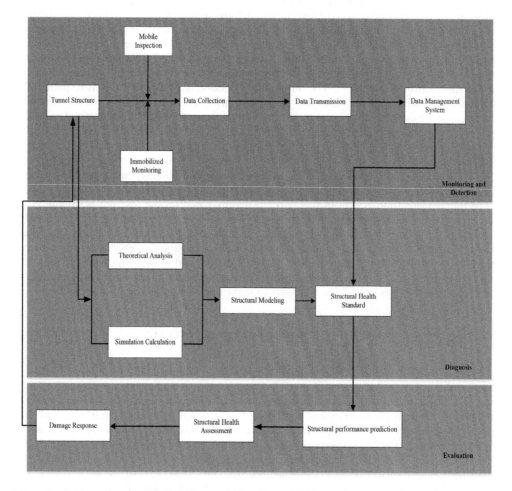

Figure 1.   Basic structure of the tunnel structure health monitoring system.

information. Table 2 is a comprehensive consideration of the immobilized sensor network and mobile inspection of two specific monitoring, detection content and indicators.

3.3 *Tunnel structure health evaluation system*

This section introduces the shield tunnel structure health evaluation system. From the perspective of shield tunnel structure health monitoring and detection data, the system maximizes the use of monitoring data. In order to reflect the change trend of tunnel structure health, this paper selects three indicators that can reflect the health trend change of tunnel structure from the 13 indicators of AHP for the prediction model of tunnel performance trend, and finally uses entropy weight model to calculate the weights and establish the tunnel health status. Figure 2 shows the overall idea of establishing the tunnel structure health evaluation model.

Table 2. Monitoring and detecting content and indicators.

Method	Content	Indicators
Automatic Monitoring	Cross-section convergence	Diameter change rate, ellipticity
	Vertical settlement	Uneven settlement, curvature
	Contact channel	Uneven settlement
	Segment seam	Opening amount, wrong amount
	Deformation joint	Opening amount, wrong amount
	Leaking water	Leakage state
	Structural surface crack	Width, length, depth
	Appearance damage	Drop powder, peel, lift, peel off
Daily Inspection	Structural surface precipitation	Rust spots, secretion, loose flowers, shelling
	Leaking water	Seepage area, leakage of sediment
	Bolt	Loose, large deformation, damage
	Lining strength	Concrete compressive strength
Special Test	Concrete carbonation	Carbonation depth
	Steel corrosion	Corrosion rate

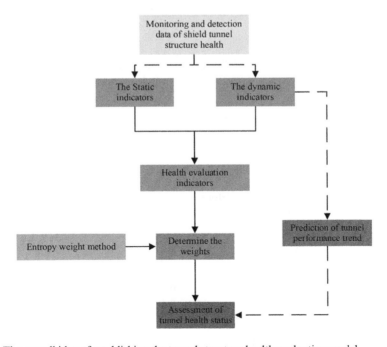

Figure 2. The overall idea of establishing the tunnel structure health evaluation model.

As is shown in Figure 2, the indicators of the tunnel health evaluation system are composed of static indicators and dynamic indicators. The data comes from the tunnel structure automatic mobile monitoring and immobilization detection. After determining the evaluation indicators, the weight of each indicator is determined by the entropy weight method. At the same time, data analysis is performed on the selected dynamic indicators to predict performance service trends. Finally, the evaluation value of the tunnel health status and the corresponding level are obtained based on the evaluation criteria of the tunnel health status.

3.3.1 *Structure health evaluation grade of shield tunnel*

The classification of healthy grade of tunnel structure section is the basis of quantitative diagnosis, which determines the choice of tunnel structure management measures. Therefore, to a certain extent, the degree of grade classification affects the reliability of diagnosis and the operability of maintenance. This paper classifies the shield tunnel structure health evaluation into four levels based on multiple domestic and foreign standard specifications and several existing results, as shown in Table 3.

3.3.2 *Health evaluation indicators and evaluation criteria*

The project of health evaluation of tunnel structure needs to be detected and monitored mainly includes four categories: structural deformation, surface disease of concrete, leakage state and material deterioration. See Table 4 for detailed detection and monitoring contents and indicators.

Table 3. Health class division and definition of shield tunnel.

Health level	Health Status	Definition
I	Normal	Performance is intact.
II	Damage	Degraded performance, a small part of the function is impaired, and has less impact on normal use.
III	Deterioration	Performance deteriorates and applicability is affected, but it does not endanger security for the time being.
IV	Danger	Serious deterioration in performance, endangering safety

Table 4. Criteria for classification of indicators.

Indicators		Level I	Level II	Level III	Level IV
Overall structural deformation	Cross-section convergence/‰	0~5	5~10	10~20	>20
	Cross-section convergence/ m^{-1}	0 ~ 1/ 15000	1/15000~1/ 5000	1/5000~1/ 2000	>1/2000
	Cross channel differential settlement/mm	0~10	10~20	20~30	>30
Connecting member deformation	Segment joint deformation/ mm	0~6	6~12	12~18	>18
	Deformation joint deformation/mm	0~10	10~20	20~30	>30
	Bolt	/	Loose	Large deformation	Damage/ break
Surface disease	Crack width/mm	0~0.2	0.2~0.3	0.3~0.4	0.4~0.5
	Damaged area/cm^2	0~10	10~30	30~100	>100
	Precipitation area/cm^2	0~10	10~30	30~50	>50
Leakage state	Leakage state/m^2	0~ 0.01	0.01~0.2	0.2~0.5	>0.5
Material degradation	Lining strength	0~0.1	0.1~0.3	0.3~0.5	>0.5
	Concrete carbonation depth/cm	0~15	15~30	30~40	>40
	Steel corrosion rate/%	0~3	3~10	10~25	>25

3.3.3 *Health evaluation index system*

Based on the analysis of the engineering practice and the causes of common diseases in shield tunnel, this paper puts forward that the damage form of shield tunnel structure is mainly divided into five categories: integral deformation of structure, deformation of connecting members, surface damage of concrete, leakage state and material deterioration.

In shield tunnel structure safety assessment index system, given that it is difficult to consider a problem at all the details, so using the Hierarchy principle of the system science, the large system decomposition-coordination principle in the theory, the problem is decomposed into several layers, each layer is composed of multiple factors, from coarse to fine, and its excellent from global to local gradually thorough analysis, will affect the shield tunnel structure health status factors a methodical, hierarchical, therefore pass class time analysis model is set up, using the Analytic Hierarchy Process (the Analytic Hierarchy Process, The thought of AHP (short for AHP) establishes the safety evaluation index system of shield tunnel structure. Figure 3 shows the hierarchical index system of shield tunnel structure safety evaluation.

3.3.4 *Prediction of tunnel structure performance trend*

In this paper, three indexes of performance trend prediction are selected to predict the trend of tunnel structure deformation, which are cross-sectional convergence, longitudinal settlement and cross-channel differential settlement respectively. Through the analysis of automatic monitoring data corresponding to these indexes, partial performance trend prediction was achieved by using EDM classical pattern decomposition method based on time series. EDM method extracts several Intrinsic Mode Function (IMF) from the original signal. Different IMF reflects the characteristics of different scales of the original signal, and the residual term reflects the basic trend of the original time series.

Essentially, the EMD method is to smooth the time series. The result is to decompose the fluctuations or trends of different scales in the time series step by step and produce a series of data sequences with different characteristic scales.

The processing of time series by EMD method is a process of continuous decomposition and filtering. The specific steps are as follows:

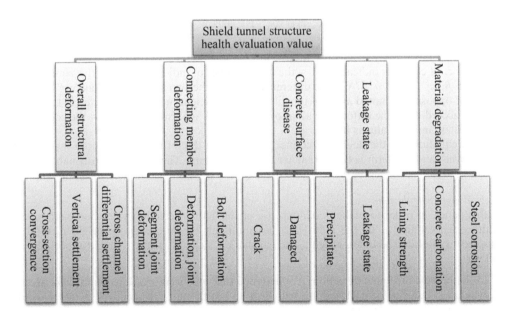

Figure 3. Shield tunnel structure health evaluation system.

Let the original time series be $x(t)$;

1. determine all maxima and minima of $x(t)$;
2. the upper and lower envelope of $x(t)$ is constructed by cubic spline interpolation according to the maximum and minimum values;
3. calculate the mean value of the upper and lower envelope of $x(t)$, and the difference between $x(t)$ and $m_{11}(t)$, $h_{11}(t) = x(t) - m_{11}(t)$;
4. replace the original time series $x(t)$ with $h_{11}(t)$, and repeat the above three steps until $h_{1k}(t)$ meets the conditions for IMF establishment. The $h_{1k}(t)$ is an IMF component, denoting $imf_1(t) = h_{1k}(t)$, $r_1(t) = x(t) - imf_1(t)$, $x(t) = r_1(t)$;
5. repeat the above four steps until $r_n(t)$ becomes a monotonic function or the difference of amplitude is less than the given threshold value S_d, and the decomposition process ends. S_d is calculated according to the results of two consecutive screening $S_d = \sum_{t=0}^{T}(|h_{(k-1)}(t) - h_k(t)|/h_k^2(t))$; including $h_k(t)$ is the difference between the current iteration, $h_{(k-1)}(t)$ is the difference between the last iteration. The value of S_d is generally defined between 0.2 and 0.3. The original time series $x(t)$ was decomposed into n the *IMF* and a residual $r(t)$, $x(t) = \sum_{i=1}^{n} imf_i(t) + r(t)$, each *IMF* reflect the characteristics of the original signal in different time scales, and the stationary time series, and the residual item $r(t)$ reflects the basic trend of original time series.

Through the analysis of automatic monitoring of high frequency data, get the basic trend of time series, compare results and degradation index standard, it concluded that the performance of the index trend prediction, so as to make the whole evaluation system in the static evaluation at the same time, trend prediction, perfect the evaluation system as well as to operational management provides the basis of a tunnel structure maintenance measures.

3.3.5 Calculate the relation degree of each index

After the detection of n tunnel sections, the sample set $A = (Q\_1, Q\_2, \ldots, Q\_n)$ was established and Q was the section. The health status evaluation grade $B = (B\_1, B\_2, B\_3, B\_4)$ is set up and B is the grade. These two sets are set up $M = (A, B)$ to set pairs, where each sample Q in set A contains 13 actual values of evaluation indexes. Based on the idea of set pair analysis, the connection degree μ between the sample interval set and the grade is:

$$\mu\_(Q, B) = S/N + (F/N) * i + (P/N) * j = a + bi + cj \tag{1}$$

where S = the number of identity indicators; F = the number of difference indicators; P = the number of contrary indicators; N = the total number of indicators; S/N = the same degree as A and B, let's call it a; F/N = the difference between the two sets, denoted as b; P/N = the degree of opposites, denoted as c; i = the difference coefficient, which is evaluated in the range of [-1,1] according to the different evaluation objects and analysis methods; j = the coefficient of opposites, and the value is fixed as -1.

From equation (1), the degree of connection contains the information of both macro and micro levels, and the term is the parameter combining them. Therefore, the reasonable determination of difference degree is the key to the analysis based on information meaning set.

3.3.6 Weight calculation

In this paper, in order to make maximum use of the actual detection value and explore the health essence reflected by the index in a large amount of data, entropy weight method is adopted to allocate the index weight. The calculation steps of entropy weight model are as follows:

1. In the process of tunnel detection, it is assumed that n sections are tested, and there are 13 evaluation indexes for each section to reflect its health status, respectively $x_i(i = 1 \sim m)$, and construct the judgment matrix R:

$$R = (r_{st})_{n \times 13} \tag{2}$$

$r_{st}$is the measured value of t-th evaluation index of the s-th evaluation object.

2. $r_{t\max}$ and $r_{t\min}$ are respectively the most satisfied and least satisfied values of different samples under the same index, then the judgment matrix R is normalized to obtain the normalized matrix Z, and the elements of Z are:

$$z_{st} = (r_{st} - r_{t\min})/(r_{t\max} - r_{t\min}) \tag{3}$$

3. Calculate the entropy value of each evaluation index H_t according to the concept of entropy:

$$H_t = -(\sum_{s=1}^{n} f_{st} \ln f_{st})/\ln 13 \tag{4}$$

Where $f_{st} = z_{st}/\sum_{s=1}^{n} z_{st}$, if $f_{st} = 0$, $H_t = 0$.

4. Calculate the entropy weight of each index:

$$w = (w_t)_{1*13}, \, w_t = (1 - H_t)/13 - \sum_{t=1}^{13} H_t \sum_{t=1}^{13} w_t, \, = 1, \tag{5}$$

5. The determination of comprehensive connection degree:

After the differential indicators of the complete part are decomposed, the comprehensive correlation degree of the F difference indicators is obtained according to equation (6).

$$\mu_{(x,B_1)} = \sum w'_i \mu_{(x_i,B_1)} \tag{6}$$

$$w'_i = w_i / \sum w_i \tag{7}$$

Where ω_i=the weight of differential index; ω_i'= the weight value of the re-normalized difference index.

6. determine the health level

By substituting equation (6) into equation (1), the expression of the final degree of correlation can be obtained. In practical engineering problems, through the decomposition of the above process, the ambiguity and uncertainty in the problem can be diluted to an acceptable range. That is, i =0 and j=-1 are used to calculate the rank correlation, and then the health grade is determined according to the maximum association criterion. However, in order to make more use of information, avoid loss of information, and determine distortion, health grade is determined by characteristic value, and the calculation formula is as follows:

$$k = \sum_{B=1}^{4} B \times \bar{\mu}_{(Q,B)} / \sum_{B=1}^{4} \bar{\mu}_{(Q,B)} \tag{8}$$

$$\bar{\mu}_{(Q,B)} = (\mu_{(Q,B)} - \min_B(\mu_{(Q,B)}))/(\max_B(\mu_{(Q,B)}) - \min_B(\mu_{(Q,B)})) \tag{9}$$

It is easy to know from the mathematical relation. According to the principle of equipartition, the classification of the characteristic value and health status of the evaluation sample corresponds to Table 5.

Table 5. Correspondence between eigenvalues and health levels.

Health level	Health status	Eigenvalues
I	Normal	[1.00, 1.75)
II	Damage	[1.75, 2.50)
III	Deterioration	[2.50, 3.25)
IV	Danger	[3.25, 4.00]

4 SAMPLE TEST

We use the following paradigm to illustrate how we use the entropy weight method to calculate the weight of different indicators and to determine the health level of each section. We take a shield tunnel in Shanghai as a sample case, which was found that there were different degrees of diseases in the tunnel structure through field investigation and detection. The detection results of the four sections are shown in Table 6.

According to the above weight determination method, after calculation, the weight of each indicator is shown in Table 7, and the health level of each section is shown in Table 8.

Table 6. Disease detection results of a road shield tunnel in Shanghai.

Index	Section 1	Section 2	Section 3	Section 4
Cross-section convergence/‰	4	16	12	8
Cross-section convergence/m^{-1}	0.0000286	0.000125	0.0004	0.0005
Cross channel differential settlement/mm	8	22	18	14
Segment joint deformation/mm	10	4	3	5
Deformation joint deformation/mm	6	12	8	6
Bolt	2	1	1	2
Crack width/mm	0.45	0.25	0.35	0.26
Damaged area/cm^2	18	20	15	20
Precipitation area/cm^2	18	40	15	35
Leakage condition/m^2	0.6	0.5	0.2	0.3
Lining strength	0.1	0.2	0.12	0.24
Concrete carbonation depth/cm	6	5	5	3
Steel corrosion rate/%	5	10	12	20

Table 7. Entropy weight method to calculate the weight of each index.

Indicators	Weight
Cross-section convergence/‰	0.0594
Cross-section convergence/m^{-1}	0.0697
Cross channel differential settlement/mm	0.0543
Segment joint deformation/mm	0.0926
Deformation joint deformation/mm	0.1305
Bolt	0.1098
Crack width/mm	0.0994
Damaged area/cm^2	0.0496
Precipitation area/cm^2	0.0805
Leakage condition/m^2	0.0653
Lining strength	0.0773
Concrete carbonation depth/cm	0.0487
Steel corrosion rate/%	0.0630

Table 8. Health level of each section.

Section	Rank values	Health level
1	1.5340	I
2	2.0636	II
3	2.0000	II
4	2.0653	II

5 CONCLUSION

(1) This paper proposes a shield tunnel structure health monitoring system combining immobilization and mobile monitoring and detection methods. The immobilized monitoring method uses embedded sensor to monitor the status of structure deformation in real time, and the mobile detection method is to scan tunneling surface disease and leakage status. The combination of the two makes the information collected by tunnel monitoring more comprehensive and extensive. At the same time, the monitoring system also has the characteristics of reasonable cost and simple engineering application.

(2) Aiming at the characteristics of high degree of refinement and strong timing of information in the real-time monitoring system and automatic inspection system of tunnels, this paper has improved the tunnel structure health evaluation system based on relevant research. The static index and dynamic index in the evaluation system are combined, which not only have the evaluation value of static tunnel structure state, but also combine the dynamic index of historical data analysis to predict future service performance, so the evaluation system is more perfect.

(3) Entropy weight method is used to determine the weight of the evaluation system, and the results are more objective.

(4) For many sensor automatic monitoring data, EDM method is used in this paper to process time series data, which can well realize partial performance trend prediction and facilitate the formulation of tunnel preventive maintenance strategy.

REFERENCES

Bhalla S, Fang YW:Zhao J, et al. 2005.Structural health monitoring of underground facilities-technological issues and challenges. *Tunneling and Underground Space Technology*20(5): 487–500.

Ding Y, Shi B, et al. 2005. Tunnel structure health monitoring system and optical fiber sensing technology. *Journal of disaster prevention and reduction engineering* 25(4):375–380.

Fan B, Zheng JQ, Zhu KY 2014. Realization of automatic deformation monitoring of tunnel construction by measuring robot. *Tunnel construction* 34(1):19–23.

Gao S, Jin Q, W.Y. Zhou, et al. 2010. Tunnel section deformation detection system based on laser ranging technology. *Measurement and control technology* 29(5):44–46.

He C, Li X, Wang B, et al. 2008. Study on health monitoring and safety evaluation of tunnel structure during operation period. *Modern Tunnel Technology* (s1):289–294.

Z.Q. Li, Xiong H.B. 2006. Research status of structural health monitoring. *Structural Engineer* 22(5):86–90.

Li X, Lin X, Zhu H, et al. 2017. Condition assessment of shield tunnel using a new indicator: The tunnel serviceability index. *Tunnelling & Underground Space Technology* 67:98–106.

Ming, L. I., Wei-Zhong, C., Jian-Ping, Y., & Zheng-Long, X. (2015). Early warning research for tunnel structure health monitoring system based on efficacy coefficient method. *Rock & Soil Mechanics* 36(2): 729–736.

Shi B, X.J. Xu, Wang D, et al. 2005. Study on distributed optical fiber strain monitoring for tunnel health diagnosis. *Journal of Rock Mechanics and Engineering* 24(15):2622–2628.

Yuan Y, Jiang X, Liu X. 2013. Predictive maintenance of shield tunnels. *Tunnelling & Underground Space Technology Incorporating Trenchless Technology Research* 38(3):69–86.

Tunnels and Underground Cities: Engineering and Innovation meet Archaeology,
Architecture and Art, Volume 9: Safety in underground
construction – Peila, Viggiani & Celestino (Eds)
© 2020 Taylor & Francis Group, London, ISBN 978-0-367-46874-3

Author Index

Printed and bound by CPI Group (UK) Ltd, Croydon, CR0 4YY

24/10/2024

01778289-0003